Medical Physics
Selected Reprints

Edited by

Russell K. Hobbie

published by
American Association of Physics Teachers

Contents

I. Basic matters

II. Radiological Physics

III. Other Medical Physics and Biomedical Engineering Areas

RESOURCE LETTER

Roger H. Stuewer, *Editor*
School of Physics and Astronomy, 116 Church Street
University of Minnesota, Minneapolis, Minnesota 55455

This is one of a series of Resource Letters on different topics intended to guide college physicists, astronomers, and other scientists to some of the literature and other teaching aids that may help improve course content in specified fields. No Resource Letter is meant to be exhaustive and complete; in time there may be more than one letter on some of the main subjects of interest. Comments on these materials as well as suggestions for future topics will be welcomed. Please send such communications to Professor Roger H. Stuewer, Editor, AAPT Resource Letters, School of Physics and Astronomy, 116 Church Street SE, University of Minnesota, Minneapolis, MN 55455.

Resource letter MP-1: Medical physics

Russell K. Hobbie
School of Physics and Astronomy, University of Minnesota, Minneapolis, Minnesota 55455

(Received 8 March 1985; accepted for publication 26 March 1985)

This resource letter provides a guide to the literature on the uses of physics for the diagnosis and treatment of disease. The letter E after an item indicates elementary level or material of general interest to persons becoming informed in the field. The letter I, for intermediate level, indicates material of somewhat more specialized nature; and the letter A indicates rather specialized or advanced material. An asterisk (*) indicates those articles to be included in an accompanying Reprint Book.

I. INTRODUCTION

Physicists sometimes use the terms medical physics and biophysics loosely and almost interchangeably. Within these fields, however, there is a distinction. In the United States the term medical physics has traditionally meant the physics used by radiologists to diagnose and treat disease. Originally, this was primarily the physics of x rays; as ultrasound was developed for diagnostic purposes, it also became part of medical physics. Recently the areas of interest to the medical physicist have expanded as still more diverse and sophisticated instruments are used for diagnosis and treatment. The American Association of Physicists in Medicine is the AIP-affiliated professional organization to which most medical physicists belong. Biophysics has been a very broad term, encompassing studies as diverse as molecular structure, physiology, biomedical instrumentation, radiologic physics, and mathematical biology. In recent years, the term biophysics has been used more narrowly to mean the study of molecular biology. Members of the Biophysical Society have interests primarily in this area.

This resource letter describes the use of physics to diagnose and treat disease in humans. This definition includes biomedical engineering as well as medical physics, but it ignores interesting applications of physics in physiology.

In many cases a textbook is the most appropriate way to begin learning about a topic in medical physics. The articles cited here should be considered representative examples rather than the state of the art or an exhaustive bibliography. One can explore further by reading the references in the articles cited, by scanning the journals listed here, or by using *Index Medicus* to find other articles on the same subject. The resource letter presents textbooks first, followed by a list of representative journals and a few representative articles in each area.

II. TEXTS

This section lists textbooks which are suitable references for a physics teacher; elementary physics texts containing biological examples have not been listed.

The reader unfamiliar with medical terms may find useful the following glossary, which was produced by the AAPM for science writers:

1. **Glossary of Terms Used in Medical Physics**, edited by R. J. Barish and P. F. Schewe (AIP, New York, 1984). (E)

Each year the Publication Committee of the AAPM prepares a bibliography of medical physics as a service to its members. It can be purchased for $3.00 for the AAPM, 335 East 45th St., New York, NY 10017.

A. General biophysics and medical physics

2. **Biophysical Science, 2nd ed.**, E. Ackerman, L. B. M. Ellis, and L. E. Williams (Prentice-Hall, Englewood Cliffs, 1979). This is a comprehensive general biophysics text. It covers sensory systems, nerve and muscle, ionizing and nonionizing radiation, molecular biology, thermodynamics, transport, and instrumentation. (I)
3. **Physics With Illustrative Examples from Medicine and Biology. Vol. 1. Mechanics. Vol. 2. Statistical Physics. Vol. 3. Electricity and Magnetism**, G. B. Benedeck and F. M. H. Villars (Addison-Wesley, Reading, 1974–1979). (I)
4. **Medical Physics**, J. R. Cameron and J. G. Skofronick (Wiley-Interscience, New York, 1978). This book has an excellent qualitative discussion of radiologic physics, as well as the physics of the senses and other medical instruments and procedures which use physics. (E)

5. **Medical Physics, Vol. 1: Physiological Physics, External Probes, Vol. 2: External Senses,** A. C. Damask (Adademic, New York, 1978). (I)
6. **Medical Physics, Vol. 3: Synapse, Neuron, Brain,** A. C. Damask and C. E. Swenberg (Academic, New York, 1984). (I)
7. **Intermediate Physics for Medicine and Biology,** R. K. Hobbie (Wiley, New York, 1978). (I)
8. **An Advanced Undergraduate Laboratory in Living State Physics,** J. P. Wikswo, B. Vickery, and J. H. Venable, Jr. (Vanderbilt University, Nashville, TN 1980). This laboratory manual describes experiments in bioelectric phenomena, diffusion, compartments, vision, and ultrasound. (I)
9. **Biomechanics of Human Motion,** M. Williams and H. R. Lissner (Saunders, Philadelphia, PA, 1962). This small book contains many clinical examples of resolution of forces and static equilibrium, including orthopedic traction apparatus. (E)

B. Radiological physics

1. General

Material on radiological physics is also found in Refs. 2, 4, 7, and 15.

10. **Absorption of Ionizing Radiation,** D. W. Anderson (University Park, Baltimore, MD, 1984). This is a comprehensive book on the interaction of photons, charged particles, and neutrons with matter, from the viewpoint of radiologic physics. It also discusses ionization, excitation, microscopic energy distribution, and biological radiation damage. (I)

2. Diagnostic radiology

11. **The Physical Basis of Medical Imaging,** C. M. Coulam, J. J. Erickson, F. D. Rollo, and A. E. James (Appleton-Century-Crofts, New York, 1981). (I)
12. **Christensen's Introduction to the Physics of Diagnostic Radiology, 3rd ed.,** T. S. Curry, J. E. Dowdey, and R. C. Murry (Lea & Febiger, Philadelphia, 1984). This is the third edition of a text for residents. The physics is elementary, but a physicist should find this is a good book from which to learn the medical aspects of radiology. (E)
13. **Medical Radiation Physics, 2nd ed.,** W. R. Hendee (Yearbook Medical Publishers, Chicago, 1979). (I)
14. **Medical Images and Displays: Comparisons of Nuclear Magnetic Resonance, Ultrasound, X-rays, and Other Modalities,** R. S. Mackay (Wiley-Interscience, New York, 1984). This book discusses all of the modalities currently used for medical imaging in a physical, nonmathematical way. There are many illustrations and examples. (I)

3. Radiation therapy

15. **The Physics of Radiology, 4th ed.,** H. E. Johns and J. R. Cunningham (C. C. Thomas, Springfield, 1983). This is the standard text on radiological physics, with an emphasis on therapy. It discusses the production of x rays, nuclear decay modes, accelerators, the interaction of radiation with matter, dosimetry, measurement of radiation, the physics of radiotherapy, treatment planning, and nuclear medicine. (I,A)
16. **The Physics of Radiation Therapy,** F. M. Khan (Williams & Wilkins, Baltimore, 1984). (I)
17. **Radiation therapy physics, 2nd ed.,** W. R. Hendee (Yearbook Medical Publishers, Chicago, 1981). (I)

4. Nuclear medicine

Some material on nuclear medicine is found in Refs. 2, 4, 7, and 15.

18. **Introductory Physics of Nuclear Medicine, 2nd ed.,** R. Chandra (Lea & Febiger, Philadelphia, 1982). (E)
19. **Nuclear Medicine: An Introductory Text,** P. J. Ell and E. S. Williams (Blackwell, Oxford, 1981). (E)
20. **Nuclear Medicine Physics, Instrumentation and Agents,** edited by F. D. Rollo (Mosby, St. Louis, 1977). (I,A)
21. **Physics in Nuclear Medicine,** J. A. Sorenson and M. F. Phelps (Grune & Stratton, New York, 1980). (E)
22. **Nuclear Medical Physics,** edited by L. Williams (CRC, Boca Raton, 1985). (I,A)

5. Ultrasound

23. **Biomedical Ultrasonics,** P. N. T. Wells (Academic, New York, 1977). (I)

6. Magnetic resonance imaging

24. **Magnetic Resonance in Medicine and Biology,** M. A. Foster (Pergamon, New York, 1984). This book describes many uses of nuclear magnetic resonance in biology, including magnetic resonance imaging. (I)

7. Radiation protection

25. **Introduction to Health Physics, 2nd ed.,** H. Cember (Pergamon, New York, 1982). (I)

C. Biomedical engineering

26. **Medical Physics and Physiological Measurement,** B. H. Brown and R. H. Smallwood (Blackwell, Oxford, 1981). (I)
27. **Introduction to Bioinstrumentation,** C. D. Ferris (Humana, Clifton, 1979). (I)
28. **Principles of Applied Biomedical Instrumentation, 2nd ed.,** L. A. Geddes and L. E. Baker (Wiley, New York, 1975). (I)
29. **Medicine and Clinical Engineering,** B. Jacobson and J. G. Webster (Prentice-Hall, Englewood Cliffs, 1977). This translation of a Swedish book introduces medical diagnosis and therapy to clinical engineers, medical physicists, physiotherapists, pharmacists, and administrators. The physics and engineering are elementary, but it is a good introduction to clinical terms and applications. (E)
30. **Non-invasive Physiological Measurements,** edited by P. Rolf (Academic, New York, 1983). (I)
31. **Electrical Safety in Health Care Facilities,** H. H. Roth and I. M. Kane (Academic, New York, 1975). (I)
32. **Clinical Engineering: Principles and Practices,** edited by J. G. Webster and A. M. Cook (Prentice Hall, Englewood Cliffs, 1979). As hospitals use more sophisticated equipment, clinical engineering is emerging as a separate discipline. The clinical engineer keeps the equipment calibrated, working, and safe. (I)
33. **Medical Instrumentation: Application and Design,** edited by J. G. Webster (Houghton Mifflin, Boston, 1978). (I)
34. **Biomedical Instruments, Theory and Design,** W. Welkowitz and S. Deutsch (Academic, New York, 1976). (I)

An extensive bibliography of biomedical engineering texts is found in

35. **"A biomedical engineer's library,"** J. G. Webster, J. Clin. Eng. 7, 67–72 (1982). (E)

III. JOURNALS

There are a number of journals devoted to biomedical engineering and medical physics.

Annals of Biomedical Engineering.

Annual Reviews of Biophysics and Biomedical Engineering. (Early volumes contained material on biomedical engineering. In recent years the publication has been devoted exclusively to molecular biophysics.)

Clinical Physics and Physiological Measurement.

IEEE Engineering in Medicine and Biology Magazine. (This magazine is analogous to *Physics Today* and contains tutorial articles at the elementary to intermediate level.)

IEEE Transactions on Biomedical Engineering. (I,A)

IEEE Transactions on Medical Imaging. (I,A)

Journal of Nuclear Medicine. (I,A)

Medical Physics. (I,A)

Physics in Medicine and Biology.

IV. RADIOLOGICAL PHYSICS

A. Basic matters

1. Photon interactions

Photon interactions and the relationship of dose to beam characteristics are still active areas of research. A few recent articles are cited here.

36. **"Energy imparted, energy transferred, and net energy transferred,"** F. H. Attix, Phys. Med. Biol. **28**, 1385–1390 (1983). (I)
37. **"Calculation of scattering cross sections for increased accuracy in diagnostic radiology. I. Energy broadening of Compton-scattered photons,"** G. A. Carlsson and C. A. Carlsson, Med. Phys. **9**, 868–879 (1982). (A)
38. **"Coherent scatter in diagnostic radiology,"** P. C. Johns and M. J. Yaffe, Med. Phys. **10**, 40–50 (1983). (I)

2. Image science

A great deal is known about the physics and psychophysics of images. While a good deal of this was developed for photography, it is also applied in radiology. There is also much information about the effect of photon statistics on image quality and the detectability of lesions.

39. **"Medical imaging systems,"** R. W. Redington and W. H. Berninger, Phys. Today **34**, 36–41 (1981). (E)
40. **"Medical imaging, vision and visual physics,"** in *Medical Radiography and Photography*, C. C. Jaffe (Eastman Kodak, Rochester, 1984), Vol. 60, No. 1. A monograph on some of the psychophysical aspects of detecting abnormal features in a radiologic image. (I)
41. **Radiological Imaging,** H. H. Barrett and W. Swindell (Academic, New York, 1981), 2 Vols. (I)
42. **"Transfer function analysis of radiographic imaging systems,"** C. E. Metz and K. Doi, Phys. Med. Biol. **24**, 1079–1106 (1979). (A)
43. **"A physical statistics theory for detectability of target signals in noisy images. I. Mathematical background, empirical review, and development of theory,"** P. R. Moran, Phys. Med. Biol. **9**, 401–413 (1982). (A)
44. **"Toward a unified view of radiological imaging systems. Part I: Noiseless images,"** R. F. Wagner, K. E. Weaver, E. W. Denny, and R. G. Bostrom, Med. Phys. **1**, 11–24 (1974). (I)
45. **"Toward a unified view of radiological imaging systems. Part II: Noisy images,"** R. F. Wagner, Med. Phys. **4**, 279–296 (1977). (I)
46. **"Overview of unified SNR analysis of medical imaging systems,"** R. F. Wagner and D. G. Brown, IEEE Trans. Med. Imag. **MI-1**, 210–213 (1982). (E)
47. **"Low contrast sensitivity of radiologic, CT, nuclear medicine, and ultrasound medical imaging systems,"** R. F. Wagner, IEEE Trans. Med. Imag. **MI-2**, 105–121 (1983). (I)

B. Dosimetry

48. **"Forty years of development in radiation dosimetry,"** J. W. Boag, Phys. Med. Biol. **29**, 127–130 (1984). (I)
49. **Fundamentals of Radiation Dosimetry,** edited by J. R. Greening. Hospital Physicists' Association Handbook No. 6. (Adam Hilger, Bristol, 1981). (I)
50. **"The theoretical and microdosimetric basis of thermoluminescence and application to dosimetry,"** Y. S. Horowitz, Phys. Med. Biol. **26**, 756–824 (1981). (A)
51. **Thermoluminescence Dosimetry,** A. F. McKinlay. Hospital Physicists' Association Handbook No. 5. (Adam Hilger, Bristol, 1981). (I)

C. Radiation biology

Radiobiology is concerned with the effects of radiation on viruses, cells, and organisms. Paradoxically, one area of research is concerned with understanding radiation carcinogenesis; another is concerned with improving the efficacy of radiation in cancer therapy.

52. **"Forty years of radiobiology: its impact on radiotherapy,"** J. J. Fowler, Phys. Med. Biol. **29**, 97–113 (1984). (I)
53. **"Radiation and the single cell: The physicist's contribution to radiobiology,"** E. J. Hall, Phys. Med. Biol. **21**, 347–359 (1976). (I)
54. **"Radiation carcinogenesis,"** H. I. Kohn and R. J. M. Fry, N. Engl. J. Med. **310**, 504–511 (1984). (E,I)
55. **"Radiation biology: The conceptual and practical impact on radiation therapy,"** H. D. Suit, Radiat. Res. **94**, 10–40 (1983). (I)
56. **Radiation Carcinogenesis: Epidemiology and Biological Significance,** (Progr. Cancer Therapy, Vol. 26), edited by J. D. Boice, Jr. and J. Fraumeni, Jr. (Raven, New York, 1984). (A)

D. Radiation protection

57. **"Radiation exposure in our daily lives,"** S. C. Bushong, Phys. Teacher **15**, 135–144 (1977). (E)
58. **The Biological Risks of Medical Irradiation,** edited by G. D. Fullerton, R. Waggener, D. T. Kopp *et al.*, AAPM Monogr. no. 5. (AIP, New York, 1980). (I)
59. **Health Effects of Low-level Radiation,** edited by W. R. Hendee (Appleton-Century-Crofts, East Norwalk, 1984). (I)
60. **"Forty years of development in radiation protection,"** F. W. Spiers, Phys. Med. Biol. **29**, 145–152 (1984). (I)

An interesting new development in radiation protection is the recognition that naturally occurring radon-222 and its decay products are considerably more concentrated in indoor air than in outdoor air. This effect is exacerbated for smokers, and droplets of radioactive tobacco tar lodge at the bifurcations of the bronchi.

61. **"Alpha radiation dose at bronchial bifurcations of smokers from indoor exposure to radon progeny,"** E. A. Martell, Proc. Natl. Acad. Sci. **80**, 1285–1289 (1983). (I)

E. Diagnostic radiology (roentgenology)

1. General

One can best learn about the clinical aspects of radiology from one of the textbooks listed in Sec. II. The articles listed here provide some of the history of the field.

62. **"Forty years of development in diagnostic imaging,"** M. J. Day, Phys. Med. Biol. **29**, 121–126 (1984). (I)
63. **"History of medical physics,"** J. S. Laughlin, Phys. Today **36**, 26–33 (1983). (E)
64. **"The early history of radiological physics: A fourth state of matter,"** I. A. Lerch, Med. Phys. **6**, 255–266 (1979). (I)

A description of various kinds of radiologic imaging can be found in

65. **"The physics of medical imaging,"** P. Moran, R. J. Nickles, and J. A. Zagzebski, Phys. Today **36**, 36–42 (1983). (E,I)

2. Computed tomography

Tomography is derived from the Greek *tomos*, meaning slice. Tomography has long been a standard radiographic technique, in which the film and x-ray tube are rotated about a point or line passing through an organ of interest, thereby blurring structures which are not close to the pivot. In computed tomography, two-dimensional slices are reconstructed from a series of projections. The technique was developed simultaneously in radioastronomy, crystallography, radiology, and nuclear medicine. Two physicists shared the Nobel Prize in medicine for this development. The first two references are their Nobel lectures.

66. **"Early two-dimensional reconstruction and recent topics stemming from it,"** A. M. Cormack, Med. Phys. **7**, 277–282 (1980). (I)
67. **"Computed medical imaging,"** G. N. Hounsfield, Med. Phys. **7**, 283–290 (1980). (I)
68. **"Overview of computerized tomography with emphasis on future developments,"** R. H. T. Bates, K. L. Garden, and T. M. Peters, Proc. IEEE **71**, 356–372 (1983). (I)
69. **Image Reconstruction from Projections: The Fundamentals of Computerized Tomography**, G. T. Herman (Academic, New York, 1980). (I,A)
70. **"Design constraints in computed tomography: A theoretical view,"** D. L. Parker, J. L. Crouch, K. P. Peschmann *et al.*, Med. Phys. **9**, 531–539 (1982). (A)

It takes several seconds to record the projections for x-ray transmission tomography. The Dynamic Spatial Reconstructor has been developed at the Mayo Clinic to study the beating heart. Multiple slices are taken simultaneously in 1/60 s. A recent paper describing the machine is

71. **"High-speed three-dimensional x-ray computed tomography: The dynamic spatial reconstructor,"** R. A. Robb, E. A. Hoffman, L. J. Sinak *et al.*, Proc. IEEE **71**, 308–319 (1983). (I)

An experimental technique which may become more important in the future involves detecting the Compton-scattered photons rather than the unattenuated beam.

72. **"Compton scatter axial tomography with x-rays: SCAT–CAT,"** L. Brateman, A. M. Jacobs, and L. T. Fitzgerald, Phys. Med. Biol. **29**, 1353–1370 (1984). (I)

3. Digital radiology

Digital radiography involves making two digital images and subtracting one from the other. In temporal subtraction, pictures of blood vessels with and without contrast media are used; in energy subtraction, images from x-ray beams of different average energy are subtracted. The former technique shows the blood vessels; the latter allows one to emphasize or obliterate the image due to a particular kind of tissue.

a. Temporal subtraction

73. **Digital Radiography**, W. Brody (Raven, New York, 1984). (E,I)
74. **"Computerized fluoroscopy in real time for non-invasive visualization of the cardiovascular system,"** R. A. Kruger, C. A. Mistretta, T. L. Houk *et al.*, Radiology **130**, 49–57 (1979). (I)
75. **Digital Radiography: A Focus on Clinical Utility**, edited by R. R. Price, F. D. Rollo, W. G. Monahan, and E. A. James (Grune and Stratton, New York, 1982). (I,A)
76. **"Intravenous digital subtraction: A summary of recent developments,"** S. A. Riederer and R. A. Kruger, Radiology **147**, 633–638 (1983). (I)
77. **"The technical characteristics of matched filtering in digital subtraction angiography,"** S. J. Riederer, A. L. Hall, J. K. Maier *et al.*, Med. Phys. **10**, 209–217 (1983). (A) This paper discusses a new technique which combines several images as a bolus of contrast material passes through the field of view. The digital filter is matched to the bolus concentration.

b. Energy subtraction

78. **"Spectral considerations for absorption edge fluoroscopy,"** F. Kelcz, C. A. Mistretta, and S. J. Riederer, Med. Phys. **4**, 26–35 (1977). (I)
79. **"Digital K-edge subtration radiography,"** R. A. Kruger, C. A. Mistretta, A. B. Crummy *et al.*, Radiology **125**, 243–245 (1977). (I)

F. Radiation therapy (radiation oncology)

Radiation therapy is the greatest area of employment of medical physicists, since many patient treatments require planning by a physicist and may require the custom fabrication of absorbers. The most common beams are high energy photons or electrons; neutrons, protons and pions are also used at some centers. A recent development is the heating of tissue, hyperthermia, as an adjuvant to radiotherapy. Brachytherapy uses radioactive sources that are implanted in tissue or placed in a body cavity to deliver ionizing radiation to the tumor.

1. General

80. **"Developing aspects of radiotherapy,"** J. F. Fowler, Med. Phys. **8**, 427–434 (1981). (I)
81. **"Radiation treatment planning,"** J. S. Laughlin, F. Chu, L. Simpson, and R. C. Watson, Cancer **39**, 719–728 (1977). (I)
82. **"Forty years of development in radiotherapy,"** W. J. Meredith, Phys. Med. Biol. **29**, 115–120 (1984). (I)
83. **Radiotherapy Treatment Planning**, R. F. Mould. Hospital Physicists' Association Handbook No. 7. (Adam Hilger, Bristol, 1981). (I)
84. **"A review of time-dose effects in radiation therapy,"** R. E. Peschel and J. J. Fischer, Med. Phys. **7**, 601–608 (1980). (I)
85. **Advances in Radiation Therapy Treatment Planning**, edited by A. E. Wright and A. L. Boyer, AAPM Monogr. no. 9. (AIP, New York, 1983). (I,A)

2. X rays and electrons

86. **"Physical aspects of supervoltage x-ray therapy,"** F. Bagne, Med. Phys. **1**, 266–274 (1974). (A)
87. **"Electron linear accelerators for radiation therapy: History, principles and contemporary developments,"** C. J. Karzmark and N. C. Pering, Phys. Med. Biol. **18**, 321–354 (1973). (I)
88. **Practical Aspects of Electron Beam Treatment Planning**, C. Orton and F. Bagne, AAPM Monogr. no. 2. (AIP, New York, 1978). (I)

3. Brachytherapy

89. **Recent Advances in Brachytherapy Physics**, edited by D. R. Shearer. AAPM Monogr. no. 7. (AIP, New York, 1981). (I)

4. Neutrons

90. **"Fast neutron radiotherapy: for equal or better?"** J. J. Broerse and J. J. Batterman, Med. Phys. **8**, 751–760 (1981). (I)
91. **"Fast neutron radiotherapy,"** P. H. McGinley, Phys. Teacher **11**, 73–78 (1973). (E)

5. Protons and pions

92. **"Planning proton therapy of the eye,"** M. Goitien and T. Miller, Med. Phys. **10**, 275–283 (1983). (I)
93. **"The physics of cancer therapy with negative pions,"** C. Richman, Med. Phys. **8**, 273–291 (1981). (I)
94. **Pion and Heavy Ion Radiotherapy: Pre-clinical and Clinical Studies**, edited by L. D. Skarsgard (Elsevier Biomedical, New York, 1983). (I,A)

6. Hyperthermia

95. **"Hyperthermia for the engineer: A short biological primer,"** G. M. Hahn, IEEE Trans. Biomed. Eng. **BME-31**, 3–8 (1984). (E)

96. **Physical Aspects of Hyperthermia**, edited by G. Nussbaum. AAPM Monogr. no. 8. (AIP, New York, 1983). (I)

G. Nuclear medicine

1. General

Diagnostic nuclear medicine techniques involve measuring the distribution of radioactive substances in various organs. The spatial resolution is not as good as in radiology, but one obtains information about *function*—the uptake and disappearance of the isotope from the organ. General texts are a good place to start. The following articles give some of the history.

97. **"Early history (1936–1946) of nuclear medicine in thyroid studies at Massachusetts General Hospital,"** R. D. Evans, Med. Phys. **2**, 105–109 (1975). (E)
98. **"Forty years of development in radioisotope non-imaging techniques,"** H. Miller, Phys. Med. Biol. **29**, 157–162 (1984). (I)
99. **"Nuclear Medicine: How it began,"** W. G. Myers and H. N. Wagner, Jr., Hosp. Prac. **9**(3), 103–113 (1974). (E)
100. **"Forty years of development in radioisotope imaging,"** N. Veall, Phys. Med. Biol. **29**, 163–164 (1984). (I)
101. **"Radioimunoassay: A probe for the fine structure of biological systems,"** R. S. Yalow, Med. Phys. **5**, 247–257 (1978). (A)

The following reference describes functional imaging, especially in cardiology.

102. **"Nuclear Medicine,"** W. J. Macintyre. Four chapters in *Proceedings of the International School of Physics, "Enrico Fermi," Course LXXVI*, edited by J. R. Greening, (North-Holland, Amsterdam, 1981). (I,A)

2. Emission tomography

X-ray transmission tomography reconstructs the two-dimensional function $\mu(x,y)$ from a series of projections of $\int\mu(s)ds$. In emission tomography, the concentration of the isotope $C(x,y)$ is reconstructed form a series of projections $\int C(s)ds$. In single photon emission computed tomography (SPECT) a gamma emitter is used. In positron emission tomography (PET) a positron emitter is used, and the two annihilation photons are detected in coincidence. Because the positron emitters have very short half-lives, this requires an accelerator at the hospital. An advantage is that natural chemicals incorporating positron emitters can often be made. A recent book covering both aspects is

103. **Computed Emission Tomography**, edited by P. J. Ell and B. L. Holman (Oxford Univ., Oxford, 1982). (I,A)

a. Single photon emission tomography

104. **"Single-photon emission computed tomography,"** G. F. Knoll, Proc. IEEE **71**, 320–329 (1983). (I)
105. **"An overview of a camera-based SPECT system,"** K. M. Greer, R. L. Jaszczak, and R. E. Coleman, Med. Phys. **9**, 455–463 (1982).
106. **"A comparison of three systems for performing single-photon emission tomography,"** M. A. Flower, R. W. Rowe, S. Webb, and W. I. Keyes, Phys. Med. Biol. **26**, 671–692 (1981). (A)

b. Positron emission tomography

107. **"Positron tomography and nuclear magnetic resonance imaging,"** G. L. Brownell, T. F. Budinger, P. C. Lauterbur *et al.*, Science **215**, 619–626 (1982). (I)
108. **"Brain function and blood flow,"** N. A. Lassen, D. H. Ingvar, and E. Skinhoj, Sci. Am. **239**(4), 62–71 (1978). (E)
109. **"Positron-emission tomography,"** M. M. Ter-Pogossian, M. E. Raichle, and B. E. Sobel, Sci. Am. **243**(4), 170–181 (1980). (I)
110. **"Positron tomography and three-dimensional reconstruction technique,"** D. A. Chessler, in *Tomographic Imaging in Nuclear Medicine*, edited by G. S. Freedman (Soc. Nucl. Med., New York, 1973), pp. 176–183. (E)
111. **"Patterns of human local cerebral glucose metabolism during epileptic seizures,"** J. Engel, D. E. Kuhl, and M. E. Phelps, Science **218**, 64–66 (1982). (I)

H. Ultrasound

Ultrasound has established a firm place in medical diagnosis. Conventional ultrasound uses reflections from impedance discontinuities between structures in the body. Doppler ultrasound can detect moving structures such as the beating fetal heart or measure the velocity of red cells in flowing blood. There are several good general texts and articles.

112. **"Medical ultrasonics,"** P. N. T. Wells, IEEE Spectrum **21**(12), 44–51 (1984). (E)
113. **Doppler Ultrasound and Its Use in Clinical Measurement,** P. Atkinson and J. P. Woodcock, (Academic, New York, 1982). (I)
114. **"Ultrasound in medical diagnosis,"** G. B. Devy and P. N. T. Wells, Sci. Am. **238**(5), 98–112 (1978). (E)
115. **"State-of-the-art of single-transducer ultrasonic imaging technology,"** R. C. Eggleton, Med. Phys. **3**, 303–311 (1976). (I)
116. **Medical Physics of CT and Ultrasound: Tissue Imaging and Characterization,** edited by G. D. Fullerton and J. A. Zagzebski. AAPM Monogr. no. 6. (AIP, New York, 1980). (I)
117. IEEE Trans. Biomed. Eng. Special Issue on Medical Ultrasound. **BME-30** (8), August, 1983. (I)
118. **"State-of-the-art in two-dimensional ultrasonic transducer array technology,"** M. G. Maginess, J. D. Plummer, W. L. Beaver, and J. D. Meindl, Med. Phys. **3**, 312–318 (1976). (I)

Computed tomographic reconstructions are also done with ultrasound, although diffraction and scattering effects make the reconstruction considerably more complicated.

119. **"Computerized tomography with ultrasound,"** J. F. Greenleaf, Proc. IEEE **71**, 330–337 (1983). (I)
120. **"Ultrasonic reflectivity imaging in three dimensions: Exact inverse scattering solutions for plane, cylindrical and spherical apertures,"** S. J. Norton and M. Linzer, IEEE Trans. Biomed. Eng. **BME-28**, 202–220 (1981). (A)

I. Magnetic resonance

Nuclear magnetic resonance is currently a very active field. It is being called magnetic resonance to avoid the radioactive connotations of "nuclear."

1. Imaging

Magnetic resonance imaging provides cross sections similar to those of x-ray transmission tomography. The quantity imaged is the proton density of the tissue or the spin–spin or spin–lattice relaxation time. Because there are many ways of pulsing the magnetic moments in the tissue nuclei, the interpretation of magnetic resonance images is considerably more complicated than for other modalities.

121. **"Diagnostic NMR,"** R. W. Redington, IEEE Trans. Med. Imag. **MI-1**, 230–233 (1982). (E)
122. **"Positron tomography and nuclear magnetic resonance imaging,"** G. L. Brownell, T. F. Budinger, P. C. Lauterbur *et al.*, Science **215**, 619–626 (1982). (I)

123. **"Nuclear magnetic resonance technology for medical studies,"** T. F. Budinger and P. C. Lauterbur, Science **226**, 288–298 (1984). (A)
124. **"The physics of proton NMR,"** R. L. Dixon and K. E. Ekstrand, Med. Phys. **9**, 807–818 (1982). (I)
125. **Technology of Nuclear Magnetic Resonance**, P. D. Esser and R. E. Johnston (Soc. Nucl. Med., New York, 1984). (I)
126. **"An introduction to NMR imaging: from the Bloch equations to the imaging equation,"** W. S. Hinshaw and A. H. Lent, Proc. IEEE **71**, 338–350 (1983). (A)
127. **"Realistic expectations for the near term development of clinical NMR imaging,"** L. Kaufman and L. E. Crooks, IEEE Trans. Med. Imag. **MI-2**, 57–65 (1983). (I)
128. **"A unified description of NMR imaging, data-collection strategies, and reconstruction,"** K. F. King and Paul R. Moran, Med. Phys. **11**, 1–14 (1984). (I)
129. **NMR Imaging in Biomedicine**, P. Mansfield and P. G. Morris (Academic, New York, 1982). (I)
130. **"NMR imaging in medicine,"** I. L. Pykett, Sci. Am. **246**(15), 78–88 (1982). (I)
131. **"Principles of nuclear magnetic resonance imaging,"** I. L. Pykett, J. H. Newhouse, F. S. Buonanno *et al.*, Radiology **143**, 157–168 (1982). (I)

There are unanswered questions about the effect of strong magnetic fields and changing magnetic fields on the body, as well as the radiofrequency power involved in magnetic resonance imaging.

132. **"Power deposition in whole-body NMR imaging,"** P. A. Bottomley and W. A. Edelstein. Med. Phys. **8**, 510–512 (1981). (I)
133. **"Nuclear magnetic resonance (NMR) in-vivo studies: Known thresholds for health effects,"** T. F. Budinger, J. Comput. Assist. Tomog. **5**, 800–811 (1981). (I)

2. Blood flow

134. **"The NMR blood flowmeter—theory and history,"** J. H. Battocletti, R. E. Halbach, S. X. Salles-Cunh, and A. Sances, Jr., Med. Phys. **8**, 435–443 (1981). "The NMR blood flowmeter—design," Med. Phys. **8**, 444–451 (1981). "The NMR blood flowmeter—applications," Med. Phys. **8**, 452–458 (1981). (I,A)
135. **"Nuclear magnetic resonance blood flow measurements in the human brain,"** J. R. Singer and L. E. Crooks, Science **221**, 654–656 (1983). (I)

V. OTHER MEDICAL PHYSICS AND BIOMEDICAL ENGINEERING AREAS

A. Electrical signals from the body

There is an electrical potential difference of about 60–90 mV across the membrane of nearly every cell. As a nerve cell conducts or a muscle cell prepares to contract, it is swept by a wave of depolarization in which the membrane potential reverses sign. The resulting potential differences on the body surface give rise to the electrocardiogram, the electroencephalogram, the electromyogram, and the electroretinogram.

The electrocardiogram detects surface potential differences due to depolarization of heart muscle cells during the cardiac cycle. Thousands of papers have been written about the electrocardiogram, including quite sophisticated approaches to solving the "inverse problem," that is, determining the volume current generator distribution from the potential distribution on the body surface. The usual clinical interpretation is based on either a fairly simple model or empirical correlations with disease. An elementary discussion, which should be read in conjunction with Ref. 7, can be found in

136. **"Improved explanation of the electrocardiogram,"** R. K. Hobbie, Am. J. Phys. **52**, 704–705 (1984). (I)

A more formal discussion, which includes material about the inverse problem, is in

137. **Bioelectric Phenomena**, R. Plonsey (McGraw-Hill, New York, 1969). (A)

Current research can be found in many journals, including IEEE Trans. Biomed. Eng.

The electroencephalogram measures the much smaller potentials due to the collective effect of all the nerve cells in the brain. The interpretation of this signal is even more empirical than the electrocardiogram.

The electromyogram measures electrical activity in skeletal muscle during contraction, or motor and sensory nerve conduction. The electroretinogram measures the potential between cornea and eyelid evoked by repeated flashes of light. Signals picked up on the scalp over the visual cortex are called visual evoked responses. An auditory evoked response is picked up by scalp electrodes in response to a repetitive noise stimulus. These latter two signals are so weak that signal averaging techniques must be used.

138. **Clinical Applications of Evoked Potentials in Neurology. (Advances in Neurology, Vol. 32)**, J. Courjon, F. Maugiere, and M. Revol (Raven, New York, 1982). These articles are very clinical, but they show how the techniques are used.
139. **Electrodiagnosis of Neuromuscular Disease, 3rd ed.**, J. Goodgold and A. Eberstein (Williams & Wilkins, Baltimore, 1983). (I, but quite clinical)
140. **"The clinical use of auditory evoked potentials,"** A. R. D. Thornton, in *Proceedings of the International School of Physics "Enrico Fermi," Course LXXVI*, edited by J. R. Greening (North-Holland, Amsterdam, 1981), pp. 384–396. (I)

B. Magnetic signals from the body

It is now possible using SQUID magnetometers to detect the magnetic fields arising from the body. These can be from either electric currents as in magnetocardiography or permanent magnetic moments. Permanent magnetism comes from inhaled or ingested iron particles. The magnetic susceptibility of the liver gives a measure of the iron stores in the body.

141. **"Magnetic fields of the human body,"** D. Cohen, Phys. Today **28**, 33–43 (1975). (E)
142. **"Magnetic susceptibility measurement of human iron stores,"** G. M. Brittenham *et al.*, N. Eng. J. Med. **307**, 1671–1675 (1982). (I) This is a "clinical" paper about using a SQUID to measure iron stores in the liver.
143. **"Ferrimagnetic particles in the lung. Part 1: The magnetizing process; Part II: The relaxation process,"** D. Cohen, I. Nemoto *et al.*, IEEE Trans. Biomed. Eng. **BME-31**, 261–285 (1984). (I)
144. **"Magnetocardiography: an overview,"** D. B. Geselowitz, IEEE Trans. Biomed. Eng. **BME-26**, 497–504 (1979). (I)
145. **"Biomagnetic Instrumentation,"** G. L. Romani, S. J. Williamson, and L. Kaufman, Rev. Sci. Instrum. **53**, 1815–1845 (1982). (I)
146. **"Proceedings of the Fourth International Workshop on Biomagnetism,"** edited by G. L. Romani and S. J. Williamson, Il Nuovo Cimento **2D**(2) March–April, 1983. (I,A)
147. **"Noninvasive magnetic detection of cardiac mechanical activity: (1) Theory, (2) Experiments,"** J. P. Wikswo, Med. Phys. 7, 297–314 (1980). (I)
148. **Biomagnetism, An Interdisciplinary Approach**, S. J. Williamson, G. L. Romani, L. Kaufman *et al.*, (Plenum, New York, 1983). (A)

C. Pacing

The cardiac pacemaker is one of the triumphs of biomedical engineering. It has provided life extension of good quality. More than 500 000 patients in the United States have permanently implanted pacemakers.

149. **"An engineering overview of cardiac pacing,"** P. P. Tarjan and A. D. Berstein, IEEE Eng. Med. Bio. Mag. **3**(2), 10–14 (1984). A number of other relevant articles appear in the same issue. (E)
150 "Control of tachyarrhythmias by electrical stimulation—Techniques and mechanisms," R. Mehra, IEEE Eng. Med. Biol. Mag. **3**(2), 29–24 (1984). (E)
151. **"Physiological stimulators: From electric fish to programmable implants,"** L. J. Seligman, IEEE Trans Biomed. Eng. **BME-29**, 270–284 (1982). (I)

A very good review article from the physician's point of view is found in

152. **"Cardiac pacing in the 1980s,"** P. L. Ludmer and N. Goldschlager, N. Engl. J. Med. **311**, 1671–1680 (1984). (I)

D. Other effects of electric and magnetic fields

Electrical stimulation is also used to enhance tissue regeneration and to control pain.

153. **"The effects of pulsed magnetic fields of the type used in the stimulation of bone fracture healing,"** A. T. Barker and M. J. Lunt, Clin. Phys. Physiol. Meas. **4**, 1–28 (1983). (I)
154. **"A review of electromagnetically enhanced soft tissue healing,"** C. B. Frank and A. Y. J. Szeto, IEEE Eng. Med. Bio. Mag. **2**(4), 27–32 (1983). (E)
155. **"Skeletal tissue electromechanics and electrical stimulation of growth and remodelling,"** A. J. Grodzinsky and L. A. Hey, IEEE Eng. Med. Bio. Mag. **2**(4) 18–22 (1983). (E)
156. **"Transcutaneous electrical nerve stimulation for pain control,"** A. Y. J. Szeto and J. K. Nyquist, IEEE Eng. Med. Bio. Mag. **2**(4), 14–18 (1984). (E)

The question of the effect of electromagnetic fields is an active area of research. A recent review of the effect of power line frequencies is found in

157. **"Power-line fields and human health,"** M. G. Morgan, H. K. Florig, I. Nair, and D. Lincoln, IEEE Spectrum **22**(2), 62–68 (1985). (I)

E. Prostheses

A wide variety of prosthetic devices are available, ranging from artificial limbs with varying degrees of sophistication to artificial sensory organs. Some limb prostheses are now being controlled by electromyographic signals picked up from the patient.

158. **"New developments in mobility and orientation aids for the blind,"** J. A. Brabyn, IEEE Trans. Biomed. Eng. **BME-29**, 285–289 (1982). (I)
159. **"Development of the Utah artificial arm,"** S. C. Jacobsen, D. F. Knutti, R. T. Johnson, and H. H. Sears, IEEE Trans. Biomed. Eng. **BME-29**, 249–269 (1982). (I)
160. **"Review of current status of cochlear prostheses,"** R. L. White, IEEE Trans. Biomed. Eng. **BME-29**, 233–238 (1982). (I)

F. Lasers and Optics

161. **"Lasers in surgery and medicine,"** A. L. McKenzie and J. A. S. Carruth, Phys. Med. Biol. **29**, 619–642 (1984). (I) This is a two-part review by a physicist and a surgeon.
162. **"Physics and ophthalmology,"** R. A. Weale, Phys. Med. Biol. **24**, 489–504 (1979). (I)

G. Phototherapy

The chief use of ultraviolet radiation has been for the treatment of psoriasis. The other widespread use of phototherapy is irradiation with blue light for the treatment of neonatal jaundice. Some experimental work is being done on phototherapy of cancer, after the tumor has been sensitized by a dye given systemically. Some of the references discuss the harmful effects of ultraviolet radiation.

163. **Ultraviolet ratiation in medicine**, B. L. Diffey. Medical physics handbook no. 11. (Adam Hilger, Bristol, 1982). (I)
164. **"Ultraviolet radiation physics and the skin,"** B. L. Diffey, Phys. Med. Biol. **25**, 405–426 (1980). (I)
165. **Biological Effects of Ultraviolet Radiation**, W. Harm (Cambridge Univ., Cambridge, 1980). (I)
166. **"Blue light and bilirubin excretion,"** A. F. McDonagh, L. A. Palma, and D. A. Lightner, Science **208**, 145–151 (1980). This article reviews phototherapy for neonatal jaundice. (I)
167. **"Dosimetry considerations in phototherapy,"** A. E. Profio and D. R. Doiron, Med. Phys. **8**, 190–196 (1981). (A)
168. **"Dye-sensitized photodynamic inactivation of cells,"** J. P. Pooler and D. P. Valenzeno, Med. Phys. **8**, 614–628 (1981). (I)

H. Dialysis

Approximately 45 000 patients in the United States are undergoing chronic dialysis for renal disease. For a discussion of some of the design problems, see

169. **"Artificial kidneys: Problems and approaches,"** C. F. Gutch, Ann. Rev. Biophys. Bioeng. **4**, 405–429 (1975). (I)

Dialysis is not without its problems; to understand the effect of dialysis on a patient, read

170. **"Iatrogenic problems in end-stage renal failure,"** H. Calland, N. Engl. J. Med. **287**, 334–336 (1972). (E)

I. Elemental analysis

171. **"An x-ray fluorescence technique to measure the mercury burden of dentists,"** P. Bloch and I. M. Shapiro, Med. Phys. **8**, 308–311 (1981). (I)
172. **X-ray Fluorescence (XRF and PIXIE) in Medicine**, edited by R. Cesareo (Acta Medica, Rome, 1982). (I)
173. **"Techniques of in vivo neutron activation analysis,"** D. R. Chettle and J. H. Fremlin, Phys. Med. Biol. **29**, 1011–1043 (1984). (I)
174. **"Techniques for trace element analysis: x-ray fluorescence, x-ray excitation with protons, and flame atomic absorption,"** R. M. Wheeler, R. B. Liebert, T. Zabel *et al.*, Med. Phys. **1**, 68–71 (1974). (I)

J. Mass spectrometry

The most common medical use of the mass spectrometer has been in conjunction with the gas chromatograph for toxicology. It has also been used for respiratory gas analysis.

175. **"Routine use of a flexible gas chromatograph mass spectrometer computer system to identify drugs and their metabolites in body fluids of overdose victims,"** C. E. Costello, H. S. Hertz, T. Sakai, and K. Bieman, Clin. Chem. **20**, 255–265 (1974). (I)
176. **"Mass spectrometer evaluation of ventillation-perfusion abnormalities in respiratory distress syndrome,"** C. E. Hunt, S. Matalon, O. D. Wangensteen, and A. S. Leonard, Ped. Res. **8**, 621–627 (1974). (I)
177. **The Medical and Biological Application of Mass Spectrometry**, edited by J. P. Payne, J. A. Bushman, and D. W. Hill (Academic, London, 1979). (I)

K. Miscellaneous

178. **"Electronics and the diabetic,"** A. M. Albisser, and W. J. Spencer, IEEE Trans. Biomed. Eng. **BME-29**, 239–248 (1982). (I)

179. **"Forty years of instrumentation for medicine,"** F. T. Farmer, Phys. Med. Biol. **29**, 139–144 (1984). (I)
180. **"Biophysical and engineering aspects of cryosurgery,"** R. D. Orpwood, Phys. Med. Biol. **26**, 555–576 (1981). (I)

VI. EFFECTIVENESS AND ECONOMICS

One must always ask whether a particular device, test, or treatment works and is less expensive than what it replaces. The effect on one patient of "high tech" treatments—dialysis and transplant for renal disease—is discussed in Ref. 170. Reference 181 below discusses the public health and social implications of a "high-tech" diagnostic technique: computerized medical imaging. References 182–184 describe the "ROC" (Receiver Operating Characteristic) technique for comparing the accuracy of different tests.

181. **"Tutorial on the health and social value of computerized medical imaging,"** H. V. Fineberg and H. E. Sherman, IEEE Trans. Biomed. Eng. **BME-28**, 50–56 (1981).
182. **"Basic principles of ROC analysis,"** C. E. Metz, Semi. Nucl. Med. **8**, 283–298 (1978) (I)
183. **"Assessment of diagnostic technologies,"** J. A. Swets, R. M. Pickett, S. F. Whitehead *et al.*, Science **205**, 753–759 (1979). (I)
184. **Evaluation of Diagnostic Systems: Methods from Signal Detection Theory**, J. A. Swets and R. M. Pickett (Academic, New York, 1982). (I)

IEEE TRANSACTIONS ON MEDICAL IMAGING, VOL. MI-2, NO. 3, SEPTEMBER 1983 105

Low Contrast Sensitivity of Radiologic, CT, Nuclear Medicine, and Ultrasound Medical Imaging Systems

ROBERT F. WAGNER

Abstract—The physical sensitivity of a medical imaging system is defined as the square of the output signal-to-noise ratio per unit of radiation to the patient, or the information/radiation ratio. This sensitivity is analyzed at two stages: the radiation detection stage, and the image display stage. The signal-to-noise ratio (SNR) of the detection stage is a physical measure of the statistical quality of the raw detected data in the light of the imaging task to be performed. As such it is independent of any software or image processing algorithms which belong properly to the display stage. The fundamental SNR approach is applied to a wide variety of medical imaging applications and measured SNR values for signal detection at a given radiation exposure level are compared to the optimal values allowed by nature. It is found that the engineering falls short of the natural limitations by an inefficiency of about a factor two for most of the individual radiologic system components, allowing for great savings in the exposure required for a given imaging performance when the entire system is optimized. The display of the detected information is evaluated from the point of view of observer efficiency, the fraction of the displayed information that a human observer actually extracts. It has been found that the human observer is able to extract more than 50 percent of the displayed information in simple images when they are presented such that the noise is easily visible; otherwise, the internal noise sources of the observer degrade observer efficiency for lesion detection to much lower values. Methods for optimizing both the detection and the display of medical image information are presented in terms of the information/radiation ratio together with brief descriptions of the measurement methodology required to assess the images.

Introduction

A DECADE ago many investigators of medical imaging systems insisted that the assessment of these systems was a highly subjective procedure not subject to rigorous analysis. We know today that this was due to limitations in their analytic approach rather than to any natural obstacles to scientific inquiry into the subject matter. In this overview we will present the rigorous contemporary approach to the physical assessment of medical imaging systems with emphasis on the quantification of the information yield per unit of radiation exposure or dose. We refer to this information/radiation ratio as the *physical* sensitivity of the imaging system. *Clinical* sensitivity and related measures for any diagnostic system have been treated by Swets *et al.* [1], [2], Metz [3], and Lusted [4].

Manuscript received November 30, 1982; revised June 2, 1983.
The author is with the Office of Science and Technology, National Center for Devices and Radiologic Health, Food and Drug Administration, Rockville, MD 20857.

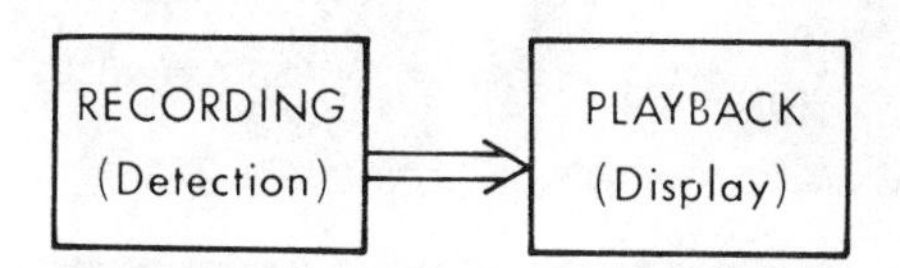

Fig. 1. Irreducible block diagram for imaging systems.

A rigorous solution to the problem of quantifying the physical performance of imaging systems was made possible by a conceptualization forced by the emergence of computer-assisted systems. It became clear that imaging is essentially a two stage process (Fig. 1) in which the first stage involves the detection or recording of the information in a radiation stream and the second stage involves the configuration of this information as an image, including any reconstruction or image processing algorithm, and actual display or playback for consumption by a human observer. The first or detection stage is assessed in terms of an ideal observer, task dependent signal-to-noise ratio (SNR), which is simply a measure of the statistical quality of the raw, i.e., unprocessed, detected data in the light of the task required of the observer. The second or display stage is evaluated with respect to the coupling of this information to the human observer, a phenomenon assessed in terms of observer or display/observer efficiency. This paper reviews these quantitative concepts in a qualitative way with applications to the major medical imaging systems in current use.

The SNR for X-Ray or Gamma-Ray Projection Images

We may deduce the form of the ideal observer SNR by studying Fig. 2, a simulation due to Rose [5] of the gradual improvement in photographic image information content obtained by increasing the number of photons utilized in making an image. Each step represents an increase by a factor between 5 and 10. We deduce that the SNR must go up as the number of photons, N, increases; in fact, as with most measurements of random phenomena, it goes as $N^{1/2}$. The other factor in the ideal SNR is object contrast C, e.g., in Fig. 2 it is easier to detect the higher contrast hair than the lower contrast nose. X-ray or radiographic images, being photon images, have the same form SNR as this photographic SNR, viz. $CN^{1/2}$. In Fig. 3 we give the X-ray attenuation coefficients for bone, muscle, and fat as a function of X-ray photon energy [6]; we see that the contrast between any two of these is a decreasing

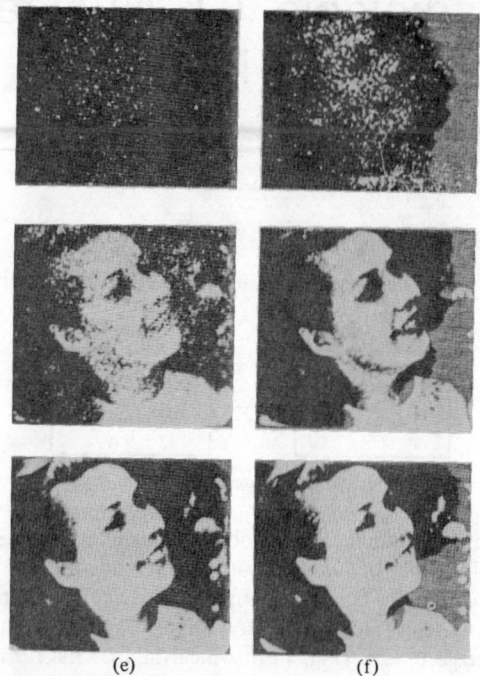

Fig. 2. Series of images showing the amount of information (SNR) conveyed by a total of N photons; N varies from 10^3 to 10^7 in steps of about a factor 5 from image to image (courtesy of A. Rose and Plenum Press.)

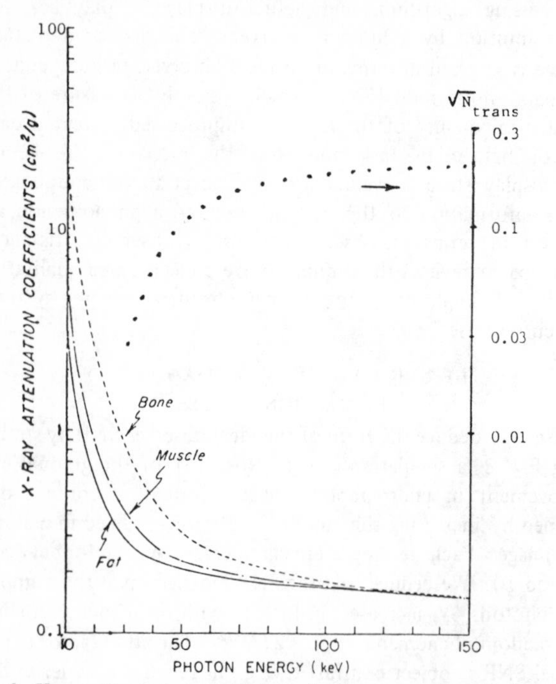

Fig. 3. X-ray attenuation coefficients for bone, muscle, and fat as a function of X-ray photon energy (Ter-Pogossian [6]); square root of number of photons transmitted through 20 cm water as a function of photon energy (dotted).

function of energy. However, the number of photons transmitted through a typical body thickness of 20 cm and detected by an ideal detector is an increasing function of energy. The square root of this number is plotted as a function of energy in Fig. 3. We might suspect that the combination of contrast

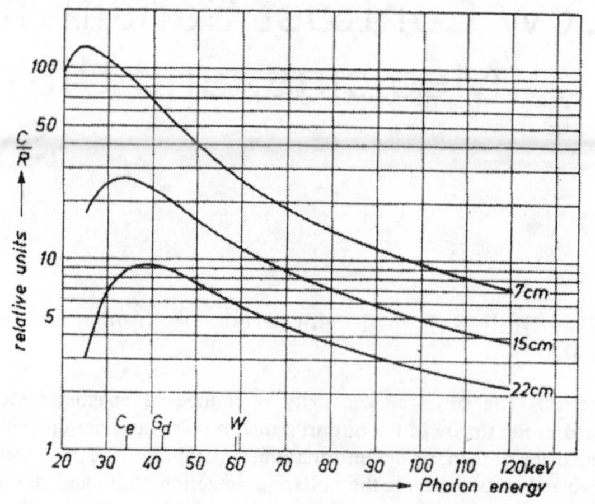

Fig. 4. Contrasts to fluctuations ratio C/R (equivalent to SNR used here) as a function of photon energy calculated by Oosterkamp [7] for detecting bone in soft tissue: equal integral dose to patient; curve parameter is patient thickness (courtesy of S. Balter and Philips Medical Systems, Inc.).

and $N^{1/2}$ in the SNR might produce an optimum imaging energy. Oosterkamp [7] first studied this question and his results for the task of detecting bone in soft tissue are shown in Fig. 4 in the form of SNR for fixed integral dose to the patient (total energy absorbed) versus X-ray energy, with body thickness as the curve parameter. For greater body thicknesses there is an optimum energy in the neighborhood of 40 keV, and this moves to lower energies in the 20 to 30 keV range, as body thickness diminishes. In Fig. 5 we move down in energy to the task of mammography and several detection tasks considered by Jennings *et al.* [8] for the average compressed breast with a thickness of about 5 cm. Here the results are presented in the form of the relative patient exposure required for a fixed SNR. The optimum energy has moved down to close to 20 keV, which is higher than the characteristic molybdenum X-ray lines at 17.5 and 19.5 keV commonly used in screen-film mammography. Jennings and colleagues showed that moving to the higher optimal energy through K-edge filtered tungsten anode beams allows for a patient exposure reduction of about a factor 2.5 without compromising imaging performance. (See, also, Muntz *et al.* [9], [10].) Motz and Danos [11] indicate that the sensitivities, or information–$(SNR)^2$/exposure–of most tasks outside of mammography can be simultaneously optimized by using beams in which photons lower than 40 keV have been removed. Such images displayed on conventional X-ray film would appear flat in contrast (see Fig. 3); but this suggestion becomes practical in a system where the display contrast is variable, such as a video or digital imaging system with an adjustable output contrast rather than the fixed contrast of film.

Conventional X-ray images integrate over a spectrum and therefore do not use all of the energy information available in the beam. The attenuation versus energy functions shown in Fig. 3 are comprised of three components: the photoelectric absorption, elastic or Rayleigh scattering, and inelastic or Compton scattering. However, Alvarez and Macovski [12] have shown that a vector space of only two dimensions is adequate

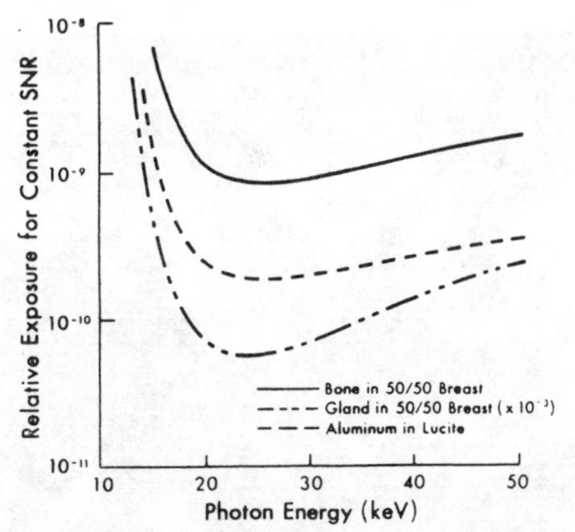

Fig. 5. Relative patient exposure for fixed SNR as a function of photon energy for several mammographic imaging tasks (Jennings *et al.* [8]).

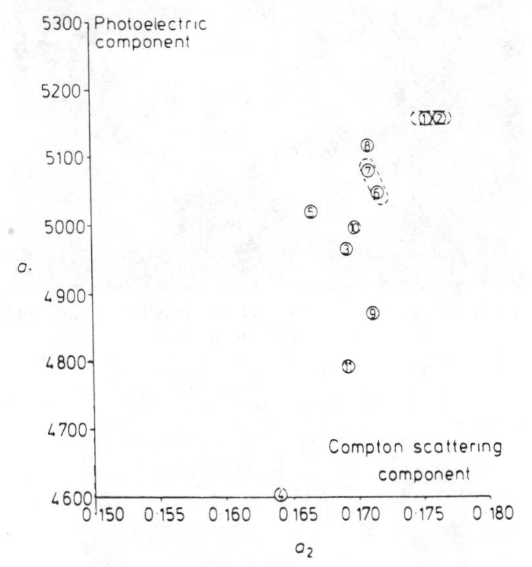

Fig. 6. Two-dimensional vector representation of human soft tissue determined from measurements made over the diagnostic range. Discrimination along a_1 axis ranges from clotted blood (top) through normal and abnormal brain tissue to water (bottom) (Alvarez and Macovski [12]).

to span these functions for biological materials of interest: one basis function is $1/E^3$, approximately the photoelectric function for these materials; the other is the Klein-Nishina cross section function for Compton scattering from the free electron. In this space many tissues that are degenerate or indistinguishable in a single conventional image can be spread out and distinguished from one another if images taken at two different energies are properly combined. Consider the two-dimensional representation of tissue vectors in Fig. 6. The projection of these vectors onto the horizontal axis is equivalent to high energy radiography in which only the Compton component is present; the projection onto various vectors or axes above the horizontal axis is equivalent to radiography at various conventional energies where contrast from differences in atomic number, as manifested in the photoelectric effect, is indistinguishable from contrast due to differences in density manifested in both absorption and scattering. By combining

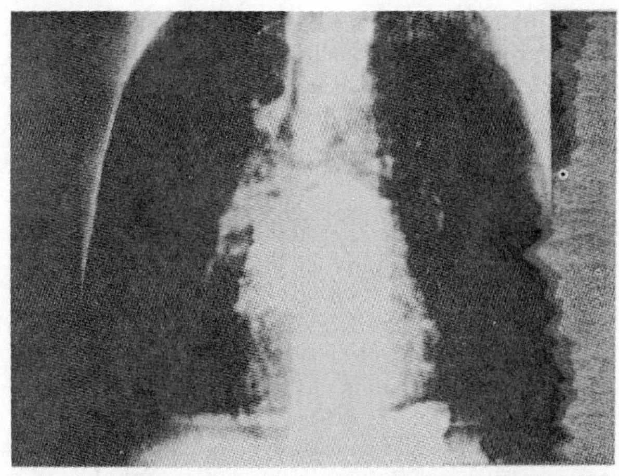

Fig. 7. Subtraction of images made at two energies and weighted such that the difference is insensitive to the presence of bone (courtesy of the Advanced Imaging Techniques Laboratory, Stanford University, and the General Electric Company [13]).

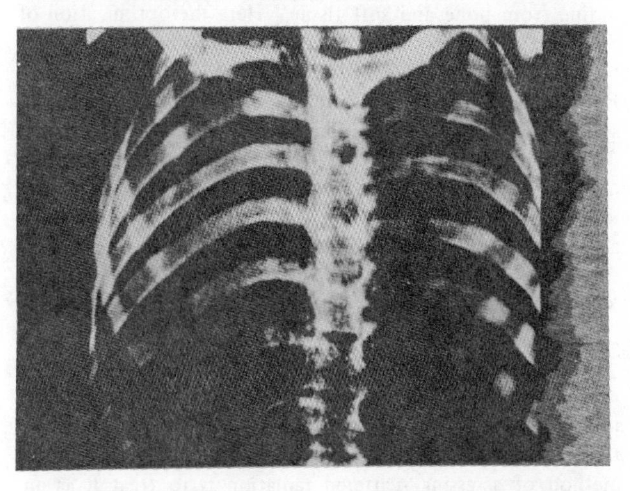

Fig. 8. Subtraction of images made at two energies and weighted such that the difference is insensitive to the presence of soft tissue (courtesy of the Advanced Imaging Techniques Laboratory, Stanford University, and the General Electric Company [13]).

two different projections by generalized subtraction techniques [13] new kinds of images become possible, such as shown in Figs. 7 and 8. The first is a subtraction of images made at two different energies and weighted such that the difference is insensitive to the presence of bone; the second is the subtraction of the images weighted such that the difference is insensitive to the presence of soft tissue. Vectors in the basis sets of the approaches mentioned here are not orthogonal. Weaver has recently completed a principal components analysis of the space of human tissue attenuation functions to find optimal orthogonal basis vector sets [14].

Now, just as above where we found an optimal energy for single energy projection radiography, investigators working in the two energy scheme have solved for the optimal energy partitioning of the available tube output or patient exposure to maximize the SNR for a given detection or discrimination task [12],[15],[16]. For example, Alvarez [12] has calculated the relative errors in the measured tissue vectors when using a 105 kVp X-ray spectrum and two-channel analysis of the detected beam. He found that the threshold setting between the two channels that minimizes the errors in the resulting tissue vectors lies within a broad neighborhood about 50 keV; that the errors in the photoelectric component and the Compton component can be simultaneously minimized; and that the photoelectric component has a substantially greater error due to the greater attenuation by the body of the low energy photons that carry this information. Another interesting result of such studies is that of Kelcz, Joseph, and Hilal [17] who indicate the superiority of using beams at two energies over using a single beam with tandem detectors where the second receives radiation not absorbed in the first. In this latter scheme, extra signal counts from one channel reduce contrast in the difference image while also serving as noise counts in the difference image because of the unavoidable overlap of the two spectra detected in this way.

Three energy techniques have been discussed for some time by Mistretta and colleagues [18]-[20] for the isolation of iodine from bone and soft tissue. Here the optimization of SNR per patient exposure or per X-ray tube output is driven by the demands on tube loading necessitated by the following: For the isolation of iodine and the cancellation of bone and soft tissue to become a well-conditioned problem, a three-dimensional energy space must be considered and three beams must be used, of which one must be lower than 33 keV, the iodine absorption edge; the latter beam is weakly penetrating and many X-ray tubes have been burnt out in the pursuit of this technique. It is limited by the current technology of X-ray tube heat loading and dissipation.

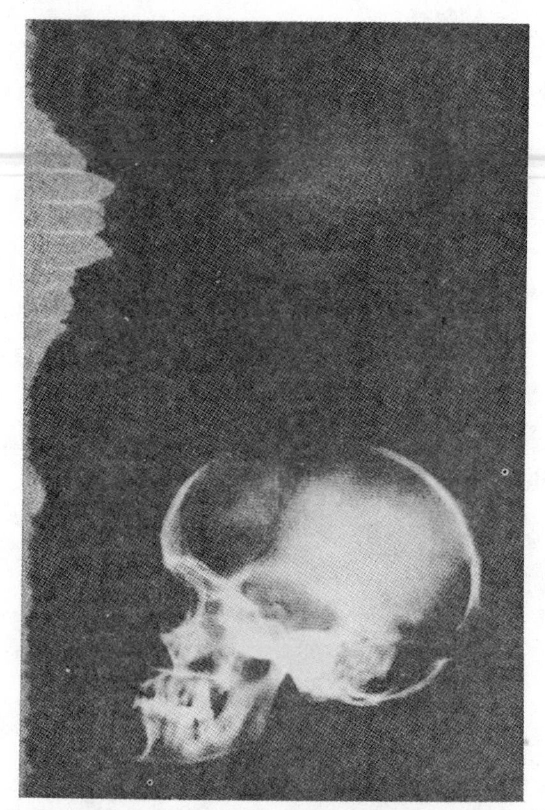

Fig. 9. Skull radiograph made without scatter reducing grid (top) and with a grid (bottom).

Scatter Rejection

In conventional radiology the amount of detected scattered radiation ranges from being comparable to the primary beam to being almost an order of magnitude greater. The traditional method of assessing scattered radiation is to treat it as an unwanted degrader of image contrast, e.g., as shown in Fig. 9(a) and (b). Although this is true it is not a fundamental point of view. Scattered radiation is more fundamentally an additional source of noise counts laid down upon the signal counts. A demonstration of this due to Wagner, Barnes, and Askins [21] is given in Figs. 10 and 11. Fig. 10, bottom, shows a phantom image made with a conventional scatter reducing grid technique. A superior scatter removal device consisting of scanning slits was used to make the image at the top of Fig. 10. The exposure required for the superior technique to achieve the same SNR as the conventional technique was calculated to be less by a factor of about 2. However, we see that the film in the "superior" case displays nothing or close to nothing. But when this film was enhanced by the autoradiographic intensification technique of Askins, the display contrast limitation of the film was overcome and the image at the top of Fig. 11 resulted: a 2.2-fold patient exposure reduction compared to the grid images at the bottom of the figures for the same (actually a bit more) SNR. The measurements and analysis of [21] indicate that a *perfect* scatter removal device could maintain a given SNR, while yielding about a three-fold patient exposure reduction compared to commercially available grids working in a high scatter situation (e.g., the pelvis); alternatively, it could be used to improve the SNR by the square root of this factor while maintaining a given level of patient exposure.

Fig. 10. Comparison of normally exposed grid image (bottom) with underexposed image made with superior scatter removal system (top) (Wagner, Barnes, and Askins [21]).

Fig. 11. Comparison of normally exposed grid image (bottom) with autoradiographically enhanced version of image made with superior scatter removal system (top) (Wagner, Barnes, and Askins [21]).

Detector Efficiency and Resolution Loss

The contemporary view of detector efficiency (inefficiency) and resolution (resolution loss) is that they are different aspects of the same phenomenon. The explanation of this requires us to introduce the concepts of detective quantum efficiency (DQE) and noise equivalent quanta (NEQ). We will begin by returning to Fig. 2(d) which was made with approximately 500 000 counts, or exposure quanta Q. Fig. 2(c) was made with approximately 100 000 exposure quanta. If the radiation stream used to make image 2(d) was incident upon an image detector with DQE = 0.20 or 20 percent, image 2(c) would result. We say that the number of noise equivalent quanta is 10^5, but the number of exposure quanta Q is 5×10^5. More fundamentally, DQE is a statistical or informational concept [22], [23]. It may be defined as the ratio $\mathrm{SNR}^2_{\mathrm{out}}/\mathrm{SNR}^2_{\mathrm{in}}$. For photon images the input SNR is $C(Q)^{1/2}$ and the output SNR is interpreted as $C(\mathrm{NEQ})^{1/2}$, where C is the input contrast. The relation $\mathrm{DQE} = \mathrm{NEQ}/Q$ appears as a ratio of mean numbers of quanta only because of the Poisson statistics of photons, i.e., the mean and variance are equal. DQE is formally identical to the concept of statistical efficiency, a measure of the fraction of the information contained in a sample that is extracted by an estimator; here, the estimator is the imaging procedure. In practice, NEQ and DQE are determined from output signal and noise measurements in the Fourier domain, or spatial frequency space, and they both are measured as a function of spatial frequency [22], [23]. The noise measurements in the form of output noise variance, in particular the variance in the form of its spatial frequency decomposition–the output noise power spectrum–is referred back through the transfer characteristic of the imaging system and interpreted in input units, which for photon imaging systems is the photon or quantum.

For a radiographic intensifying screen/film combination the output noise in density units is referred back to the exposure axis through the transfer characteristic consisting of the modulation transfer function (MTF), i.e., the blur function in Fourier or spatial frequency space, and large area characteristic curve (density versus log exposure) with slope γ, which are shown schematically in Fig. 12 from Sandrik and Wagner [24]. We emphasize that this is not a model; it is simply a scaling or referral of output noise to input quantities by the fundamental laws of propagation of error. The NEQ(f) spectrum obtained this way for three intensifying screen/film systems is given in Fig. 13. The falloff with frequency means that the number of photons available to image large areas is not available to image edges and detail. We see that typical intensifying screens yield an image equivalent to that produced by 4×10^4 quanta/mm^2 (corresponding to a density of 1.0 on the film). Whereas the exposure quanta required with Hi-Plus, Par, and Detail screens are in the ratio 1:2:8, the NEQ are in the ratio of 1:1:2, i.e., image performance does not necessarily track with system speed. When these results are normalized to required input quanta Q, the DQE(f) spectrum of Fig. 14 results. These curves show that Hi-Plus is more efficient in the low spatial frequencies, corresponding to large area effects, and Detail is more efficient in the high spatial frequencies, corresponding to edges and detail. These functions constitute a rigorous specification of detector efficiency and "resolution" loss with one continuous function. A characteristic bandwidth of this function, or in ordinary space a characteristic length, corresponds to an average region in space over which signal counts are spread upon background counts by the system blur, thus increasing the noise, and decreasing the SNR. A rigorous specification of this averaging is given in [24], with typical values for the characteristic length ranging around 0.25 mm.

In the last decade X-ray phosphors have been developed from rare-earth elements used for color TV CRT screens. These phosphors are intrinsically brighter than the conventional calcium tungstate X-ray phosphors, allowing them to be used with finer grained film. But they also have greater X-ray attenuation properties, leading to increased DQE. Since the absorption edge of these phosphors falls between 35 and 50 keV, their actual DQE gain is a sensitive function of X-ray beam quality and so the collection of data for NEQ and DQE curves is an elaborate task that is currently incomplete. However, several points on the DQE curves for the rare-earth systems can be deduced from the literature [25]-[27]. Rare-earth screens can achieve about 1.4 to 1.8 times the low frequency DQE of Hi-Plus screens while maintaining comparable MTF; they can achieve about twice the low frequency DQE of Par screens, and at least 2.5 times that of detail screens, maintaining comparable MTF. In principle, they can excel by more than these values in the higher spatial frequencies since they can be paired with finer grained films. Work is in progress to refine these statements.

Sandrik and Wagner [24] showed how to use the NEQ and DQE concepts to carry out an optimization of magnification radiography of fine detail in an object. Results of output SNR^2 per effective patient exposure are presented in Fig. 15. At a

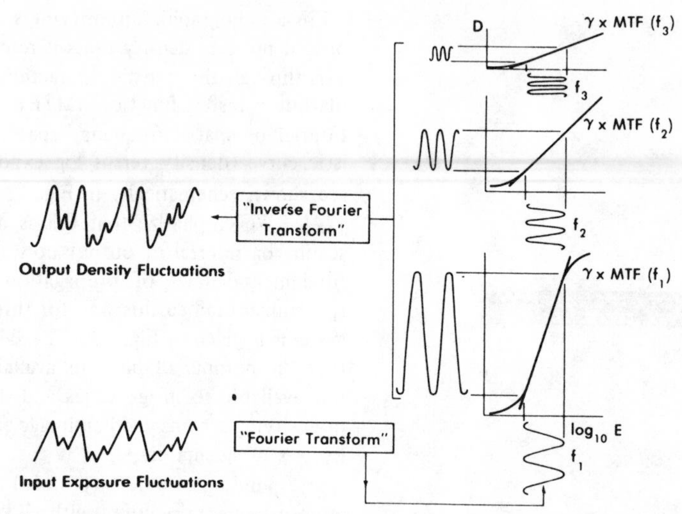

Fig. 12. Three spatial frequencies, $f_1 < f_2 < f_3$, of the X-ray input pattern are shown being operated on independently in frequency space by the combined film (γ) screen (MTF) characteristics for small signal transfer superimposed to produce the output density pattern (Sandrik and Wagner [24]).

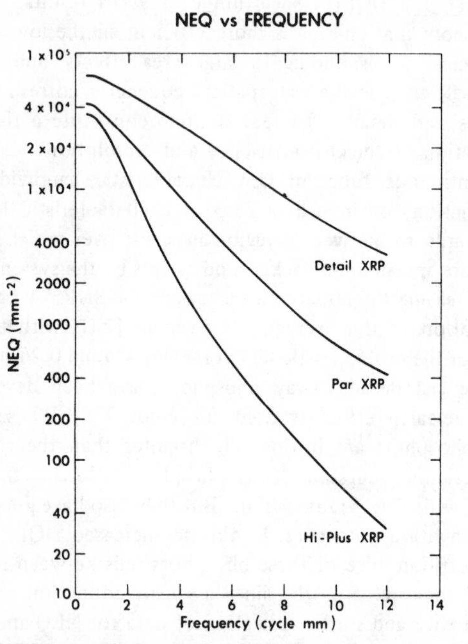

Fig. 13. NEQ versus frequency for three screen-film systems (Sandrik and Wagner [24]).

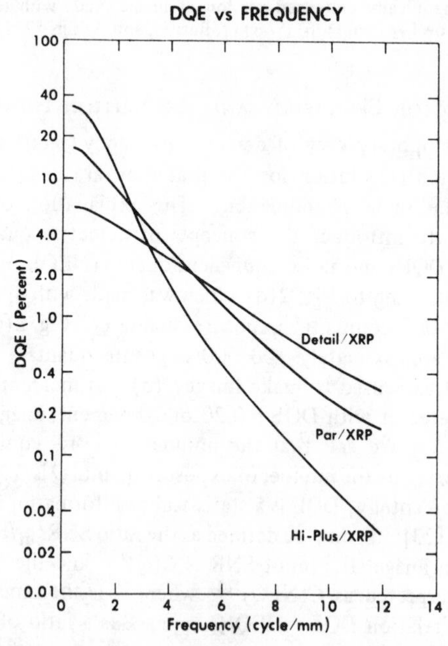

Fig. 14. DQE versus frequency for three screen-film systems (Sandrik and Wagner [24]).

radiographic magnification of about 2X the information out per unit of patient radiation is almost twice that at contact (magnification equal to unity) for a Hi-Plus screen with a 0.3 mm focal spot.

Finally, these concepts can be used to study system latitude or dynamic range, i.e., SNR as a function of exposure level. In Figs. 16 and 17 we present the NEQ(f) and DQE(f) spectra for a Par speed screen/film system as a function of film density level or exposure level, with spatial frequency as the parameter [28]. Ordinarily, we would expect that as input or exposure quanta Q are increased, the picture should become richer in equivalent quanta NEQ. We see from Fig. 16 that this is only true over a small range. In the mid-density range the increase of exposure quanta is apparently effective only to increase the film darkening and display contrast, but not to increase the SNR in the form of NEQ. This means that the DQE, shown in Fig. 17, falls off with exposure in the mid-density range. Screen/film systems are popularly assumed to have a latitude or dynamic range of about two orders of magnitude. However, if we define dynamic range as the range over which NQE has greater than 50 percent of its maximum value we see that these systems have a dynamic range more

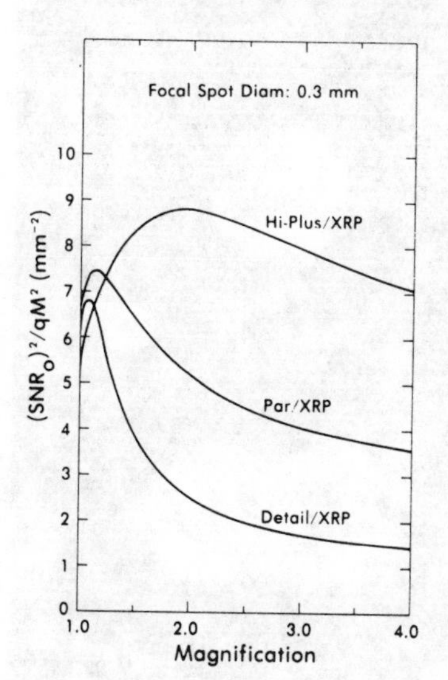

Fig. 15. Variation of SNR^2 in the object (patient) plane as a function of magnification with a 0.3 mm focal spot normalized to patient entrance exposure qM^2 (Sandrik and Wagner [24]).

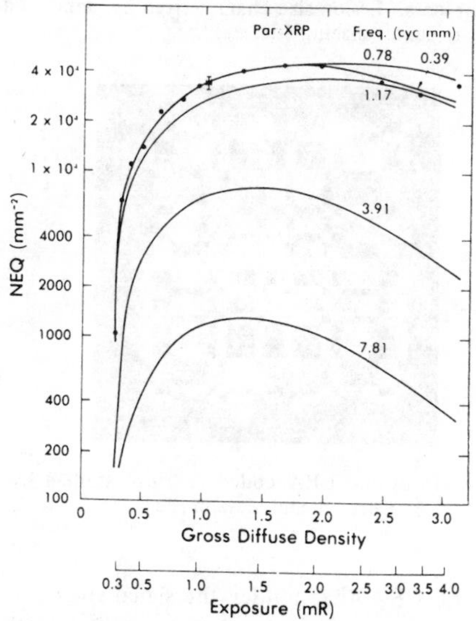

Fig. 16. NEQ versus gross diffuse density and exposure for Par/XRP screen/film system; curve parameter is spatial frequency (Sandrik, Shuping, and Wagner [28]).

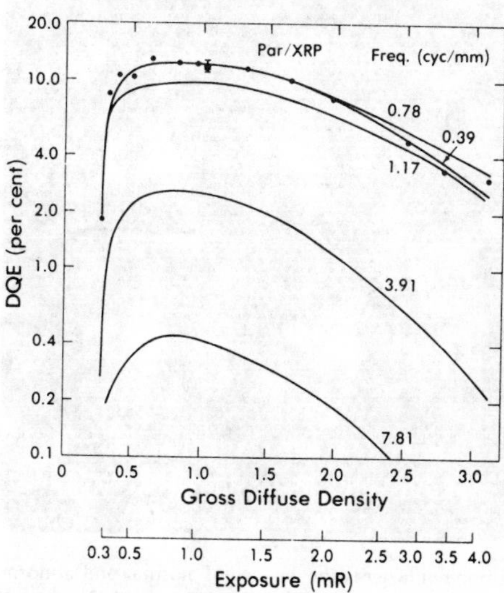

Fig. 17. DQE versus gross diffuse density and exposure for Par/XRP screen/film system; curve parameter is spatial frequency (Sandrik, Shuping, and Wagner [28]).

like one order of magnitude. The limitation is due to the falling off in the number of developable grains per incident X-ray photon at both ends of the range—film grain noise is dominating X-ray quantum noise.

Analogous results for High-Plus screens are incomplete, but again we can infer from published data that the High-Plus low frequency DQE curves would follow those of Par, but at about twice the level. They would not be able to keep up this gain at higher frequencies because of the lower MTF. Their dynamic range at the higher frequencies is then even smaller than an order of magnitude. A well-designed rare-earth combination should be able to improve greatly on this latitude in the mid-frequency range.

The measurements required for the above analysis were the film characteristic curve, the screen/film MTF, and the corresponding noise power spectrum. The Bureau of Radiological Health has been involved in a program to develop consensus measurement methodology for these measurements. The program began as a collaboration with the University of Chicago [27] and then the Los Alamos National Laboratory [29] (see, also, [30]), and has spread to about a dozen laboratories worldwide. These references document the emergence of the field of radiographic image performance measurements as an objective metrology with an absolute scale of performance that is portable between laboratories. The portability arises from calibrations discussed in [29]. The need for such measures—extending to higher spatial frequencies—has been emphasized by Hanson [31] in an analysis that shows that higher frequencies overwhelmingly dominate lower frequencies in the performance of higher order discrimination tasks.

The Nuclear Medicine Dilemma

In nuclear medicine imaging a radionuclide emitter is introduced into the patient and is preferentially taken up by the organ of interest, as for example in Fig. 18; it is then imaged essentially in the manner of a pinhole camera where a lead collimator plays the role of the pinhole aperture. Investigators of such systems are faced with the apparent difficulty that using a small aperture allows for fine image resolution but poor collection efficiency of the emitted radiation; a large collecting aperture gives improved collection efficiency at the price of poorer resolution. They were inspired by coded aperture techniques in gamma-ray astronomy to attempt to have the best of both worlds: a large aperture for high collection efficiency; and fine substructure in the aperture to encode high-frequency or fine detail information from the

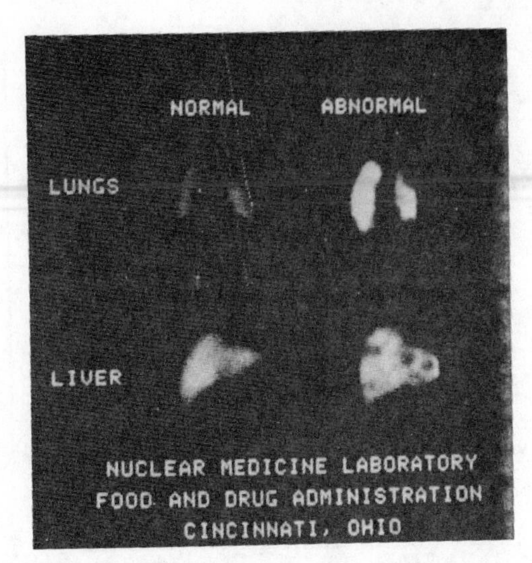

Fig. 18. Radionuclide emission images of normal and abnormal lungs and liver (courtesy of R. J. Jennings and FDA Nuclear Medicine Laboratory).

object to be imaged [32]. Among the apertures suggested were the rotating slit, the Fresnel zone plate, the annulus, random pinhole array, and various so-called uniformly redundant arrays (URA). These patterns have sharply peaked autocorrelations so that the aperture function itself can serve as the mask for decoding the resulting image–when the decoding mask is walked over or convolved with the image it will generate spikes when it senses its own shape, the coded shadow of a point source.

The concept of "resolution" used by the analysts of coded aperture systems has never been rigorously defined. Wagner, Brown, and Metz [33] suggested a Rayleigh-like discrimination task as a measure of resolution. An ideal observer is presented with a noisy image (raw data) and asked to determine whether it resulted from a single "star" or lesion or two such objects near coalescence, as shown in Fig. 19. This discrimination task has all the features required of a definition of "resolution" since the task can be studied as the size of the lesion and the interlesion distance are varied. In [33] the authors studied the detection theoretic SNR or ideal observer SNR for such a discrimination task when the scene is imaged through the coded aperture shown in Fig. 20 and when it is imaged through a simple large open aperture. The coded aperture of Fig. 20 is one cycle of a URA pattern currently receiving much attention; the large open aperture studied had the same open area as the URA–since the URA is roughly 50 percent transmitting (50 percent open area), the simple collimating aperture selected was a square with one side equal to $1/\sqrt{2}$ the side of the URA. It was found that the URA had a collection efficiency advantage, measured in SNR^2 per collection time, of about an order of magnitude over the simple collimator when the lesions or stars were point sources, as in some gamma-ray astronomy. For large lesions the URA was comparable to or inferior to the simple collimator. Contrary to claims of other investigators there was almost no middle ground available, i.e., as soon as the objects depart from point sources they cast

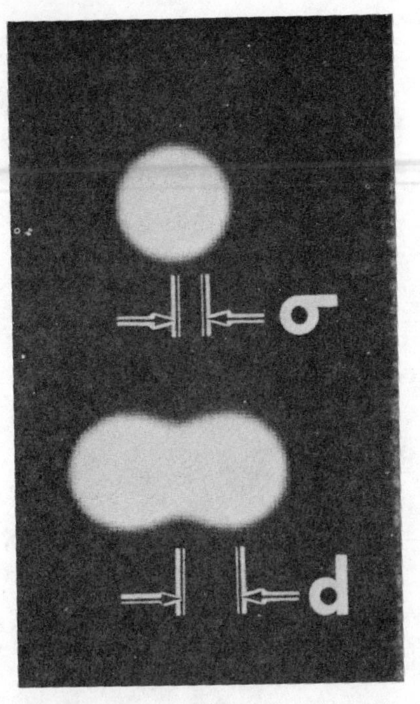

Fig. 19. Ideal observer discrimination task: to determine whether the data correspond to a single lesion or star, or to the coalescence of two lesions or stars. Lesion size characterized by standard deviation σ, and center-to-center spacing 2-D.

Fig. 20. One cycle of the URA coded aperture studied by Wagner, Brown, and Metz in [33].

shadows of the URA that render the signal from one star as additional noise overlap for the detection of the other star. We concluded that–for discrimination tasks like the above–it made more sense to simply open up the collimator than to use a coded aperture, unless the object consists of point sources as in some gamma-ray astronomy. Furthermore, since a pinhole in practice almost never has an efficiency advantage over a parallel-hole collimator [34], our argument supports the latter conventional application *a fortiori*.

Kaufman *et al.* [35] have recently reviewed the significant improvements in nuclear medicine gamma camera resolution and scatter elimination available from high-energy-resolution semiconductor detectors. This technology also gives positional information free of the distortion that results from analog readout methods in scintillation cameras. It is this distortion that becomes the textured noise in scintillation camera images that has been measured by Shosa *et al.* [36]. According to Kaufman and to Shosa, the presence of this textured noise effectively reduces the large area or low frequency DQE to about 0.6 at current counting levels, and to lower values at higher counting levels, and is in their view responsible for the lack of incremental improvement from increases in count levels in scintillation camera images. The germanium camera studied by these authors would therefore represent a great advance on several fronts if it could be fabricated over a reasonably large format at a reasonable expense.

Images Reconstructed from Projections–Computed Tomography

The first major medical imaging advance to come from the evolution of high speed digital electronics was the computed tomographic (CT or CAT) scanner. Low contrast soft tissue differences in X-ray attenuation were visualized for the first time *in vivo*, for example, normal and abnormal tissues and fluids in the brain (Fig. 21). The radiological physics aspects of computed tomography are almost identical to those of conventional radiography. However, the imaging is unconventional in that only a cut, usually 1 cm thick, through the patient is imaged from many different angles ($\geqslant 200$) with a narrow beam, and a computer algorithm is used to reconstruct from these projection images a cross-sectional view in the plane of the cut or tomogram (Gk. *tomos*: cut or slice). The image in each projection is blurred by the X-ray tube focal spot and the detector apertures with their respective collimation. These apertures blur the signal but not the noise, which remains Poisson noise uncorrelated from measurement bin to bin. At the next stage the computer algorithm contributes three kinds of filtration which blur both the signal and the noise (Fig. 22): 1) backprojection, 2) the associated frequency ramp filter for removing the $1/r$ spread function that would otherwise result from backprojection of image points, and 3) an apodization or high-frequency rolloff filter to soften the appearance of image noise. Wagner, Brown, and Pastel [37] presented a construction based on these points for obtaining the hardware–focal spot and detector–blur by combining two measurements: the total system MTF, which contains all four blurrings, and the system noise power spectrum, which contains only the software blurring. The construction is shown in Fig. 23. By dividing the noise power spectrum into the properly weighted total system MTF^2, the MTF^2 of the hardware results. The hardware MTF is simply the frequency space representation of the blur function in the original projections. Only this blurring is fundamental–it determines the region over which signal counts are spread upon background counts with the concomitant increase in noise, decrease in SNR, discussed above for conventional radiography. The size of this region ranged from 1 to 2 mm for the first and second generation head scanners. We call this the "terminal blur" since it cannot be removed by image processing, although this has been attempted periodically.

When this construction and division are carried out on an absolute scale they yield the NEQ(f) spectrum for CT that is totally analogous to the NEQ(f) spectrum for conventional radiography discussed above. The absolute scale for the ordinate

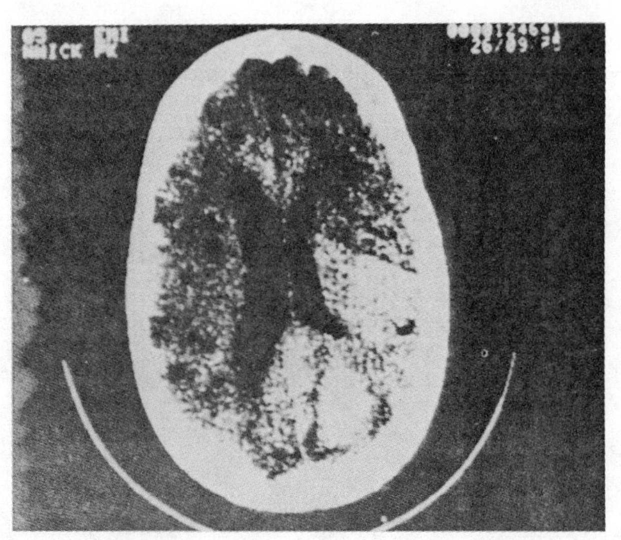

Fig. 21. Brain scan from EMI CT5005, second generation CT scanner.

of the second generation system of Fig. 23 is $1.0 \equiv 1.8 \times 10^8$/cm along the projections. (NEQ for CT is given as a linear density since CT averages over the cut direction; since the latter is usually about 1 cm, the aereal density is about the same.) In Fig. 24 we give the NEQ spectrum for the first generation CT system; the absolute scale for the ordinate here is $1.0 \equiv 1.4 \times 10^8$/cm. We see therefore that the second generation represented both an increase in NEQ level and an increase in NEQ bandwidth–this was achieved without an increase in dose. Finally, we point out that these CT scans (head mode) were made with a total of almost 10^{10} effective detected counts. It is this great number that allows CT to make the subtle gray matter/white matter distinction in images of the brain.

Experimental studies of the contrast required for threshold visibility as a function of lesion size have been carried out by Cohen and DiBianca [38] and by Cohen [39] for CT scanners. An example of a contrast-detail study of a moderate and a high "resolution" third generation system operating at two dose levels is given in Fig. 25. Notice again that dose and resolution are not coupled together. Small diameter, high contrast detectability is only a function of the apertures, which are smaller for the model 8800 than for the 7800. Large diameter, low contrast detectability is only a function of the exposure or dose level. We may relate these experimental findings to NEQ concepts. It is the low-frequency portion or absolute level of the NEQ spectrum that is the principal determinant of large area, low contrast detectability; this level depends on the number of detected quanta and is thereby related to dose. On the other hand, the NEQ bandwidth is a principal determinant of high contrast detail detectability; this depends on the focal spot and detector apertures. In principle (but not inexpensively), these can be improved with technology. All of these points can be summarized by saying, counter to opinions still prevailing, that high contrast, small diameter resolution in the plane of the CT cut is not necessarily tied to exposure or dose.

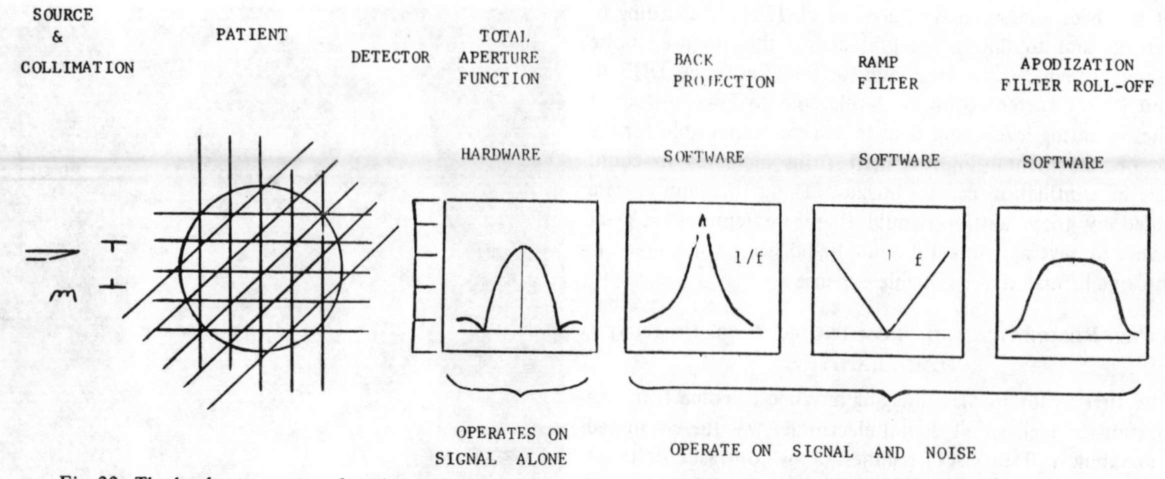

Fig. 22. The hardware aperture function, and three software filters used in CT; the former acts only on the signal, the latter on the signal and noise allowing for the isolation of the aperture function by the measurements of Fig. 23.

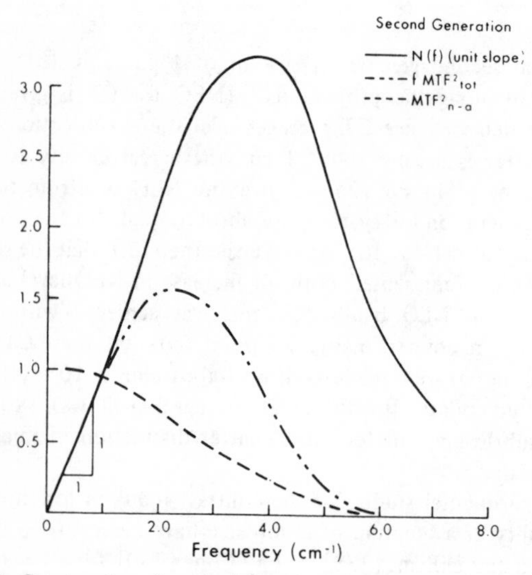

Fig. 23. Construction for obtaining the aperture or nonalgorithmic MTF^2; the total system MTF^2, weighted for two dimensions, is divided by the normalized noise power spectrum (NPS) (from Wagner, Brown, Pastel [37]). Absolute scale $1.0 \equiv 1.8 \times 10^8$/cm (cf. Fig. 24).

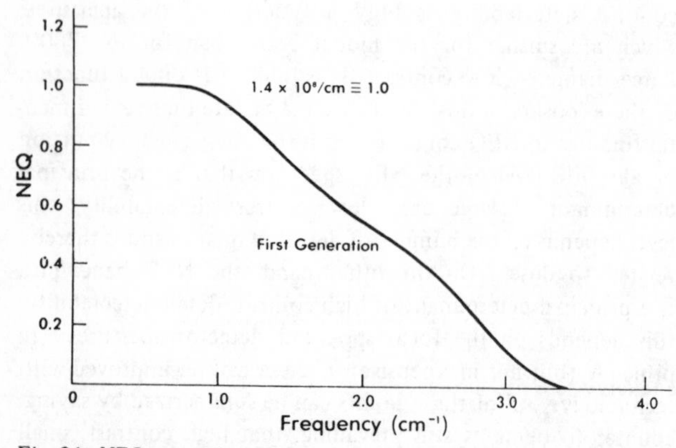

Fig. 24. NEQ versus frequency spectrum for first generation CT system (from Wagner, Brown, Pastel [37]) analogous to Fig. 13 above for screen/film systems.

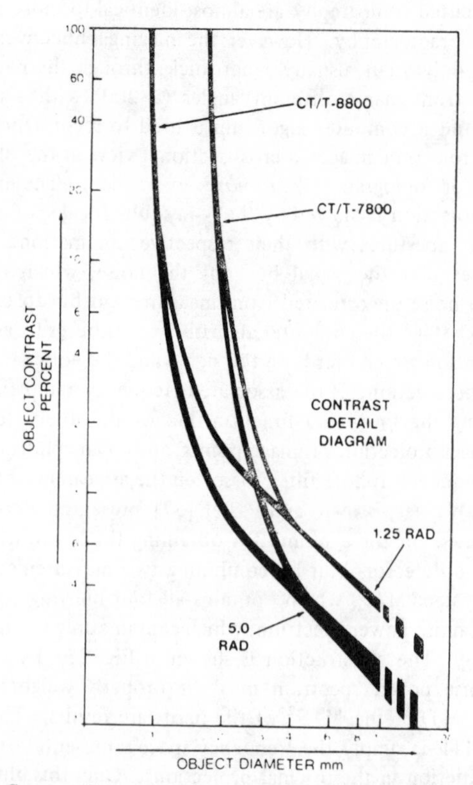

Fig. 25. Contrast-diameter diagrams for two CT systems operating at two phantom dose levels (rows of disks, not isolated disks, were used) (courtesy of the General Electric Company).

The detective quantum efficiency of CT can be determined in a manner similar to that for conventional images. In [37] it was found that the large area or low-frequency noise equivalent quanta represented approximately 65 percent of the exposure quanta for the EMI Mark I first generation CT, i.e., the low-frequency DQE = 0.65, or the large area information in the image represented 65 percent of the information incident on the detectors. This was remarkable in the face of promises from the mathematical community working on CT algorithms

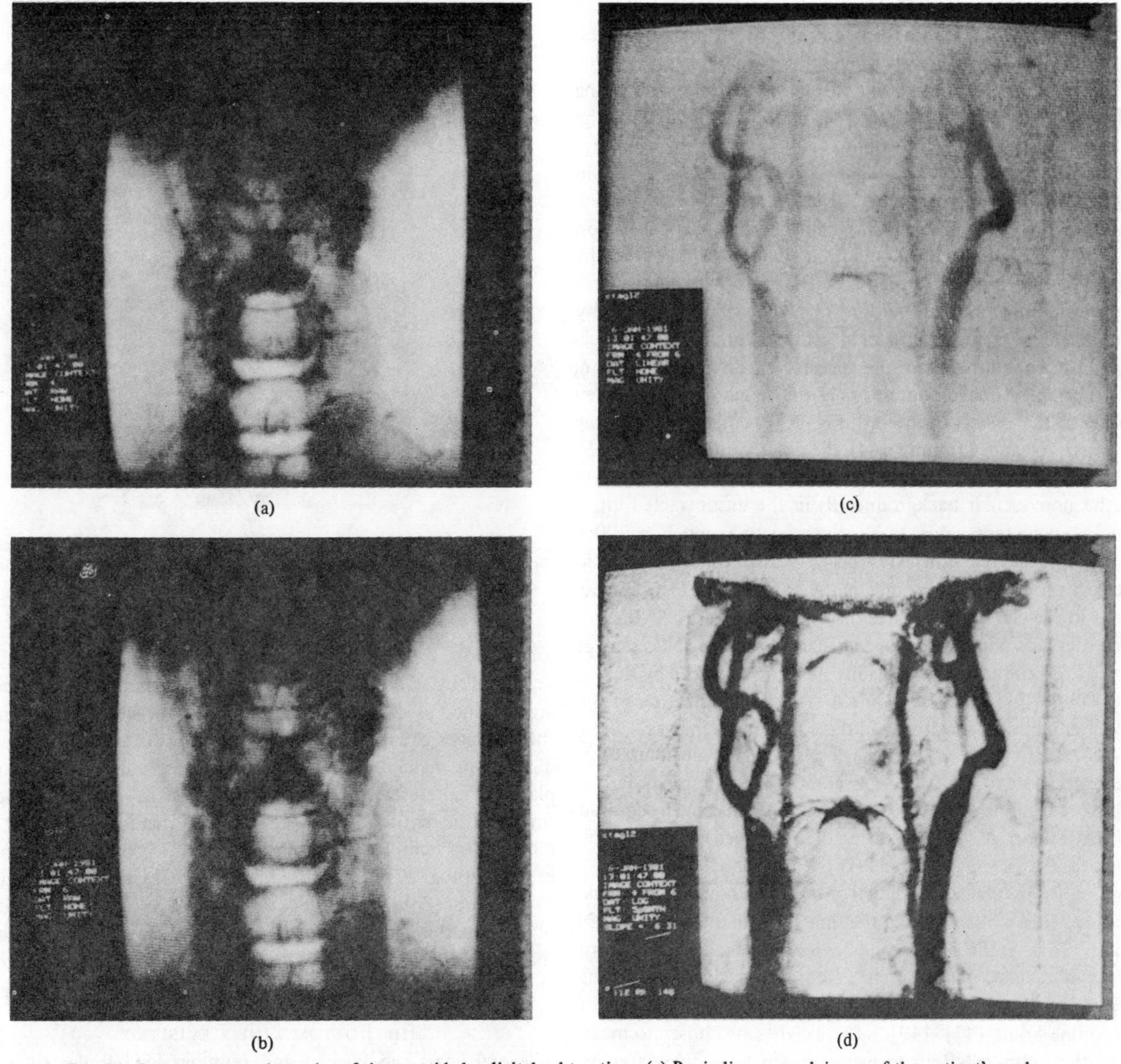

Fig. 26. Intravenous angiography of the carotids by digital subtraction. (a) Pre-iodine or mask image of the patient's neck. (b) Post-iodine-injection image. (c) Image resulting from the subtraction of (a) from (b). (d) Contrast enhanced version of (c). (Courtesy of T. W. Ovitt, University of Arizona.)

that given further support they could reduce CT dose by an order of magnitude. In fact, at that time it *was* possible through improvements in *hardware* to reduce CT dose further since the beam collimation was poor. In the second generation system only about $\frac{1}{3}$ of the downstream beam was intercepted by the detectors. The dose efficiency of second generation CT was therefore estimated to be approximately $0.65 \times 0.33 = 0.22$. An independent estimate by Hanson [40] placed the efficiency between 0.12 and 0.17. Since that time low dose collimation has been introduced, and in principle it has become possible to carry out CT at better than 50 percent total dose efficiency. These are indirect arguments that the contribution of the CT algorithm to dose inefficiency is negligible. A direct argument has been given by Hanson [41], building on the argument of Tretiak [42]; Hanson had earlier pointed out that filtered backprojection preserves the information in the projections [40]. This was implicit in the CT information analysis of [37], but never made explicit.

Additional Cutaway and Difference Images

In addition to its use of a great number of quanta, CT achieves its success by isolating the cut of interest from over- and underlying structure. Conventional radiographs contain a compression or integration of all the information along the direction of the beam, as well as an integration over the energy spectrum and the exposure time. One of the more significant achievements of the last decade has been the invention or reinvention of techniques for differentiating and isolating portions or cuts of information along dimensions where it is usually integrated. We have already seen above how the *energy* information that is usually integrated can instead be isolated by working with two beams at different energies, with the resulting soft tissue only, or bone only images.

The most active area among the various difference imaging schemes being pursued currently is digital radiography in the form of subtraction angiography, or intravenous angiography. An example of the type of images obtained is presented in Fig. 26, taken from the work of Ovitt *et al.* [43]. A mask image in Fig 26(a), i.e., the image formed before the injection of iodine, is digitally subtracted from an image formed after the injection Fig. 26(b). (Because of the sensitivity of the procedure the injection can be done intravenously rather than by the much more invasive arterial catheterization.) If the patient is cooperative and the images remain in registration, the subtracted image contains only the signal from the added iodine in the vessels of interest, Fig. 26(c) and (d). According to Mistretta and Crummy [44] the vascular iodine signal (S) might typically be 1 part in 20 of the range of fluctuation of the nonvascular background (B) in the unsubtracted intravenous angiogram, and can be increased to as much as 200:1 in a well-registered subtracted image. Mistretta calls this ratio the S/B ratio. Its importance in the context of a highly structured image clearly dominates that of the stochastic SNR's discussed in this paper. Some generalization of the SNR concept to the S/B ratio or the conspicuity measure advanced by Revesz and Kundel [45] will be required as a characterization of the increased physical sensitivity in these difference images.

Real-time or *fluoroscopic* imaging is being revolutionized by the conversion to digital data handling. The temporal derivative of the iodine signal through vasculature has been acquired by Mistretta *et al.* [46] by using the (averaged) frame(s) previous to the (averaged) frame(s) of interest as the mask in a continuous sequence of digital subtractions of fluoroscopic images. This mode is called the time interval difference (TID) mode. This simple subtraction has been generalized to various forms of recursive and matched filtering that use a weighting of a string of many fluoroscopic frames during the passage of the iodine bolus [47]-[49]. The weights are chosen to maximize the iodine SNR in images summed from these frames. These methods all suppress the dc or steady component to remove nonvascular background (but not necessarily misregistration artifacts [50]). They allow for an optimum use of the radiation exposure during the *fluoroscopic* sequence to produce one or a few high quality archival radiographs from a sequence of low quality noisy frames. Recursive and matched filter fluoroscopic schemes may have a dose utilization efficiency in excess of 60 percent and do not require high quality video. On the other hand, the digital *radiographic* schemes discussed previously make a sequence of a dozen or so high quality digital radiographs while archiving only a few. The dose utilization with respect to the archiving may be on the order of 15 percent. Furthermore, the high quality of the digital radiographs requires a high quality video chain.

Finally, we point out that *conventional* (analog) tomography may be extended by the power of digital acquisition and calculation. In conventional tomography the source and image receptor are moved counter to each other in such a way that all planes except the plane of the fulcrum of the source/receptor motion are blurred out. In tomosynthesis, a technique suggested by Grant [51] and others, a number of images are obtained at different source and image receptor positions; these images can then be added together after shifting them digitally or manually with respect to a fulcrum at an arbitrary plane of interest. This tomographic plane can be walked through the entire depth of the body thereby generating a more or less complete family of tomograms. An early example is given in Fig. 27. A measure of the increase in sensitivity due to this procedure would require knowledge of the variability of tissue attenuation through the body. It is this variation that serves to null out lesions of interest or serves as unwanted clutter to their detectability in conventional projection imaging.

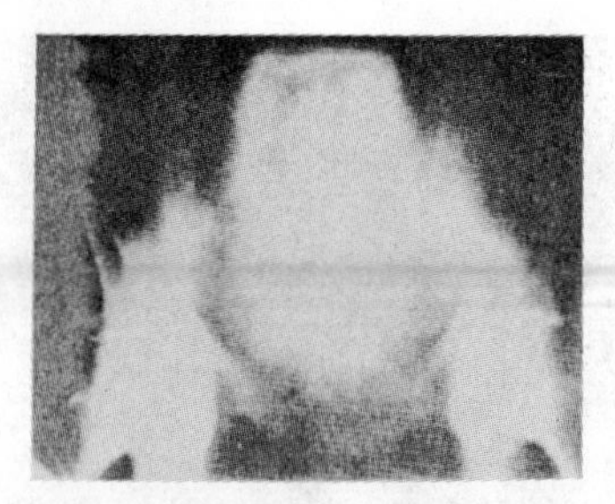

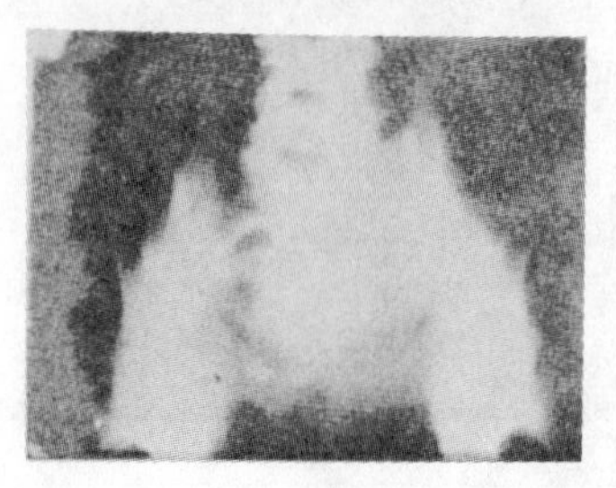

Fig. 27. Two from a family of tomograms at different depths in the pelvis synthesized from 20 images made at different projection angles (Grant [51]).

The Bottom Line Revisited

Several years ago Wagner and Jennings [52] indicated in a "bottom line" study that most of the links of the radiographic system were subject to further exposure and dose efficiency improvements by factors of order 2 while retaining the contemporary standard of "image quality." They include a factor two by changing from conventional calcium tungstate intensifying screens to rare-earth screens, a factor two from improved scatter rejection schemes, a factor two from optimizing the X-ray spectrum or energy used, a factor two from using the optimal geometrical configuration to minimize the SNR degradation due to aperture or blurring effects, a factor of about 1.1 to 1.3 by using low attenuation support materials made using the advances in carbon fiber material technology, and perhaps a factor two or less from display optimization to be mentioned below. (All of these factors have been discussed above with the exception of savings from the carbon fiber materials, which are discussed in [53].) It was pointed out that a factor between 5 and 10 of this possible 20-40 fold improvement might be collected by working within the constraints of the then current technology. Recently, Muntz *et al.*

[54] have carried out a multiparameter optimization of mammography and indicate that the exposure efficiency of current low-dose systems can be improved by about a factor 4 through an optimal configuration, spectrum, and scatter rejection using *current* screen/film technology. Soon the digital radiographic technology will offer the possibility of working at the higher end of this scale for general radiography since it has none of the display contrast limitations of film. We have just seen that some of the digital radiographic subtraction schemes remain inefficient in their clinical application but that the fluoroscopic implementations of them may cancel out these inefficiencies. That would leave no fundamental obstacles to a total optimization of the use of X-radiation for the formation of medical images–aside from the realistic questions of cost and convenience. These questions could evaporate with the evolution of the technology.

Ultrasound *B*-Scans

During the last twenty-five years a number of technical developments promoted the evolution of the early one-dimensional *A*-scans–which consisted of returned signals from pulsed RF ultrasound displayed as a voltage versus time trace, into the *B*-scan image–which results from envelope detection of the RF from a given line and the display of many lines as a two-dimensional gray scale image. In Fig. 28 we give a typical modern *B*-scan ultrasound (US) image of a normal liver. There has been interest in the texture of these images and the question of whether the statistics of this texture carries "tissue signature" information. There has been even more interest in ultrasound attenuation measurements and their spectral content as markers signaling tissue character. And more recently investigators have been studying the possibility of using these images to obtain absolute volume backscatter coefficients that might be used in the same way as the attenuation coefficient information available from CT scanners for classifying the state of health or disease of tissue. ([55] updates these projects.) None of these concepts has yet led to a commercial development but quantitative US is almost certain to have some clinical role in the near future.

Wagner, Smith, Sandrik, and Lopez [56] have recently demonstrated that the texture in a *B*-scan of a scattering phantom is the result of coherent scatter and interference of US waves totally analogous to the speckle in coherent radar and laser imaging. If the particles in the scattering medium are many and fine on the scale of the wavelength then the texture of the speckle (its correlation properties or second-order statistics) is characteristic only of the transducer used to insonify and sample the medium. However, if the particles cluster on the scale of the wavelength, the authors indicate that the clustering will broaden the speckles and therefore allow quantitative texture analysis. Also, Fellingham-Joynt and colleagues have been successful at using the speckle amplitude and correlation properties to pick up quasi-periodic structure in the scattering medium [57], [58].

Smith, Wagner, Sandrik, and Lopez [59] used their statistical analysis of the *B*-scan speckle to analyze the detectability of focal lesions in these images. Previously, Smith and Lopez [60] had designed a phantom for studying detail visibility in *B*-scans. Scans of the phantom generated the images shown in Fig. 29, from which they developed contrast-diameter visibility threshold curves such as that shown in Fig. 30 (analogous to Fig. 25). These curves give the threshold contrast in dB versus circular lesion size in mm at the threshold determined by human observers.

When the detection task was analyzed in terms of ideal observer signal detection and ideal observer SNR, a new definition of ultrasound contrast C_ψ was suggested which for low contrasts is simply the mean fractional change in backscatter intensity between lesion and surround. Just as in the X-ray case, the SNR then took the form $\mathrm{SNR} = C_\psi N^{1/2}$, where N is now the expected number of speckle spots over the lesion area and is proportional to that area. Also, just as in the X-ray case, the threshold contrast-detail curve for the ideal observer

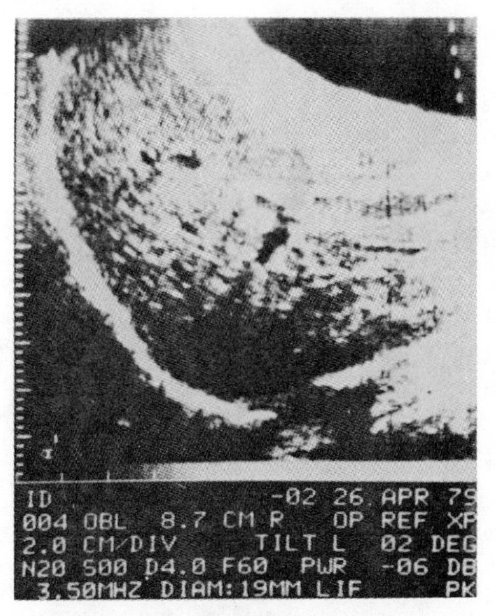

Fig. 28. Modern ultrasound *B*-scan image of a normal liver.

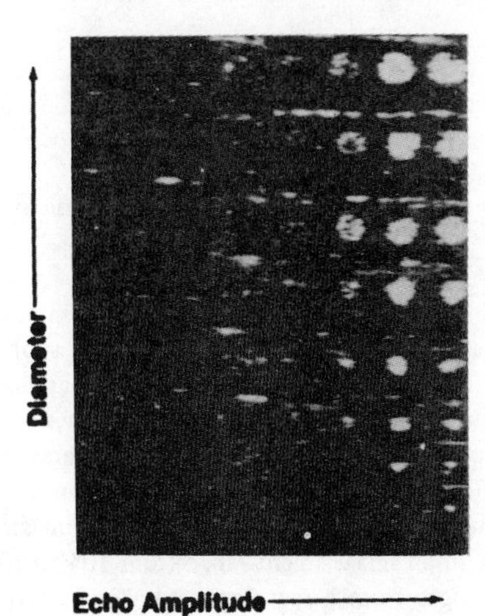

Fig. 29. Scans of the Smith and Lopez contrast-detail phantom [60]: diameter decreases from top to bottom of figure; contrast decreases from outside to center of image.

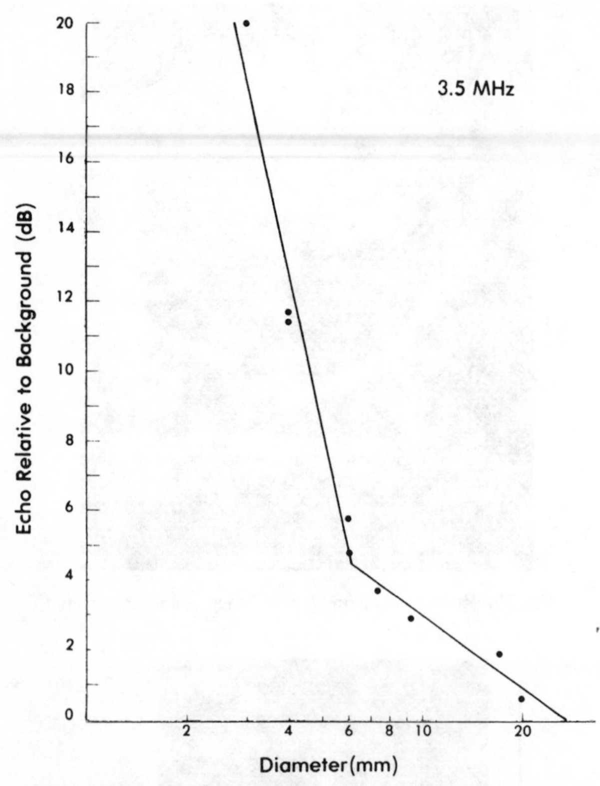

Fig. 30. Contrast-diameter visibility threshold curve from an ultrasound B-scan phantom image such as Fig. 29.

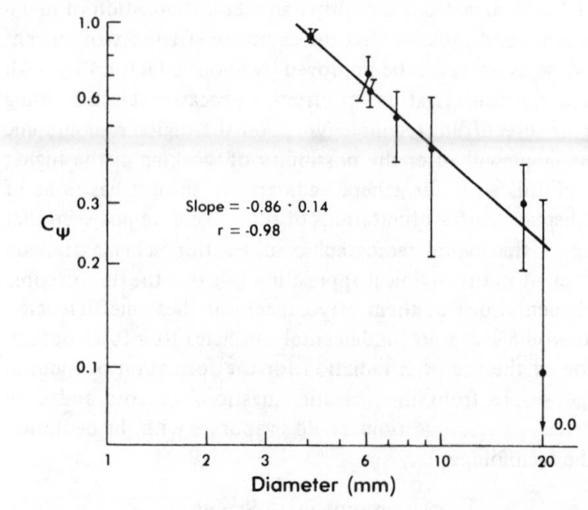

Fig. 31. Contrast-diameter data from Fig. 30 replaced in terms of C_ψ, the contrast that arises in the detection theoretic analysis of B-scan images.

was calculated to have a slope -1 when plotted in terms of the new contrast. This happens since the speckle spot serves the same role for sampling the ultrasound field as does the photon for sampling an X-ray image. In both cases the SNR goes as the root of this number over the lesion area, and so lesion contrast at threshold goes inversely to lesion diameter. However, whereas in photon projection and CT imaging the sample number in an image might range from 10^6 to 10^{10}, in ultrasound it ranges from a few dozen to about a thousand. The speckle samples are resolvable, being comparable in size to the resolution cell of the imaging system. This means that lesions must be large or contrast very great, or a number of independent images must be compounded for lesions to be detectable.

When the human observer data were replotted in terms of the new contrast the straight line of Fig. 31 resulted. The slope is consistent with the value for the ideal observer, and the human observers were detecting lesions when they stood out from the background by 2-3 standard deviations of the background speckle noise. This result means that the lesion detectability with the Smith and Lopez phantom is limited by the fundamental physics of the coherent speckle formation and cannot be much improved by image processing. However, scattering phantoms with weaker concentrations of scatterers have been made and then detectability has been limited by scanner electronic sensitivity. The regimes of speckle limited and instrument limited detection have not yet been completely mapped out and will depend on frequency since the phantom or patient penetration is a decreasing function of US frequency. A principal result of this work, then, was the extension of the ideal observer SNR approach developed in connection with X-ray imaging to a fundamentally different modality, further unifying the approach to the evaluation of medical imaging systems. This investigation continues and will be extended to several classes of tissue-information-bearing speckle which we do not discuss here.

Observer Studies

Contrast-detail diagrams analogous to those presented above in the CT and US cases have been studied by Burgess, Humphrey, and Wagner [61] in the context of conventional X-ray noise, and by Wagner, Brown, and Pastel [37] and Wagner [62] for CT noise. In the former study the observer threshold corresponded to signals that were 3-4 standard deviations of the noise and in the latter case to signals that were almost 7 standard deviations of the noise. We believe that these differences are more characteristic of the observers' criteria than they are of the different imaging systems. These findings raised several fundamental questions concerning observer studies intended for contrast-detail analysis: 1) how reproducible is the observer threshold and is it a function of the imaging parameters (size, shape, contrast, texture)? 2) how portable are these thresholds, i.e., can one observer communicate his criteria to another? The thresholds mentioned above, viz. 3, 4, 7, were *SNR's in the image* for signals considered at threshold by the observer; then, 3) can the observer's threshold be measured in the observer system, i.e., at the decision level of the observer, on the same scale as the SNR in the image? Until these questions are answered the contrast-detail observer method of assessing imaging systems may not have the statistical power required for discerning small to modest differences in imaging system performance [63]. Metz, Brown, and Wagner [64] have begun to study this problem. The only result to date is the conclusion that the higher the level of signal required by the observer as his threshold criterion, the better the observer precision in determining the contrast-detail diagram. In the final section we address the question of the absolute scale of human observer performance.

Observer Efficiency: The DQE of the Display/Observer Link

Burgess, Wagner, Jennings, and Barlow [65] have used two alternative forced choice (2AFC) experiments to measure the SNR in the human observer/display system. In these experiments a specified lesion was first shown to the observer who then looked at two images, one of which contained the signal near threshold at a specified location (signal specified, position known exactly); the observer designates the image containing the lesion, or for discrimination tasks the image containing the lesion at higher contrast. Through hundreds of repeated trials per lesion type the observers' percent correct score is accurately measured and converted to a SNR by inverting the appropriate error function. This SNR can be compared with the SNR of the ideal observer performing the same task, which can be calculated from the signal and noise characteristics or by programming a computer to perform as an ideal observer (maximum likelihood, correlation detector, or matched filter). The ratio of the squares of these SNR's is the observer efficiency, formally the same concept as DQE above, but now the DQE of the display/observer coupling [65]. A summary display of the experimental results of Burgess *et al.* with signals embedded in white noise is given in Fig. 32. This is a plot of signal energy (integrated squared contrast) required for 76 percent correct decisions, as a function of noise power spectral density. Also shown are the results expected for the ideal observer and the ideal observer with an internal source of noise. At the highest noise levels the requirements of the human observer are about twice those of the ideal observer, and this corresponds to an observer efficiency of 50 percent. At the lower noise levels the human observer requirements are much greater than those of the ideal observer, e.g., near zero noise the human observer still requires significant signal contrast because of the internal fluctuations or internal noise of the eye-brain system—this is the well-known contrast threshold of the human observer. These results can be interpreted to indicate that the display to the human observer should be at a sufficient contrast that the noise is easily visible to him. Only then does he have the possibility of extracting most of the information from the image. In fact, at high noise levels and for certain spatial frequencies near the maximum frequency response of the eye-brain, we find efficiencies in excess of 80 percent. Another interpretation of these results is that simple contrast enhancement is an important form of image enhancement (as found above in the scatter reduction study); also, attempts to aid the observer to overcome the masking effects of noise by image processing may have some payoff, but it will not be great.

A simple demonstration can be given to show the plausibility of the results just given. Consider the two images in Fig. 33 [66]. An ideal observer will find the noise in the image on the left to interfere greatly with his perception of the image content—the noise and the signal spectra overlap. The ideal observer will find that the noise in the image on the right does not mask the image information at all—the noise spectrum is two octaves above the signal spectrum, with no overlap—the ideal observer can completely separate this noise from the signal by a simple filtering operation. Are we ideal observers? Certainly we have great trouble with the noise in the image on the left. We have some trouble with the noise on the right—we have less if we remove our galsses, shake our heads, or squint. It is therefore reasonable to conclude that we are inferior to the ideal observer, but perhaps not by much. We continue to find high human observer efficiencies after tens of thousands of observations and Burgess has extended this work to CT-like images with a wide range of lesions and target types.

An additional interesting aspect of the images in Fig. 33 should be pointed out. The pixel variance in the image on the

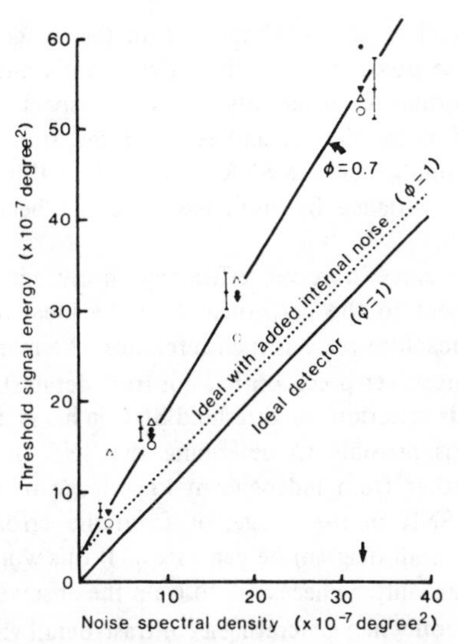

Fig. 32. Signal energy required for 76 percent correct decision in the two alternative forced choice experiments of Brugess *et al.* [65]. Results for the ideal detector with and without an internal source of noise are shown.

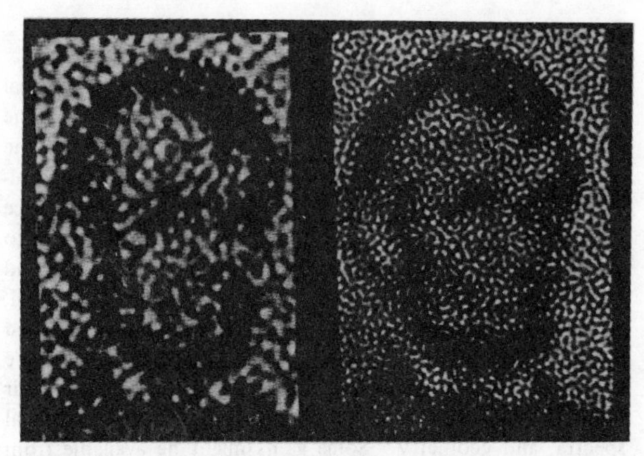

Fig. 33. Photograph of a statesman with noise overlapping the signal spectrum (left), and with noise totally outside the signal spectrum (right) (courtesy of AAAS and L. Harmon [66]; copyright, 1973, AAAS).

left is equal to the pixel variance in the image on the right. Their noise power spectra differ significantly and couple to or mask information differently. It is this aspect of the masking power of noise that is addressed in the statistical decision theory from which the SNR's sketched in this paper derive. The pixel variance by itself has no direct bearing on signal detectability.

Finally, since observer efficiency measurements are done with respect to the performance of the ideal observer, they have an absolute scale as their reference. We mentioned above that the observer precision in a contrast-detail study is a function of his criterion, or threshold SNR in his system. It therefore seems possible to determine the SNR in the observer system either from independent knowledge of his efficiency and the SNR in the image, or from the error bars in the contrast-detail diagram he generates. If this works in practice it will then only be necessary to train the observer to maintain this criterion when generating a contrast-detail visibility curve. Then, with sufficient phantom images, this technique should be adequate to the task of distinguishing systems whose physical performance is similar, or to the null task of demonstrating them to be equivalent.

Conclusions

The use of statistical decision theory to obtain the ideal observer SNR for a given lesion detection task serves as the basis of a rigorous solution to the problem of assessing the sensitivity of medical imaging systems. It allows us to calculate the ideal performance, that is, the best performance allowed by nature with a given level of radiation exposure to the patient. By measurements within this context we find how far short of the ideal performance actual systems fall, and by implication how much the radiation may be reduced if the system is engineered up to the ideal performance. We find that this is sizeable for radiologic systems due to their modest DQE, low scatter rejection efficiency, suboptimal spectra, and geometry. Some gains might be available from image processing, especially simple contrast enhancement; the gains from noise processing will not be great since the human observer already functions in excess of 50 percent efficiency when viewing noisy images.

This approach is fundamental and rigorous and has been extended from X-ray to CT and to US systems. We are currently extending it to three-dimensional imaging systems, including transmission and emission systems and the various modes of nuclear magnetic resonance (NMR) imaging.

Acknowledgment

I gratefully acknowledge the invitation by R. H. Schneider, the Director of the Office of Science and Technology at the National Center for Devices and Radiological Health, to address these questions. Acknowledgment and gratitude are due to many collaborators, especially those listed in the references, for their contributions to the work briefly described here; likewise to the staff of the Medical Physics Branch of these laboratories under D. G. Brown for their support; and to the Acoustics Branch of the same laboratories for collaborative association through S. W. Smith. I am grateful to B. K. Kramer for expertly processing this manuscript and to D. F. Elbert for the high quality illustrations.

References

[1] J. A. Swets, R. M. Pickett, S. F. Whitehead, D. J. Getty, J. A. Schnur, J. B. Swets, and B. A. Freeman, "Assessment of diagnostic technologies," *Science*, vol. 205, pp. 753-759, 1979.

[2] J. A. Swets and R. M. Pickett, *Evaluation of Diagnostic Systems: Methods from Signal Detection Theory*. New York: Academic, 1982.

[3] C. E. Metz, "Basic principles of ROC analysis," *Seminars Nucl. Med.*, vol. VIII, pp. 283-298, 1978.

[4] L. B. Lusted, "Signal detectability and medical decision-making," *Science*, vol. 171, pp. 1217-1219, 1971.

[5] A. Rose, *Vision: Human and Electronic*. New York: Plenum, 1973.

[6] M. M. Ter-Pogossian, *The Physical Aspects of Diagnostic Radiology*. New York: Harper and Row (Hoeber), 1967, p. 172.

[7] W. J. Oosterkamp, "Monochromatic X-rays for medical fluorscopy and radiography," *Medicamundi*, vol. 7, pp. 68-77, 1961.

[8] R. J. Jennings, R. J. Eastgate, M. P. Siedband, and D. L. Ergun, "Optimal X-ray spectra for screen-film mammography," *Med. Phys.*, vol. 8, pp. 629-639, 1981.

[9] E. P. Muntz, M. Welkowsky, E. Kaegi, L. Morsell, E. Wilkinson, and G. Jacobson, "Optimization of electrostatic imaging system for minimum patient dose or minimum exposure in mammography," *Radiology*, vol. 127, pp. 517-523, 1978.

[10] E. P. Muntz, E. Wilkinson, and F. W. George, "Mammography at reduced doses: Present performance and future possibilities," *Amer. J. Roentgenology*, vol. 134, pp. 741-747, 1980.

[11] J. W. Motz and M. Danos, "Image information content and patient exposure," *Med. Phys.*, vol. 5, pp. 8-22, 1978.

[12] R. E. Alvarez and A. Macovski, "Energy-selective reconstructions in X-ray computerized tomography," *Phys. Med. Biol.*, vol. 21, pp. 733-744, 1976.

[13] L. A. Lehman, R. E. Alvarez, A. Macovski, and W. R. Brody, N. J. Pelc, S. J. Riederer, and A. L. Hall, "Generalized image combinations in dual kVp digital radiography," *Med. Phys.*, vol. 8, pp. 659-667, 1981.

[14] J. B. Weaver and A. L. Huddleston, "Attenuation parameters used to characterize tissues in multiple energy CT and digital radiography" (abstr.), *Med. Phys.*, vol. 9, p. 637 1982.

[15] S. J. Riederer and C. A. Mistretta, "Selective iodine imaging using K-edge energies in computerized X-ray tomography," *Med. Phys.*, vol. 4, pp. 474-481, 1977.

[16] P. S. Yeh, A. Macovski, and W. Brody, "Noise analysis in isolation of iodine using three energies," *Med. Phys.*, vol. 7, pp. 636-643, 1980.

[17] F. Kelcz, P. M. Joseph, and S. K. Hilal, "Noise considerations in dual energy CT scanning," *Med. Phys.*, vol. 6, pp. 418-425, 1979.

[18] F. Kelcz and C. A. Mistretta, "Absorption edge fluoroscopy using a 3-spectrum technique," *Med. Phys.*, vol. 3, pp. 159-168, 1976.

[19] T. L. Houk, R. A. Kruger, C. A. Mistretta, S. J. Riederer, C.-G. Shaw, J. C. Lancaster, and D. C. Flemming, "Real time digital K-edge subraction fluoroscopy," *Invest. Radiol.*, vol. 14, pp. 270-278, 1979.

[20] S. J. Riederer, R. A. Kruger, and C. A. Mistretta, "Limitations to iodine isolation using a dual beam non-K-edge approach," *Med. Phys.*, vol. 8, pp. 54-61, 1981.

[21] R. F. Wagner, G. T. Barnes, and B. S. Askins, "Effect of reduced scatter on radiographic information content and patient exposure: A quantitative demonstration," *Med. Phys.*, vol. 7, pp. 13-18, 1980.

[22] J. C. Dainty and R. Shaw, *Image Science*. New York: Academic, 1974.

[23] R. Shaw, "Evaluating the efficiency of imaging processes," *Rep. Prog. Phys.*, vol. 41, pp. 1103-1155, 1978.

[24] J. M. Sandrik and R. F. Wagner, "Absolute measures of physical image quality; Measurement and application to radiographic magnification," *Med. Phys.*, vol. 9, pp. 540-549, 1982.

[25] R. F. Wagner and K. E. Weaver, "Prospects for X-ray exposure reduction using rare-earth intensifying screens," *Radiology*, vol. 118, pp. 183-188, 1976.

[26] R. F. Wagner, "Fast Fourier digital quantum mottle analysis with

application to rare-earth intensifying screen systems," *Med. Phys.*, vol. 4, pp. 157-162, 1977.

[27] K. Doi, G. Holje, L.-N. Loo, H.-P. Chan, J. M. Sandrik, R. J. Jennings, and R. F. Wagner, "MTF's and Wiener spectra of radiographic screen-film systems," U.S. Dep. HHS, Publ. FDA 82-8187, U.S. GPO, Washington, 1982.

[28] J. M. Sandrik, R. E. Shuping, and R. F. Wagner, to be published.

[29] J. M. Sandrik, R. F. Wagner, and K. M. Hanson, "Radiographic screen-film noise power spectrum: Calibration and intercomparison," *Appl. Opt.*, Oct. 1982, in press.

[30] J. M. Sandrik and R. F. Wagner, "Radiographic screen-film noise power spectrum: Variation with microdensitometer slit length," *Appl. Opt.*, vol. 20, pp. 2795-2798, 1981.

[31] K. M. Hanson, "Variability in the ideal observer," in *Proc. Soc. Photo-Optical Instrum. Eng., SPIE, Medicine XI*, Bellingham, WA, 1983.

[32] R. G. Simpson, and H. H. Barrett, "Coded-aperture imaging," in *Imaging for Medicine*, vol. 1, S. Nudelman and D. D. Patton, Eds. New York: Plenum, 1980.

[33] R. F. Wagner, D. G. Brown, and C. E. Metz, "On the multiplex advantage of coded source/aperture photon imaging," in *Proc. Soc. Photo-Optical Instrum. Eng., SPIE*, Bellingham, WA 1981, vol. 314, pp. 72-76.

[34] H. H. Barrett and W. Swindell, *Radiological Imaging: The Theory of Image Formation, Detection and Processing*, vol. 1. New York: Academic 1981, p. 174.

[35] L. Kaufman, D. Shosa, L. Crooks, and J. Ewins, "Technology needs in medical imaging," *IEEE Trans. Med. Imaging*, vol. MI-1, pp. 11-16, 1982.

[36] D. Shosa, L. Kaufman R. Hattner, and W. O'Connell, "Measurement of the texture contribution to image noise in scintigrams," in *Proc. Soc. Photo-Optical Instrum. Eng., SPIE, Medicine VIII*, Bellingham, WA, 1980, vol. 233, pp. 134-136.

[37] R. F. Wagner, D. G. Brown, and M. S. Pastel, "The application of information theory to the assessment of computed tomography," *Med. Phys.*, vol. 6, pp. 83-94, 1979.

[38] G. Cohen and F. A. DiBianca, "The use of contrast-detail-dose evaluation of image quality in a compute tomographic scanner," *J. Comput. Assist. Tomogr.*, vol. 3, pp. 189-195, 1979.

[39] G. Cohen, "Contrast-detail-dose analysis of six different computed tomographic scanners," *J. Comput. Assist. Tomogr.*, vol. 3, pp. 197-203, 1979.

[40] K. M. Hanson, "Detectability in computed tomographic images," *Med. Phys.*, vol. 6, pp. 441-451, 1979.

[41] ——, "On the optimality of the filtered backprojection algorithm," *J. Comput. Assist. Tomogr.*, vol. 4, pp. 361-363, 1980.

[42] O. J. Tretiak, "Noise limitations in X-ray computed tomography," *J. Comput. Assist. Tomogr.*, vol. 2, pp. 477-480, 1978.

[43] T. W. Ovitt, "Intravenous angiography using digital subtraction methods," in *Digital Subtraction Arteriography: An Application of Computerized Fluoroscopy*, C. A. Mistretta, A. B. Crummy, C. M. Strother, and J. F. Sackett, Eds. Chicago, IL: Year Book Medical, 1982, pp. 31-35.

[44] C. A. Mistretta and A. B. Crummy, "Diagnosis of cardiovascular disease by digital subtraction angiography," *Science*, vol. 214, pp. 761-765, 1981.

[45] G. Revesz, H. Kundel, and M. Graber, "The influence of structured noise on the detection of radiological abnormalities," *Ivest. Radiol.*, vol. 9, pp. 479-486, 1974.

[46] R. A. Kruger, C. A. Mistretta, T. L. Houk, W. Kubal, S. J. Riederer, D. L. Ergun, C. G. Shaw, J. C. Lancaster, and G. G. Rowe, "Computerized fluoroscopy techniques for intravenous study of cardiac chamber dynamics," *Invest. Radiol.*, vol. 14, pp. 279-287, 1979.

[47] R. A. Kruger and J. A. Nelson, "Digital angiography using band-pass filtration," *Diagnos. Imaging*, vol. 4, pp. 24-29, 1982.

[48] S. Riederer, A. Hall, and N. Pelc, "Matched filtering in digital fluorography" (abstr.), *Med. Phys.*, vol. 9, p. 608, 1982.

[49] R. A. Kruger and P.-Y. Liu, "Digital angiography using a matched filter," *IEEE Trans. Med. Imaging*, vol. MI-1, pp. 16-21, 1982.

[50] C. A. Mistretta, private communication, Jan. 4, 1983.

[51] D. G. Grant, "Tomosynthesis: A three-dimensional radiographic imaging technique," *IEEE Trans. Biomed. Eng.*, vol. BME-19, pp. 20-28, 1972.

[52] R. F. Wagner and R. J. Jennings, "The bottom line in radiologic dose reduction," in *Proc Soc. Photo-Optical Instrum. Eng., SPIE*, Bellingham, WA, 1979, vol. 206, pp. 60-66.

[53] R. E. Shuping, T. R. Fewell, R. A. Phillip, R. R. Gross, and C. K. Showalter, "Dose reduction potential of carbon fiber material in diagnostic radiology," in *Proc. SPIE*, Bellingham, WA, 1980, vol. 233, pp. 264-268.

[54] H. Jafroudi, E. P. Muntz, H. Bernstein, and R. J. Jennings, "A multiparameter optimization of mammography," in *Proc. Soc. Photo-Optical Instrum. Eng., SPIE, Medicine X*, Bellingham, WA, 1982.

[55] M. Linzer, Ed., (conf. abstr.), *Ultrason. Imaging*, vol. 4, pp. 171-199, 1982.

[56] R. F. Wagner, S. W. Smith, J. M. Sandrik, and H. Lopez, "Second-order statistics of speckle in ultrasound *B*-mode scans of a uniform scattering phantom," *IEEE Trans. Sonics Ultrason.*, June 1983.

[57] L. G. Joynt, S. E. Green, P. J. Fitzgerald, D. S. Rubenson, and R. L. Popp, "*In vivo* ultrasound characterization of mural thrombi using amplitude distribution analysis" (abstr.), *Ultrason. Imaging*, vol. 3, p. 191, 1981.

[58] F. G. Sommer, L. F. Joynt, B. A. Carroll, and A. Macovski, "Ultrasonic characterization of abdominal tissues via digital analysis of backscattered waveforms," *Radiology*, vol. 141, pp. 811-817, 1981.

[59] S. W. Smith, R. F. Wagner, J. M. Sandrik, and H. Lopez, "Low contrast detectability and contrast/detail analysis in medical ultrasound," *IEEE Trans. Sonics Ultrason.*, June 1983.

[60] S. W. Smith and H. Lopez, "A contrast-detail analysis of diagnostic ultrasound imaging," *Med. Phys.*, vol. 9, pp. 4-12, 1982.

[61] A. E. Burgess, K. Humphrey, and R. F. Wagner, "Detection of bars and discs in quantum noise," in *Proc. Soc. Photo-Optical Instrum. Eng., SPIE*, Bellingham, WA, vol. 173, pp. 34-40, 1979.

[62] R. F. Wagner, "Imaging with photons: A unified picture evolves," *IEEE Trans. Nucl. Sci.*, vol. NS-27, pp. 1028-1033, 1980.

[63] L.-N. Loo, K. Doi, M. Ishida, and C. E. Metz, "An empirical investigation of variability in contrast-detail diagram measurements," in *Proc. Soc. Photo-Optical Instrum. Eng. Medicine XI*, 1983, vol. 419, to be published.

[64] C. E. Metz, D. G. Brown, and R. F. Wagner, unpublished analysis and computer models.

[65] A. E. Burgess, R. F. Wagner, R. J. Jennings, and H. B. Barlow, "Efficiency of human visual discrimination," *Science*, vol. 214, pp. 93-94, 1981; see also *J. Appl. Photog. Eng.*, vol. 8, pp. 76-78, 1982, for further discussion of this work.

[66] L. D. Harmon and B. Julesz, "Masking in visual recognition: Effects of two-dimensional filtered noise," *Science*, pp. 1194-1197, 1973.

Reprinted from *Physics in Medicine and Biology*, **29**, (2), pp. 97–113, by kind permission of The Institute of Physics (UK). ©1984

The Eighteenth Douglas Lea Lecture

40 years of radiobiology: its impact on radiotherapy†

Jack F Fowler

Gray Laboratory of the Cancer Research Campaign, Mount Vernon Hospital, Northwood, Middx HA6 2RN, England

Abstract. The contributions of radiobiology to radiotherapy in the past 40 years are reviewed, taking the work of Gray and Read on bean roots with high LET radiation as the starting point. The main impact has been strategic and didactic because methods of measuring parameters in patients so as to affect individual treatments have not yet been developed. Many patients are now being treated with radiotherapy who would have been considered unsuitable 40 years ago and the radiobiological reasons for these changes are discussed. Improvements that have occurred in radiotherapy are reviewed and some projections about future potential improvements are made.

1. Historical introduction

Forty years ago, when the HPA was born, Douglas Lea was writing his epoch-making book at Cambridge, Gray and Read were producing their pioneering results at Northwood, using bean roots, and Robert Stone had just finished treating his series of patients with fast neutrons from the cyclotron at Berkeley, California (Lea 1946, Gray and Read 1942, 1943, 1944, 1948, Stone 1948). Lea's book provided the launching pad for a generation of radiobiologists. Gray and Read's work provided, ten years later, the main rationale for radiotherapy with high LET beams; an increase in RBE and a decrease of OER with increasing LET (Gray *et al* 1953). Figure 1 shows Gray and Read and the wooden hut at Mount Vernon Hospital in which they built, in 1936, the first neutron generator designed for radiobiological studies, a 400 kV D–D Cockroft–Walton generator (Read 1982).

> 'We believed that it was easier to get money for salaries than for equipment and the choice of £500 was based on the supposition that we would do as much construction work as we could. We built our own oil diffusion pumps and I wound the autotransformer and several small transformers. We sawed up angle iron and built it into frameworks to hold the apparatus and Gray chipped channels in the concrete floor and cemented water pipes in them. We ground up to 2 inches off the ends of four glass cylinders which formed the bodies of the rectifier column and the ion tube. As a result it was not until February 1938 that evidence was obtained of the generation of neutrons. We had begun on January 1st 1936. This delay was more serious because there were not many years left before the war broke up our partnership.'

Perhaps, if Gray and Read had been able to build their neutron generator faster, or if the war had not interrupted so much work, we should be dealing with fewer radiobiological uncertainties now.

† This lecture was given at the HPA 40th Anniversary Conference in Newcastle upon Tyne, September 1983.

0031-9155/84/020097+17$2.25

(*a*)

(*b*) (*c*)

Figure 1. (*a*) The first purpose-built high-LET radiobiology laboratory in the world at Mount Vernon Hospital, 1935. (*b*) Hal Gray. (*c*) John Read.

Hal Gray had been appointed to Mount Vernon Hospital in 1933, from Rutherford's Cavendish Laboratory at Cambridge, to become one of the first five or six hospital physicists.

Gray and Read developed dosimetry and then went on to do radiobiological experiments. Meanwhile, Ernest Lawrence had built his cyclotron. His brother John Lawrence, MD, was interested in its application to cancer treatment. Paul Aebersold, a physicist, measured neutron doses in 'n-units' and with John Lawrence carried out at Berkeley irradiations of rats bearing tumours. The tumours were probably more responsive because of graft-vs-host immunity than would be thought acceptable now, but their work led to Stone's clinical trial of neutrons from 1937 until it was stopped by the war in 1942. This was the scene 40 years ago.

The recent progress in high LET radiotherapy has been reviewed elsewhere (Fowler 1982).

2. Has radiobiology helped radiotherapy?

There are major changes in the strategy of radiotherapy, compared with 40 years ago, which are due to radiobiology. Of course the introduction of megavoltage radiotherapy has had a major effect, and without it the concepts of radiobiology could not have had the impact that they have had, as summarised in table 1.

Table 1. Steps in radiotherapy strategy.

Step	Radiobiological reason
Prophylactic irrad. of node spread areas	$\frac{3}{4}$ of full dose will sterilise small nodules
Surgery + RT in combination	Both can be less than full radical
Boost doses	Highest dose required in centre of tumour
Hyperfractionation	Small doses per F cause less late damage

Even now there have not been test procedures developed which enable the radiobiological response of tumours or normal tissues *in individual patients* to be measured in time to affect their treatment. This may come with the use of flow cytometers and monoclonal antibodies against proliferating cells but the impact of radiobiology so far has been strategic and didactic (Alper 1979).

Some of the new strategies in radiotherapy that depend upon radiobiological concepts are outlined in table 1. In the 1940s Ralston Paterson's methods had a major influence on radiotherapy in many countries (Paterson 1948). There are several important aspects which have changed as a result of radiobiological knowledge.

(1) No shrinking field or boost-dose techniques were used; the initial large treatment volume was treated to full tolerance dose.

(2) The use of somewhat reduced doses to treat areas of possible nodal spread was, if used at all, controversial.

(3) No special emphasis was placed on attempting higher doses for larger tumours. These three points are of course related.

(4) Treatment fields were square or rectangular; no edge or corner trimming or shaping was used.

(5) Radiation alone was the general rule. Little priority was given to the planned combination of radiation and surgery. The multidisciplinary approach, with joint clinics, developed much later. The use of radiotherapy planned together with surgery, so that neither need be as radical as when used alone, is still developing.

(6) Results of treatment were presented in terms of absolute survival at a fixed period (e.g., 5 years). Relative or actuarial survival was introduced later, together with the importance of local control and normal-tissue complications as biological indicators for better or worse treatments. 'Quality of life' is still being developed so as to be a measurable factor.

(7) Fractionated schedules were based on Baclesse and Coutard's 'daily doses for several weeks', sometimes with modifications towards fewer and larger fractions for convenience.

The changes in strategy that have occurred as a result of radiobiological discoveries are outlined in table 2.

Table 2. Patients now accepted for radiotherapy.

Type	Radiobiological reason
Hodgkin's disease	Lymphoid cells have small repair capacity
'Radioresistant' tumours	Range of D_0 seemed small at first
Tumours which shrink slowly	Shrinkage rate depends on cell kinetics more than on cell kill
Malignant melanoma	Big shoulder, so large doses per fraction (?)

3. Hodgkin's disease

The first example in table 2 shows how physics is intimately related with radiobiology in any advances in radiotherapy. Hodgkin's disease is, at first sight, an unlikely disease to be well treated by external radiation because the pathways of lymphatic spread are far from geometrically simple. The first step was to use the poorly penetrating 250 kV x-rays for half-trunk or whole-trunk 'baths' (Peters and Middlemiss 1958). This could not have been successful if lymphoid cells were not unusually radiosensitive: their cell survival curves have small shoulders and relatively small values of D_0. The 5-year survival of patients rose from 5% to 35% (table 3), a notable beginning. The next

Table 3. Improvement in 5-year survival from Hodgkin's disease.

Date	Treatment	5-year survival (%)
1940	Considered inevitably fatal	5
1960	250 kV x-rays	35
1970	Supervoltage mantle with exploratory surgery and lymphangiography	73
1980	Supervoltage plus chemotherapy	80

step was to apply supervoltage radiotherapy to the lymph chains only, avoiding lungs, kidneys, and much of the intestines. For this step, the introduction of the 'mantle' and the 'Y', the major developments in physics of ^{60}Co and linear accelerators were essential (Kaplan 1962). The cooperation of surgeons, pathologists, and diagnostic radiologists in defining the pathways of spread was also essential. The 5-year survival then rose from 35 to 73% (table 3). Later, the addition of chemotherapy has increased it to about 80%. Throughout, the collaboration was pushed along by determined and scientifically aware radiotherapists. From 5% to 80% survival in 40 years is a spectacular improvement.

4. Survival curves for mammalian cells

The second development in table 2 depends upon the obtaining of radiation cell survival curves for mammalian cells both *in vitro* (Puck and Marcus 1956) and *in vivo* (Hewitt and Wilson 1959, Till and McCullock 1961). It was noted in those days that the values of D_0 and n appeared to fall within a narrow range for several types of mammalian cell. Although more emphasis is placed today on differences than on similarities, this apparently narrow range of cellular radiosensitivity encouraged radiotherapists to try again, and harder, with tumours that had been believed to be radioresistant. The later concepts of radioresistance due to hypoxic cells, or to cells in late S or long G1, or to repair of potentially lethal damage, are still being used today. For example, Paterson listed carcinoma of the prostate, sarcoma of soft tissue, and carcinoma of the rectosigmoid region under the heading 'radiotherapy usually contra-indicated'. This listing arose from the earlier experience of treating very advanced tumours to the low doses necessitated by large volumes, and with 250 kV x-rays. It was the awareness of cell survival curves that persuaded pioneering radiotherapists to treat sarcoma of bone and carcinoma of the prostate with radiation, as reviewed by Suit (1983). It was the growing availability of supervoltage beams that enabled them to succeed. Tumour volume now appears a more critical factor than unidentified variations in cellular radiosensitivity. 'Radiotherapy alone for small tumours or for larger tumours if combined with surgery has been effective for nearly all tumour types' (Suit 1983). It is logical to use surgery to remove the primary tumour and then radiotherapy to eliminate smaller deposits of cancer. Both modalities can then be less than fully radical. This rationale has also led to the appreciation of the importance of the remarkably good dose distributions that can be achieved with heavy ion beams (Raju 1980), and also with dynamic conformational therapy, or 'arcing-tracking', using conventional linear accelerators (Brace *et al* 1981, Chin *et al* 1981). There are at present exciting physical developments in progress to achieve these good dose distributions (Fowler 1981).

4.1. *Cell survival curves and log cell kill*

With multiple fraction irradiation the log cell kill increases in proportion to the number of fractions. Neglecting the effect of proliferation between fractions, if a full radical dose kills 9 logs of cells (survival 10^{-9}), then two-thirds of that dose will kill 6 logs of cells, etc. Therefore, if a full radical dose will eliminate primary tumours of the order of centimetres diameter, two-thirds of a full radical dose should eliminate metastatic nodules of millimetres diameter. On this basis we can calculate that the prophylactic doses to regions of potential metastatic should depend upon how large the metastases are likely to be. For subclinical nodules (mm) the doses should be larger than for purely microscopic metastases ($\sim 100\ \mu m$). This strategy has been skilfully deployed by radiotherapists in some body sites (Fletcher 1973). It also forms the basis of shrinking field techniques, or the boosting of the central part of a target volume to a high dose.

4.2. *Dose–response curves*

The logarithmic nature of cell killing curves has consequences for the accuracy of dose delivery. If there were no heterogeneity in the cell population—which there always is in tumours—then an increase in dose by $2.3D_0$† would change the cell kill by one $\log_{10}$. In principle, this should increase the probability of local control from say 7%

† D_0 is the dose which reduces cell survival by 1/e on the logarithmic portion of a cell survival curve.

to 70%, although heterogeneity of cellular radiosensitivity or of radiation dose would reduce the increase. In practice, the change of local control with a ten per cent change in total dose varies from 10% (bladders, Morrison 1975) to 30% or more (larynx, Stewart and Jackson 1975). This steep response is an obvious encouragement to high accuracy in dose delivery in radiotherapy (figure 2).

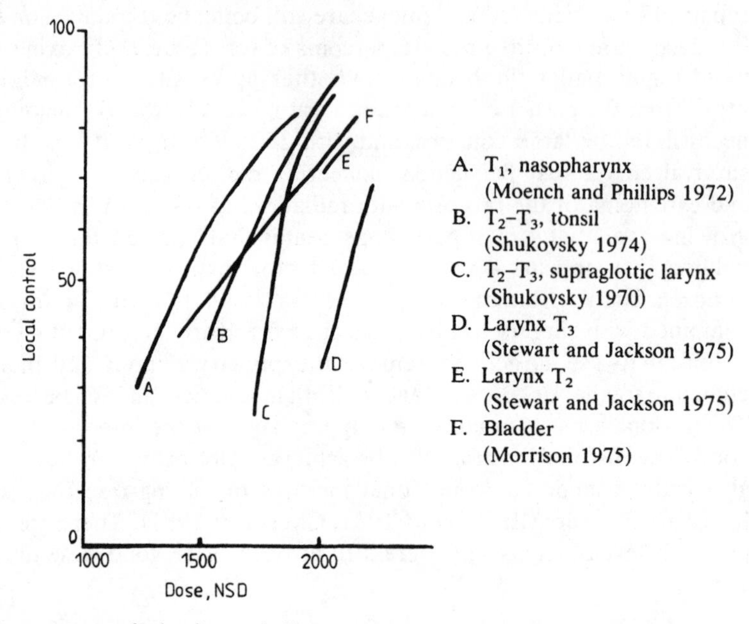

Figure 2. Clinical dose-response curves for local control of human tumours; plotted against nominal standard dose (NSD) as a way of normalising (early) normal tissue reactions when different total doses were used with different fractionation schedules (Dische (1978) q.v. for full references). (Reproduced by courtesy of the author and *Br. J. Radiol.*)

5. Tumours that shrink slowly

The third item in table 2 concerns tumours that shrink slowly.

The development of autoradiography by Howard and Pelc (1952) had an enormous impact on the whole biology of tissue development and proliferative cell renewal. Steel (1977) explained that the volume doubling time of tumours may be many times larger than the mean cell cycle time both because of large cell loss factors and because only a small portion of cells was proliferating (the growth fraction, Mendelssohn 1963). The rate of shrinkage during or after any treatment depends little on the amount of cell killing (provided that more than two decades of cells are killed). Shrinkage rate after treatment depends mainly on the rate of cell loss from the tumour, which is related to the rate of cell loss before treatment. Therefore tumours which shrink fast are those with a high natural rate of cell loss, and those which shrink slowly are tumours with a small cell loss factor (Denekamp 1972). Since the rate of shrinkage is no indication of the effectiveness of cell kill (Thomlinson 1982), there is no longer any reason to avoid the treatment of slowly shrinking tumours. 'They are not uniquely radioresistant' and such tumours include desmoids, carcinoma of the prostate or rectosigmoid and sarcoma of soft tissue. All of these have now been shown to respond successfully to radiotherapy even though they respond slowly (Suit 1983).

6. High LET radiotherapy

Robert Stone's (1948) neutron treatments at Berkeley (1937–1942) resulted in such severe late complications that no further attempts were made for 20 years. Some experiments with pigs at Hammersmith Hospital, and basic experiments with cells *in vitro*, revealed a major reason for this: that RBE increases greatly as dose per fraction is decreased. ·This finding enabled another clinical trial using neutrons to be started at Hammersmith. The rationales for high LET therapy have been recently reviewed elsewhere (Fowler 1983c). It is notable that neutron therapy clinical trials up to now have employed mostly neutron beams with penetration little better than 250 kV x-rays, except for a very few institutes with better penetration but with fixed horizontal beams. It is therefore present strategy to try neutron beams of penetration similar to 4–6 MeV photon linear accelerators with moveable isocentric beams. These are now being installed in about four clinical centres, including Clatterbridge near Liverpool. There are proposals for clinical trials based on radiobiological principles (Fowler 1983b, c). It should be noted that the cost per 6-week course of radiotherapy does not rise in linear proportion to the capital cost of a radiotherapy machine, but only about one-third as rapidly. Starting from Greene's figure of £137 per standard course of treatment (Greene 1983), the installation of a neutron cyclotron costing ten times as much as a conventional linear accelerator would increase the cost per course only to about £550 (Figure 3). Such sums can easily be exceeded in a course of anti-cancer drugs. Therefore we should not stop investigating high technology radiotherapy because it looks more expensive than present-day radiotherapy.

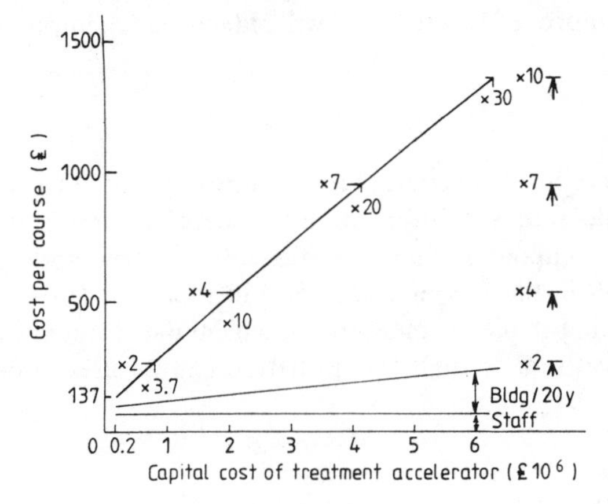

Figure 3. Costs of radiotherapy for 30-fraction treatment (or equivalent) as a function of capital cost of the accelerator. Included: radiographer, physics and technician staff; depreciation over 15 y for accelerator and simulator and 20 y for building; planning computer; electricity; 800 patients per year (Greene 1983). If more than one beam channel were assumed the costs could go down proportionately. The costs of medical staff, nurses, diagnostic tests and in-patient care are not included because they would apply to cancer treatments other than radiotherapy.

7. Highly localised doses

Heavy charged particles provide both biological advantages and excellent dose localisation (Raju 1980, Fowler 1981). Dose distributions that are approximately as good as those from negative pi mesons can be obtained by dynamic conformational therapy, i.e., arcing and tracking, using conventional photon linear accelerators controlled by microprocessors (Brace *et al* 1981, Chin *et al* 1981, Fowler 1983b). Dr Anthony Green (1959) proposed this method over 20 years ago. I welcome the currently increasing interest of accelerator manufacturers and users in this method of improving radiotherapy.

8. Is local control of tumours important?

It has sometimes been asked whether better radiotherapy is worthwhile, or whether the patients achieving local control will simply die later from their metastases. Suit (1982) has shown that better local control would indeed improve survival, especially in cancers of the intestine, uterine cervix, and head and neck. A radiobiological experiment in our laboratory showed a similar result for mouse tumours which metastasised. The response of the primary tumour increased more steeply with local X-ray dose than with systemic cyclophosphamide, as expected for the local response. But the proportion of mice rendered free of metastases also increased more steeply with the local X-ray dose than with the systemic drug dose; this was unexpected (Chu and Fowler 1982). However, consideration of the reverse result shows the obvious conclusion. If a primary tumour is *not* eliminated completely, by adequate radiotherapy or surgery, it will recur and will metastasise later. An even more definite result was reported in human patients by Anderson and Dische (1981). The proportion of patients exhibiting later metastases was strongly correlated with local control of cancer of the uterine cervix. This result is not trivial, because the long-term survival of patients was also strongly correlated with local control. It must be concluded that, in at least some common types of cancer, improved local control will indeed lead to improved survival.

9. Hypoxic cells

Some successes have been reported in trials of hyperbaric oxygen, but not in all (Fowler 1983a, b). Electron-affinic radiosensitisers of hypoxic cells have had little clinical success so far, but better compounds than misonidazole are now undergoing Phase I clinical trials: Ro-03-8799 in the UK and SR-2508 in the USA (Fowler 1983b). In addition, attempts are being made to measure which human tumours have hypoxic cells in them, so that sensitisers or high LET radiation can be used more specifically for them (Fowler 1983a, b).

10. Fractionated schedules

Cell proliferation kinetics has led to a much better understanding of the role of overall time in radiotherapy than was provided by the old 'cube root law' or by a single exponent of time as in NSD or CRE (Ellis 1969, Kirk *et al* 1971). Overall time has little effect on the total isoeffect dose for late-reacting tissues, because by definition they proliferate very little in the six weeks or so of radiotherapy. For early-reacting tissues, on the other hand, there can be a large effect of overall time, but only after a delay equivalent to the pretreatment turnover time of critical cells in the tissue at risk (table 4). For example, acute reactions in mouse skin require large increments of dose starting two weeks after fractionated daily irradiation, in rats after three weeks (Fowler 1984) and in pigs or human patients after four weeks (Turesson and Notter 1984). The increments can be as large as 100 cGy per week for acute skin reactions.

Table 4. (a) Starting time of proliferation observed in fractionated irradiations

Tissue	Mouse rat, pig	Time (days)	Comment (no. of fractions)	First author†
Jejunum	M	2–2.5	Crypt colonies (10–20f)	Tucker
Intestinal death	M	4–10	Small fields (2f)	Williams
Skin acute	M	12	Feet (9–14f)	Denekamp
	R	20	Feet (20f)	Moulder
Spinal cord				
C5–T2	R	56–112	White necr. (2f)	V d Kogel
	R	>112	Vascular (2f)	V d Kogel
C2–C4	R	20–30	Myelopathy and LI glial cells (2f)	Hornsey
L2–L5	R	6–27	(2f)	
	R	4–27	(5f)	V d Kogel
	R	11–30	(10f)	
Kidney	M	25–40	(2f)	Williams
	M	4–14	(9f)	Glatstein
	M	18–28	(14f)	Glatstein
	P	25–40	(2f)	Hopewell

(b) Extra dose required to compensate for proliferation (cGy).

Overall time (weeks)	Skin (rats)	Cervical spinal cord (rats)	Kidney (mice)
0–1	0	0	0
1–2	0	0	0
2–3	300	0	0
3–4	1000	0	0
4–5	1000	0	250 (weeks 4–7)
5–6	1000	0	
6–7	1000	0	
	Moulder and Fischer (1976)	Van der Kogel (1979)	Williams and Denekamp (1984b)

† For details of references see Fowler (1984).

The obvious conclusion is that the prolongation of overall times will spare early but not late reactions. This is one source of the dissociation between late and early damage to normal tissues in radiotherapy. Tumours, however, proliferate fast, even if their volume growth rate is slow because of high cell loss. Therefore prolongation spares tumours from cell kill, and does not spare late reactions—a totally undesirable result. The strategic conclusion is that prolonged overall times, gaps in treatment, and delays in starting treatment might be more dangerous than previously thought, at least for tumours that grow rapidly (Steel 1977). Efforts to shorten overall times must not be made by using fewer and larger fractions, because this leads to more severe late injury (Withers *et al* 1982, Fowler 1983c, 1984). The latter conclusion was drawn first from animal experiments but there is now ample evidence of the trend in clinical radiotherapy. Therefore multiple fractions per day (MFD) is a logical development in radiotherapy, and is indeed being tested in clinical trials in a number of centres. I shall describe this development as an up to date example of the effect of radiobiological principles on radiotherapy.

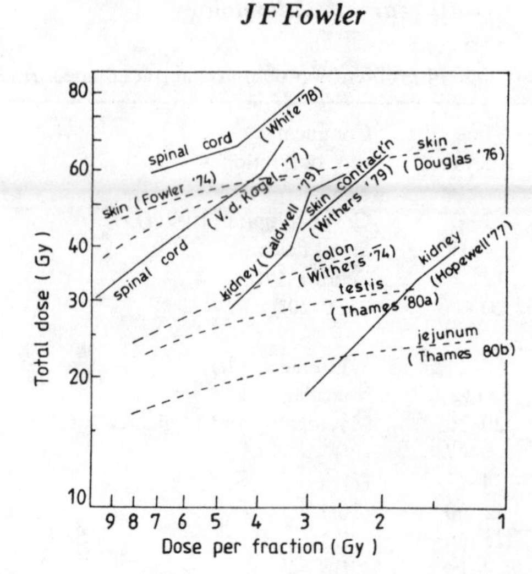

Figure 4. Total dose required to produce a given iso-effect in normal tissues as a function of dose per fraction. 2 Gy corresponds to 30 fractions, 4 Gy to 10 fractions, and 6 Gy to 6 fractions approximately. The full lines are for late effects. The dotted lines are for early-reacting tissues (Withers *et al* (1982), q.v. for full references). (Reproduced by courtesy of author, Raven Press, New York and Elsevier, Amsterdam.)

10.1. The dependence of isoeffect dose on dose per fraction

The realisation that the dependence of total isoeffect dose on the size of each dose fraction is different for late and early injury is an important recent application of animal radiobiology to radiotherapy (figures 4 and 5). During the past ten years or

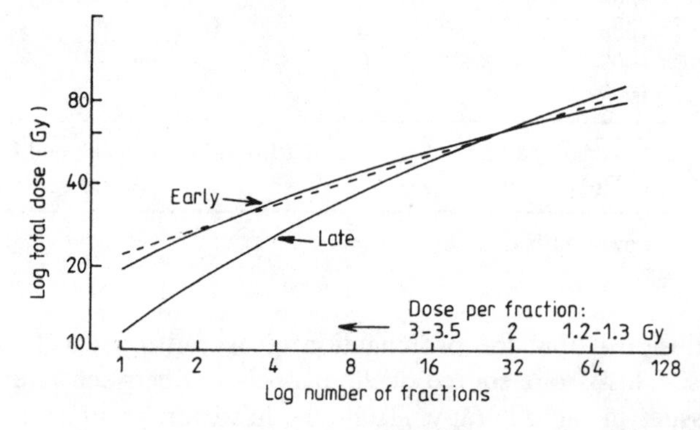

Figure 5. Schematic plot of total doses required in radiotherapy to produce clinically acceptable normal-tissue reactions, based on figure 3 (Fowler 1983c). (Reproduced by courtesy of Elsevier, Amsterdam.)

so a number of new assays for late normal-tissue damage in experimental animals has been developed. Among these tissues are spinal cord (Van der Kogel 1979), skin contraction (Wiernik *et al* 1977), bladder (Stewart *et al* 1978), lung (Travis *et al* 1980), colorectum (Terry and Denekamp 1983), kidney (Williams *et al* 1982), many but by no means all of these being from Gray's old laboratory at Mount Vernon Hospital. From multi-fraction isoeffect experiments using typically 1, 2, 4, 8, 16, 32 and (in a few experiments) 64 fractions, ideally given within a shorter overall time than the 'starting times' for time-dependent incremental doses in table 4, the shape of the dose-response curves for the critical tissues can be calculated. It is convenient, and sufficiently accurate for radiotherapy sized fractions, to express them in the form:

$$\text{effect} = \alpha d + \beta d^2 \tag{1}$$

where d is the dose per fraction and it is assumed that the intervals are sufficient to allow full repair on the 'repairable' term β. The ratio of α/β expresses the relative size of single-hit damage to 'repairable' damage. At a dose per fraction of $d = \alpha/\beta$ grays, these components of damage are obviously equal. The ratio α/β can be determined if the isoeffect doses of several different fractionation schedules are known. The ratio α/β indicates the degree of curvature on the dose response curves, large values representing rather straight curves and small values rather 'curvy' curves. The reason for our interest in α/β values for different tissues is shown in table 5. Early-reacting tissues tend to have high values, whilst late-reacting tissues have mostly low values. There is little overlap, although some might be expected around 6 Gy as further values are obtained.

Table 5. Ratio of linear to quadratic terms from multifraction experiments.

	α/β (Gy)
Early reactions	
Skin	9–12
Jejunum	6–10
Colon	10–11
Testis	12–13
Callus	9–10
Late reactions	
Spinal cord	1.0–4.9
Kidney	1.5–2.4
Lung	2.4–6.3
Bladder	3.1–7

It is not yet known exactly why the dose–response curves are straighter for rapidly than for slowly proliferating tissues, but a plausible suggestion has been made by Joel Bedford (personal communication). Cells in mitosis have steep, straight survival curves but cells in a long G_1 phase have the bending survival curves characteristic also of cells in late S phase (Sinclair and Morton 1966). If the tissue response curves reflect the shape of cell survival curves, a logical explanation is then available. This explanation links back to the 80-year-old 'law' of Bergonie and Tribondeau, that tissues which are proliferating fast appear more readiosensitive than those which proliferate slowly. Although we would now question whether 'radio responsive' would not be a better term, the general observation now seems to be coming true for small doses per fraction.

Ratios of α/β can be readily calculated by plotting the reciprocal of the total isoeffective dose against dose per fraction for several fractionation schedules for a given normal-tissue response (figures 6 and 7, Douglas and Fowler 1976). If equation (1) is valid for that tissue, a straight line can be drawn through the points. Each set of experiments provides its own check of whether the data fit equation (1), i.e., if the points lie close to a straight line. The intercept on the zero dose axis is '$\alpha \div$ damage' and the slope is '$\beta \div$ damage'. The ratio α/β can obviously be obtained readily, but absolute values require that a clone-counting assay should be available for that tissue and relevant to the isoeffect studies.

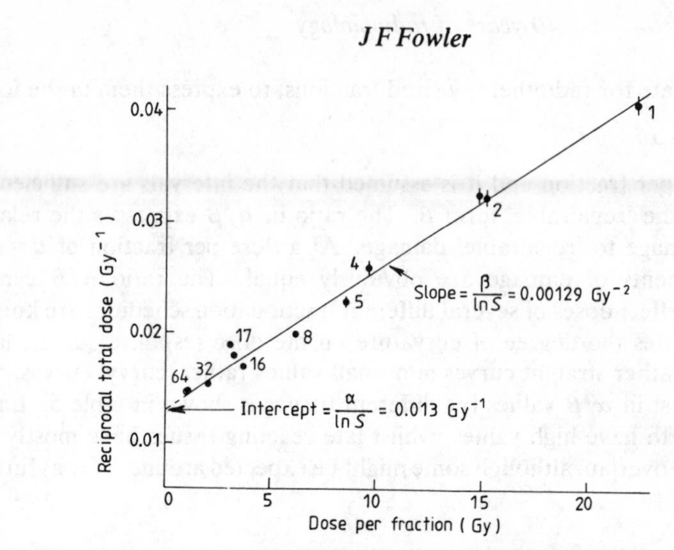

Figure 6. Reciprocal total dose as a function of dose per fraction for fractionation schedules which all yielded the same average early reaction in skin of the mouse foot. The figures beside the points indicate the numbers of fractions. The overall times were 8 days or less (Douglas and Fowler 1976). $\alpha/\beta = 10.1$ Gy. (Reproduced from Fowler (1983c) by courtesy of Elsevier, Amsterdam.)

10.2. Applications in radiotherapy

There are three practical applications in radiotherapy of this model.

(1) At a dose per fraction of 0.1 α/β the total dose will be 10% different from the α component alone. This is a just detectable difference, in clinical trials, and corresponds to Withers' (1977) 'flexure dose', the dose-per-fraction at which a just detectable sparing of damage is still present as the dose-per-fraction is decreased. It is the dose per fraction which radiotherapy might have to go down to, if equation (1) holds, to obtain the maximum benefit of the sparing of late injury by hyperfractionation. The significant point is that it corresponds to very low doses, as small as 20–40 cGy

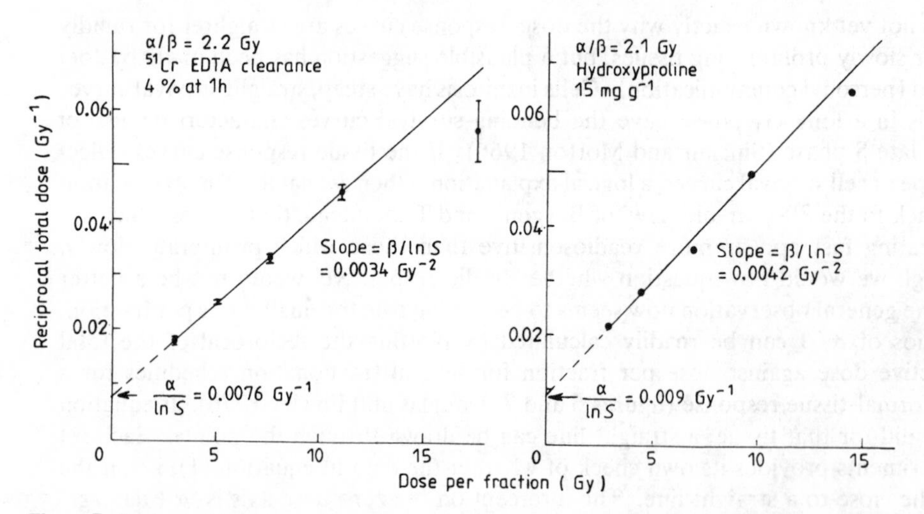

Figure 7. Reciprocal total dose plot, as in figure 5 but for two end-points 6–9 months after local irradiation of mouse kidneys. Left: kidney function measured by clearance of EDTA. Right: proportion of collagen (hydroxyproline) per dry weight of kidney at sacrifice. The points are for 1, 2, 4, 8, and 16 fractions (right to left) all yielding the same end-point (Williams *et al* 1982).

per fraction, before the additional sparing of late damage becomes negligible. The importance of treating every field on every treatment day is emphasised. For example, a thin shield in front of the kidney for all 30 fractions would be better than no shield and then stopping the irradiation of the kidney when a fixed proportion of the total dose was reached. When we remember that overall times should not be prolonged, (because tumours proliferate too much), then the necessity for multiple fractions per day is obvious. The limitation will be the minimum intervals needed for adequate repair between fractions. Although 4–8 h intervals are being used now, further work on this point is necessary.

(2) Total doses for a new fractionated schedule can now be readily calculated from those in a known schedule, with a new discrimination between tissues which NSD or CRE have not employed. The effect of overall time must be allowed for separately with the aid of table 4 or better data when they become available.

$$\text{Effect of schecule 2} = \text{Effect of schedule 1}$$

$$n_2(\alpha d_2 + \beta d_2^2) = n_1(\alpha d_1 + \beta d_1^2)$$

$$n_2 d_2(\alpha + \beta d_2) = n_1 d_1(\alpha + \beta d_1)$$

$$\frac{\text{Total dose 2}}{\text{Total dose 1}} = \frac{n_2 d_2}{n_1 d_1} = \frac{(\alpha/\beta + d_1)}{(\alpha/\beta + d_2)} \qquad (2)$$

It is obvious that the change of total dose with dose-per-fraction is larger if α/β is small (i.e., for late effects) than if it is big (i.e., for early effects) as indicated in figures 4 and 5. (Withers *et al* 1982, Barendsen 1982, Fowler 1984). Before such an approach is adopted widely, further results are required at lower doses per fraction for normal tissues. It is emphasised that equation (2) depends upon the validity of equation (1) for each fraction; upon allowing long enough intervals for the β term to recover to its original value; and upon making a separate allowance for the effects of overall time.

(3) The choice between *hyperfractionation* and *accelerated fractionation* can be made on a rational basis. *Hyperfractionation* is the use of more and smaller fractions than currently employed, such as 60 fractions of 115 cGy given twice a day in the usual six weeks. The object is to spare late injury, or more precisely, to put up the total dose so that equal late complications occur but more damage accrues to faster-proliferating tissues especially tumours. *Accelerated fractionation* is the use of the same number and size of fractions as usually, but given twice a day in half the usual overall time (Svoboda 1978). The object is to prevent too much tumour cell proliferation.

Obviously, the faster the tumour cells proliferate, the greater the advantage of accelerated fractionation. For slowly growing tumours, hyperfractionation should be the best strategy: late damage is spared, so that increased doses could be given to keep late complications at the same incidence. Calculations of the relative amounts of damage in late-reacting tissues, (assuming $\alpha/\beta = 3$ Gy), and in early-reacting tissues including tumours (assuming $\alpha/\beta = 10$ Gy) indicate that a break-even cell-number doubling time of about five days applies when either two or three fractions per day are used (Thames *et al* 1983). Table 6 shows that some types of human tumour do appear to have such rapid proliferation rates. There is no cause for complacency in our present overall times of six weeks or so.

There would be more severe early normal-tissue damage if accelerated fractionation were widely used. The present way of avoiding this—by prolongation—is wrong. The

Table 6. Doubling times for human tumours.

No	Type	Volume doubling time, T_D (days) Median (Range)	Cell number doubling time, T_{pot} (days)	Cell loss factor (%)
56	Colorectal Ca	90(60–170)	3.1	96%
27	Sq. cell Ca, Hd and Nk	45 (33–150)	6.8	85
55	Undiff. bronch. Ca	90 (40–160)	2.5	97
8	Malig. Melanoma	52 (20–150)	14	73
101	Sarcomas	39 (16–78)	23	40
11	Lymphomas	22 (15–70)	16	29
4	Childhood tumours˙	20	3.6	82

Steel (1977). (Reproduced by courtesy of the author and Oxford University Press.)

total dose should be reduced a little instead, and it has to be determined whether the new schedule is more efficient than the standard schedule even with its slightly reduced dose. This remains to be tested in clinical trials. Further efforts need to be made to measure the cell proliferation rates in individual human tumours, so that such non-standard schedules could be applied to patients who would benefit most.

This application of radiobiology to therapy has been described in some detail because its practical implications are still being worked out, including multiple fractions per day.

It is worth adding that the sparing of late radiation complications, obtainable with hyperfractionation as explained above, is also obtainable by continuous low dose rate irradiations. Then only the α component of the dose-response curve would be effective. In addition, interstitial or intracavitary irradiations at 40–60 cGy per hour can deliver a full treatment dose in about seven days. This is commendably shorter than overall times of six weeks. These are two radiobiological reasons why interstitial and intracavitary radiotherapy can give good results, as well as the localised distributions of physical dose which were also emphasized in earlier sections.

11. Conclusions

Other points emerge from other applications, as listed in table 7, where radiobiological reasons for choosing a particular strategy of radiotherapy for a particular type of tumour are summarised. These have been reviewed recently by the author (Fowler 1983a, b).

Table 7. Possible future selection of modality.

Tumour	Modality	Reason
Slow	High LET	Resist to X-rays in G_1
Fast	Accel. fract.	Fast prolif. in tumour
Hypoxic	Radiosensitisers or high LET	If measurement of hypoxic cells possible
Anaemic	Radiosensitisers or high LET	(Bush; Dische, Watson and Sealy; Taskinen)
Close to spinal cord or kidney	Hyperfract.	Sparing of late injury by small doses per fraction
Large	Hyperthermia	Lower blood flow, so H/T more effective

As mentioned above, methods are being developed in various laboratories to measure the number of hypoxic cells in individual tumours. When such methods are available for individual human tumours, both radiosensitisers of hypoxic cells and high-LET radiation can be used more selectively than at present. There is evidence that certain subgroups of patients would benefit from methods of overcoming hypoxic-cell radioresistance, but the selection of patients needs to be more specific in future.

Table 8. Prospects for the improvement of radiotherapy.

1. Localisation of disease (CT, NMR, MCA)
2. Localisation of treatment (conformational tracking: charged particles)
3. Hyperthermia
4. Vulnerable capillaries in tumours
5. Hypoxic cell sensitisers; measurement of hypoxic cells
6. High LET: resistance of slowly proliferating cells to low LET and hypoxic resistance
7. Minimising late injury (hyperfractionation)
8. Accelerated fractionation: measurement of proliferation rates in tumours
9. Interdigitating chemotherapy
10. Targetted antibodies (β, α, boron)
11. Radioprotective drugs
12. Clinical trials: two dose levels

Table 8 lists the growing points in radiotherapy which I believe are important, in a very rough order of priority. At the top are the developments concerning good localisation of dose, based on better localisations of malignant cells of course. Most of the others are discussed in detail above or in the reviews referred to. Hyperthermia is an important and rapidly developing modality, which should be used with radiotherapy, not alone.

Finally, figure 8 illustrates the pattern of irrational evaluation that has all too often been followed in the development of new modalities for clinical application. Eventually, each method achieves the quasi-equilibrium position it deserves, until further developments alter the equilibrium.

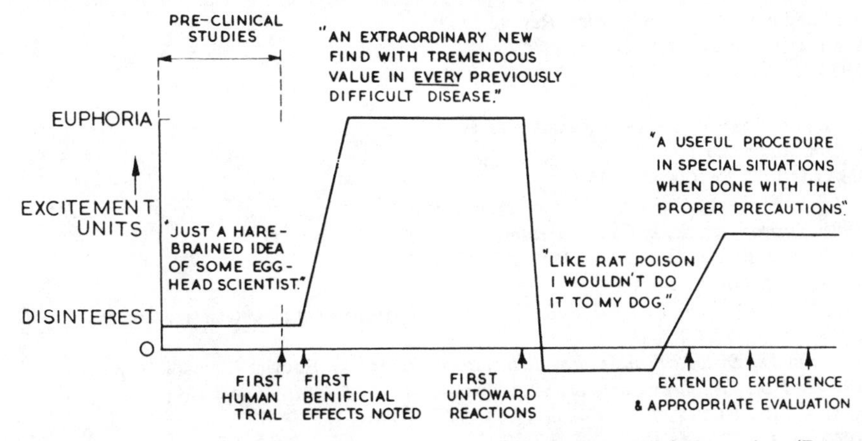

Figure 8. The irrational pattern of evaluation of new methods applied to clinical practice. (Reproduced from Fowler (1977) by courtesy of Pergamon Press.)

Acknowledgments

It is a pleasure to acknowledge the encouragement from, and the stimulating discussions with, many colleagues and friends in many laboratories, most but not all of them mentioned in the list of references. I appreciate the honour of being asked to give the Lea Memorial Lecture at the 40th Annual Meeting of the HPA at Newcastle upon Tyne.

Résumé

40 ans de radiobiologie: son impact sur la radiothérapie

Partant du travail de Gray et Read sur les racines de fèves dans les rayonnements à TEL élevé, l'auteur passe en revue les apports de la radiobiologie à la radiothérapie durant les 40 dernières années. L'impact principal de la radiobiologie sur la radiothérapie a été plutôt stratégique et didactique, car les méthodes de mesure des paramètres sur les malades, permettant d'intervenir au niveau des traitements individuels, n'ont pas encore été mises au point. De nombreux malades sont soumis aujourd'hui à une radiothérapie qui aurait été récusée il y a 40 ans; les raisons radiobiologiques de ces changements sont discutées. L'auteur passe en revue les améliorations de la radiothérapie et envisage les progrès potentiels pour le futur.

Zusammenfassung

40 Jahre Strahlenbiologie: ihr Einfluß auf die Strahlentherapie

Ein Überblick wird gegeben über die Beiträge der Strahlenbiologie zur Strahlentherapie in den letzten 40 Jahren, wobei die Arbeit von Gray und Read über Bohnenwurzeln, bestrahlt mit hohem LET als Ausgangspunkt dient. Der Haupteinfluß war planerischer und didaktischer Natur, da bisher noch keine Verfahren zur Messung von Parametern im Patienten zur Beeinflussung der individuellen Behandlung entwickelt wurden. Viele Parienten werden jetzt strahlentherapeutisch behandelt, die man vor 40 Jahren noch als ungeeignet für eine solche Therapie betrachtete. Die strahlenbiologischen Gründe für diese Veränderungen werden diskutiert. Ein Überblick wird gegeben über die Verbesserungen in der Strahlentherapie und Vorstellungen über mögliche zukünftige Verbesserungen werden aufgezeigt.

References

Alper T 1979 *Cellular Radiobiology* (Cambridge University Press)
Anderson P and Dische S 1981 *Int. J. Radiat. Oncol. Biol. Phys.* **7** 1645
Barendsen G W 1982 *Int. J. Radiat. Oncol. Biol. Phys.* **8** 1981
Brace J A, Davy T J, Skeggs D B L and Williams H S W 1981 *Br. J. Radiol.* **54** 1068
Chin L M, Kijewski P K, Svensson G K and Bjärngard B E 1981 *Int. J. Radiat. Oncol. Biol. Phys.* **7** 61
Chu A M and Fowler J F 1982 *Cancer Res.* **42** 1943
Denekamp J 1972 *Eur. J. Cancer* **8** 335
—— 1973 *Br. J. Radiol.* **46** 381
Dische S 1978 *Br. J. Radiol.* **51** 888
Douglas B G and Fowler J F 1976 *Radiat. Res.* **66** 401
Ellis F *Clin. Radiol.* **20** 1
Fletcher G H 1973 *Br. J. Radiol.* **46** 1
Fowler J F 1977 *Int. J. Radiat. Oncol. Biol. Phys.* **3** 351
—— 1981 *Nuclear Particles in Cancer Treatment* (Bristol: Adam Hilger)
—— 1982 *Int. J. Radiat. Oncol. Biol. Phys.* **8** 2207
—— 1983a *Radiother. Oncol.* **1** 1
—— 1983b in *Proc. 7th Int. Congr. of Radiation Research, Amsterdam 1983* (Amsterdam: Martinus Nijhoff Publishers) pp. 13–22
—— 1983c in *The Biological Basis of Radiotherapy* (Amsterdam: Elsevier Biomedical) pp 181 and 261
—— 1984 in *Proc. 11th Gray Conference, Glasgow—1983 Br. J. Cancer Suppl.* (in press)
Gray L H and Read J 1942 *Br. J. Radiol.* **15** 11, 39, 72, 320
—— 1943 *Br. J. Radiol.* **16** 125
—— 1944 *Br. J. Radiol.* **17** 271

Gray L H and Read 1948 *Br. J. Radiol.* **21** 5
Gray L H, Conger A D, Ebert M, Hornsey S and Scott O C A 1953 *Br. J. Radiol.* **26** 638
Green A T 1959 *Proc. R. Soc. Med.* **52** 344
Greene D 1983 *Br. J. Radiol.* **56** 189
Hewitt H B and Wilson C W 1959 *Br J. Cancer* **13** 69
Howard A and Pelc S R 1952 *Heredity Suppl.* **6** 216
Kaplan H S 1962 *Radiology* **78** 553
Kirk, J, Gray W M and Watson E R 1971 *Clin. Radiol.* **22** 145
Lea D E 1946 *Actions of Radiations on Living Cells* (Cambridge University Press)
Mendelsohn M L 1963 in *Cell Proliferation* ed. L F Lamberton and R J M Fry (Oxford: Blackwell) p. 190
Morrison R 1975 *Clin. Radiol.* **26** 67
Paterson R 1948 *The Treatment of Malignant Disease by Radium and X-rays* (London: Edward Arnold)
Peters M V and Middlemiss K C H 1958 *Am. J. Roentgenol.* **75** 553
Puck T T and Marcus P I 1956 *J. Exp. Med.* **103** 653
Raju M R 1980 *Heavy Particle Radiotherapy* (New York: Academic Press)
Read J 1982 in *Silver Jubilee Book of the Gray Laboratory* ed. J Denekamp (Northwood: Cancer Research Campaign) p. 96
Sinclair W K and Morton R A 1966 *Radiat. Res.* **29** 450
Steel G G 1977 *Growth Kinetics of Tumours* (Oxford: Clarendon) pp. 191 and 202
Stewart F A, Michael B D and Denekamp J 1978 *Radiat. Res.* **75** 649
Stewart J G and Jackson A W 1975 *Laryngoscope* **85** 1107
Stone R S 1948 *Am. J. Roentgenol. Radiat. Ther.* **59** 771
Suit H D 1982 *Cancer* **50** 1227
—— 1983 *Radiat. Res.* **94** 10
Svoboda V H J 1978 *Br. J. Radiol.* **51** 363
Thames H D, Withers H R, Peters L J and Fletcher G H 1982 *Int. J. Radiat. Oncol. Biol. Phys.* **8** 219
Thames H D, Peters L J, Withers H R and Fletcher G H 1983 *Int. J. Radiat. Oncol. Biol. Phys.* **9** 127
Thomlinson R H 1982 *Clin. Radiol.* **33** 482
Terry N H A T and Denekamp J 1982 *Br. J. Radiol.* **55** 942
Till J E and McCulloch E A 1961 *Radiat. Res.* **14** 213
Travis E L, Down J D, Holmes S J and Hobson B 1980 *Radiat. Res.* **84** 133
Turesson I and Notter G 1984 *Int. J. Radiat. Oncol. Biol. Phys.* **10** (in press)
Van der Kogel A J 1979 *Late Effects of Radiation on the Spinal Cord* (Univ. of Amsterdam Thesis, T N O Rijswijk)
Wiernik G, Hopewell J W, Patterson T J S, Young C M A, Foster J L 1977 in *Radiobiological Research and Radiotherapy* Vol I (Vienna: IAEA) p 93
Williams M V, Soranson J and Denekamp J 1982 *Br. J. Radiol.* **55** 944
Withers H R 1977 *Radiat. Res.* **71** 24
Withers H R, Thames H D and Peters L J 1982 in *Progress in Radio Oncology* Vol II ed K H Kärcher, H D Kogelnik and G Reinarts (New York: Raven Press) p 287

The physics of medical imaging

Images computed from x-ray absorption, from positron emission, from nuclear magnetic resonance and from reflections of ultrasound are showing tissues and details unimagined 25 years ago.

Reprinted with permission from *Physics Today*, **36**, pp. 36–42,

Paul R. Moran, R. Jerome Nickles and James A. Zagzebski

The growth in the applications of physics to medicine is perhaps best illustrated by the developments in medical imaging. When AAPM was founded, medical imaging consisted primarily of x-ray fluoroscopy (direct images) and radiography (exposures on film). Although these techniques still dominate the work of a radiology department, there are a number of new ways of making images of the interior of the living body. (See, for example, the cover of this issue and figure 1.) Many of these new imaging methods depend on computers to perform the enormous amounts of data-analysis required—computed tomography or nmr imaging, for example, would not be possible without computers—and in many other cases computers serve to provide enhanced images that may allow more accurate diagnoses.

In this article we will discuss several of these new medically useful imaging techniques. Digital subtraction angiography and computed tomography both make more sophisticated use of the information available from x rays. Nuclear magnetic resonance was a well-known research field in 1958; in the last few years it has become useful in medical imaging. The techniques of computed tomography have made it possible to make images of the distribution of radioactive tracers in the living body. The use of ultrasound to form images, in its infancy in 1958, is now a fairly common diagnostic tool.

While some of the contributions that physicists have made to medical imaging are eminently newsworthy, other, more mundane contributions—such as the development of quality-control devices to improve medical images—may have a more significant effect on health care. Thus, for example, unnecessary x-ray exposure has received considerable attention in the press; however, a poor x ray that causes a radiologist to miss an early diagnosis of a breast cancer may cost the patient her life, and the amount of radiation is of minor concern.

Digital subtraction angiography

One of the drawbacks of conventional film radiography is that material other than bone shows up poorly. Even when one introduces an absorbing material into the organ or vessel one wants to examine (to enhance the contrast with surrounding material) the additional absorption may be hard to see against the background of bone or other tissues. The conventional technique also fails to use all the information that may be available in the transmitted x-ray beam, such as the absorption spectrum of the material and time variations in its behavior.

The technique of "subtracting the background" is well known in other areas of physics. It can be applied to diagnostic x-ray images to provide information not available from single broadband projection images. For example, from an image obtained using an absorbing marker one can subtract an image obtained without the marker, thus making clear the contribution of the absorbing material. Alternatively, one can subtract images obtained at two different energies from each other to bring out the presence of materials whose absorption changes between these two energies—that is, materials that have an absorption edge in that energy range.

Because the subtraction technique is most commonly used to examine blood vessels and related structures, and because the images are generally obtained using digitized fluoroscopy apparatus, the technique is referred to as digital subtraction angiography.

The subtraction of background features from an image allows the clinician to work with much smaller amounts of absorbing material. One common application (shown in figure 2) is the examination of arteries; iodine injected into the bloodstream is the absorbing medium. The sensitivity to changes in absorption allows one to inject the iodine into a vein and follow its progress into the arterial system as the blood is pumped out of the heart again after passing through the lungs. Not only is venous injection much less invasive than arterial injection, but the concentration of iodine is typically twenty times lower than in direct angiography. Typically, one follows the flow of the injected iodine as it makes its first pass through the arterial system by making a sequence of images—about 1 to 30 images per second—and subtracting the pre-injection background from each. Because the iodine signals are weak, the image must be amplified, and because noise is amplified as well, the system designers must take care to provide for adequate x-ray exposure, and to use tv cameras and amplifiers with high signal-to-noise ratios. Depending on the degree of patient motion one can expect, one may also use signal-integration techniques to reduce the effects of noise.

Digital subtraction angiography is just one example of a generalized technique[1] in which data obtained under slighly different conditions are subtracted to obtain clinically useful information about the attenuation x rays. In its passage through the body (or whatever else is being examined) the intensity I of the transmitted beam is a function of its energy, of time, and of position. By examining the changes in the function $I(E,t,\mathbf{r})$ with respect to small changes in E, t, or $\mathbf{r}$, one is, in effect, looking not at I itself but at its partial derivatives—or rather, at the first few terms in the Taylor series for the function, because the intervals one uses in practice can never be truly infinitesimal.

We have already mentioned energy subtraction, which may be used to provide images selectively displaying particular organs or tissues. Iodine, for example, has an abrupt discontinuity in its absorption spectrum at 33 keV (its K-edge). Subtraction images

Paul Moran, Jerome Nickles and James Zagzebski are in the department of medical physics at the University of Wisconsin in Madison; Moran is currently visiting professor of radiology at Bowman Gray School of Medicine, Lake Forest University, Winston-Salem, N.C.

0031-9228 / 83 / 0700 36–07 / $01.00

formed from data obtained with nearly monoenergetic beams with average energies above and below 33 keV will emphasize tissues containing iodine—whether naturally or artificially introduced—while suppressing the effects of bone and other tissues. Because one is not relying on a change in iodine concentration with time, one can allow the iodine to reach a temporary equilibrium distribution before one obtains the image. Heavily filtered beams from conventional x-ray sources as well as beams from monoenergetic sources have been used to investigate this technique, and it is currently being studied with finely tuned synchrotron radiation.[2]

First-order spatial derivatives yield not only the data for traditional image-processing techniques such as edge enhancement, smoothing, and various operations on the spatial-frequency content of the image, but also for more recently introduced techniques of computed tomography.

Image processing of single films is severely limited by the fact that many variables have already been averaged over in the production of the image. They are also constrained by the fact that the trained eye and brain are already an excellent image-processing system, so that one must take care not to degrade subtle features that an observer may use to analyze an image. Because of this, computer processing of conventional radiograms has had little impact on clinical practice. However, there is considerable information in a single x-ray film that can be brought out by appropriate processing. Spatial filtration, for example, can eliminate large variations in overall brightness in different areas of the image; alternately, it can pick out features having a particular spatial frequency. A group at Wisconsin has used the latter in an attempt to show coronary arteries (high spatial frequencies) against the left ventricle (low spatial frequency); the technique, in fact, involves a second-order derivative, in that the spatial filtration is performed on a subtraction angiogram of the sort we described earlier. So far this technique has not proved as useful as the more invasive cardiac catheterization techniques.

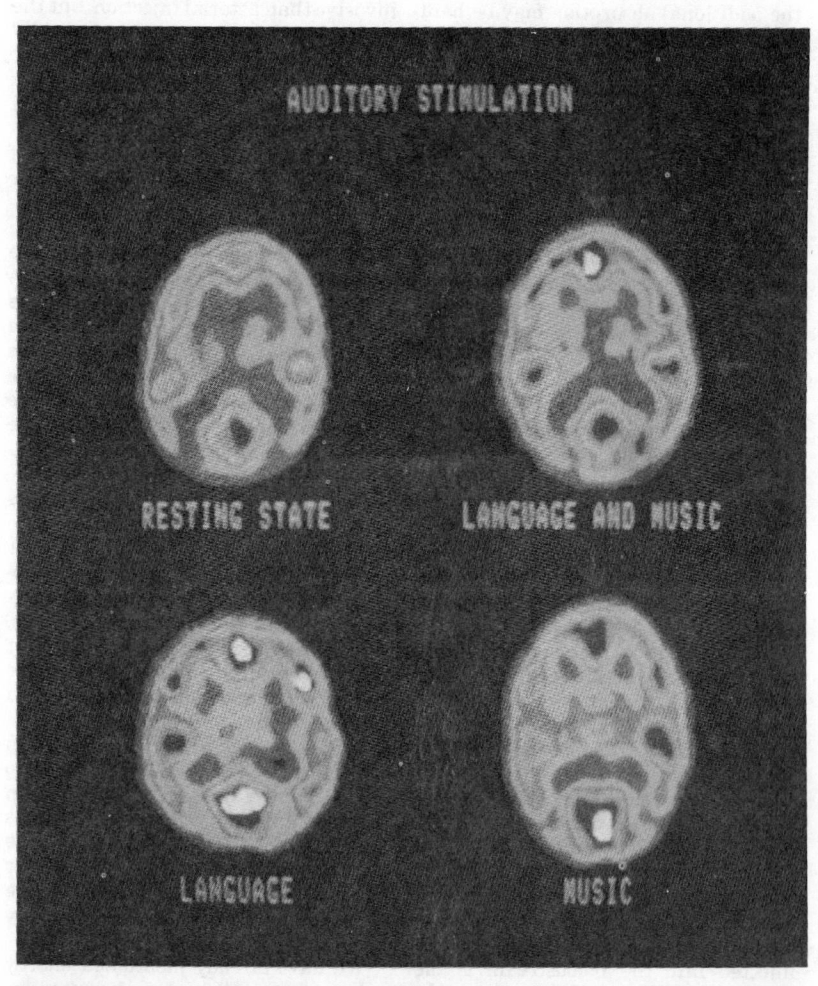

Variations in brain metabolism as a function of stimulation. These positron-emission tomograms show the distribution of F^{18}-labeled 2-deoxyglucose in a transverse slice of a living brain with various kinds of auditory stimulus. (The eyes are open for all four images.) In the resting state the ears are plugged; in the other images the stimulation consists of a segment of the radio program "The Shadow" (language) and a Brandenburg Concerto by Johann Sebastian Bach (music). The frontal lobes are indicated by the vertical arrows; the auditory cortex (temporal lobes) are indicated by horizontal arrows and the visual cortex (occipital lobes) is at the bottom of the images. (Courtesy of Michael Phelps, UCLA.) Figure 1

Other second-order techniques have also been investigated. These include:

▶ hybrid subtraction—the derivative of the intensity with respect to time and energy

▶ tomographic subtraction angiography—the derivative of the intensity with respect to spatial variables and time

▶ tomographic energy subtraction—the derivative of the intensity with respect to energy and depth

▶ multiple K-edge subtraction—the second derivative of the intensity with respect to energy

Each of these techniques combines some of the advantages of the separate subtractions involved.

The hybrid subtraction technique, for example, was introduced[3] to reduce the effects of patient motion in digital subtraction angiography. Instead of taking single exposures before and after injecting the absorbing material, one takes a pair of exposures at widely separated energies. Subtraction of the energy pair produces images in which no tissues except the carriers of the opaque medium show up; subtraction of the resulting time-separated images then removes the remaining effects of bone and displays clearly the effects of the injected contrast material.

Some of the arteries carrying the injected iodine may well overlap in the two-dimensional image obtained in a digital subtraction angiogram; this overlap often makes diagnosis difficult. One possible solution[4] to this problem is to use tomographic techniques in conjunction with subtraction angiography. To obtain such three-dimensional information one can either move the source rapidly or use multiple sources to illuminate the patient from several different angles. The resulting data can then be processed to provide images of tissues containing iodine viewed from several different angles, giving the clinician an effectively three-dimensional view of the arteries.

Energy subtraction combined with tomography can provide information on the chemical composition of various regions of the body. One can also obtain information about materials by making use of the different variations with energy of the photoelectric and Compton-effect contributions of x-ray scattering. The ratio of these two contributions at widely separated energies can be used to get some idea of the effective atomic number. In a tomogram, one can thus get more information about the chemical composition of each region than from just the absorption data at a single energy.

The first-order techniques, specifically computed tomography (equivalent to depth subtraction) and digital subtraction angiography (time subtraction) have thus far clinically had the most importance. Their impact is probably due less to the computer processing, which can enhance the display of data, than it is due to the interesting data that can be generated by subtraction processes in general. By combining data from several images obtained under slightly different conditions, one can exploit much more fully the physical, geometrical and physiological information obtainable from what would otherwise be "just an x ray."

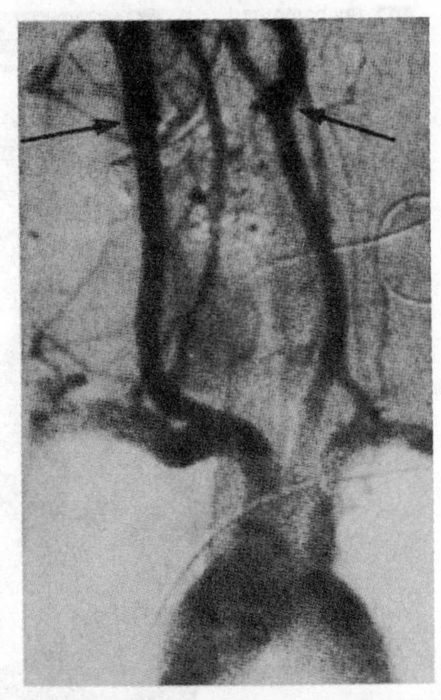

Digital subtraction angiogram of the carotid arteries that provide blood to the brain. The photo shows an x-ray image of the aorta branching, among others, into the two carotid arteries (indicated by arrows). The images are read off a fluoroscope by a tv camera, digitized and then subtracted. Note the absence of bones or other structures in the resulting image. Figure 2

Computed tomography

Computed tomography consists of producing an image of a slice (*tomos* in Greek) of a three-dimensional object by combining a large set of scans through the object at different angles, each scan producing the equivalent of strip of a standard x-ray image. The information from all these scans is then used to reconstruct the two-dimensional variation of x-ray absorption through the entire slice.

The basic techniques of producing and detecting the x-rays is over 30 years old. What is new is the information-processing capacity. Allan Cormack and Godfrey Hounsfield—as well as others—independently developed[5] the mathematical techniques for reconstructing tomograms from the integrated absorption of beams passing through the absorber. Hounsfield and Cormack won the Nobel Prize for Physiology and Medicine in 1979 for this work (see PHYSICS TODAY, December 1979, page 19). The commerical development of scanners for computed x-ray tomography followed rapidly, and most major hospitals by now have access to CT scanners.

Modern x-ray CT scanners in general incorporate a large number (500 to 1500) highly collimated detectors mounted on a circular gantry that surrounds the patient in the region from which one wants the tomogram. An x-ray tube is also mounted on a gantry and rotates around the patient, producing some 500 to 2000 views, each a different angular projection of the patient, in the course of a 360° rotation. Each view produces about 500 measurements of the x-ray transmission across the subject's diameter, and a typical view takes about 5 to 10 millisec. The average power input to the x-ray tube required for this speed is about 20 to 30 kW The x-ray beam is collimated axially and is variable between 1.5 and 10 mm; the thinner the illuminated slice, the better is the axial resolution. To remove the effects of variations in x-ray output and detector sensitivity, the individual measurements are normalized to reference detector readings and scaled to a reference material. The result is the total attenuation along the path the x-ray beam takes through the sample.

One way to understand the computations involved is to consider each scan as consisting of parallel rays. The Fourier transform of a single angular view is then, it turns out, basically the two-dimensional Fourier transform of the x-ray attenuation density in the slice, for values of the wavevector perpendicular to the view angle. (See the articles by William Swindell and Harrison Barrett, PHYSICS TODAY, December 1977, page 32, and by Rowland Redington and Walter Berninger, PHYSICS TODAY, August 1981, page 36.) The complete scan thus produces a complete Fourier transform, for wavevectors in all directions. (It is, essentially, the Fourier transform given in polar coordinates.) To reconstruct the tomogram, the data for a complete set of views are appropriately filtered ("apodized") and entered into a Fourier inversion algorithm. Other geometries of illumination and other ways of looking at the problem lead to different computational schemes, with other integral transforms replacing the Fourier transform.

The output of the calculations is typically displayed on a tv monitor divided into an array of 512×512 pixels, the brightness at each point being the result of an inversion computation that involves on the order of a million computer operations. The display is typically supported with image-array buffers and electronics so that the operator can select the mid-level brightness and the contrast range of the display. The computations are sufficiently sensitive that the x-ray quantum statistics limit the precision. Image noise, at spatial resolutions of 5 line pairs per centimeter, is typically less than $\frac{1}{2}\%$ the attenuation of an equal volume of water. The range of attenuations from air to very dense bone is about 8000 times as much. Figure 3 shows a computed tomogram.

Images for nmr

The frequency and decay rate of an nmr signal depend on the local magnetic—and hence chemical—environment of the nucleus being probed. One can thus produce images—of the interior of a body, for instance—that display the variations in nmr signals and thus display anatomical features. As in x-ray computed tomography, the measured data are given as a Fourier transform of the spatial image desired.

The initial work for producing nmr images was also done in the early 1970s. (See PHYSICS TODAY, May 1978, page 17.) Paul Lauterbur announced[6] the underlying principles of nmr "zeugmatography" in 1973, about the same time that the work on x-ray computed tomography was taking place. However, while CT scanners are now in place at many hospitals and are being actively used in diagnoses, nmr scanners are only now moving into research medical centers for clinical evaluation. The chief reason for the difference is probably historical: X-ray technology has had a long history in medicine, and computed tomography can be viewed as a more efficient use of information for which the interpretation was clear, the clinical efficacy was obvious and the engineering and technological problems were known and soluble. For nmr no such history exists, and clinicians must now begin to establish a tradition of interpreting nmr images.

Because nmr signals are intrinsically weak, data must be collected over long times to obtain reasonable signal-to-noise ratios. Thus, for example, while it takes about 2 seconds to complete an x-ray scan of a head with sufficient detail to study the morphology of the ventricles in the brain, a proton-nmr scan requires several tens of minutes to produce data for a comparable image. (Protons produce the largest intrinsic nmr signal and are the most abundant nuclei in the body; other nuclei, with smaller gyromagnetic ratios and smaller abundances, would require even longer times for scans.)

The physical appearance of the apparatus to produce an nmr scan is very similar to that used for x-ray computed tomography: A large circular gantry surrounds the patient, with the axis of the gantry usually along the long axis of the patient. The gantry houses coils to produce a strong static axial magnetic field, coils to produce gradients in this field, and an rf coil to transmit the excitatory pulses and to pick up the nmr signal.

The strong static, homogeneous base magnetic field—ranging from 0.07 T up to 1.5 T in some applications—produces a net equilibrium nuclear magnetization. From a semiclassical point of view, one can think of the nuclei as precessing when they are displaced from the equilibrium; the resonant frequency—the Larmor frequency—depends on the magnetic moment of the nuclei and on the magnetic field at the nucleus. The decay of the precession depends on local dephasing processes as well as thermodynamic relaxation processes. When one applies a short pulse of rf at the Larmor frequency, the spins are made to precess, and one can observe the resulting "ringing" as an induced emf at the Larmor frequency, which can be picked up by the same tuned coil that transmitted the excitation pulse. The signal is amplified and demodulated by phase-coherent single-sideband detectors, which use the exciting oscillator for the reference signal.

To obtain information about spatial position in the nmr signal, one adds a gradient to the applied field. The spatial variation in the field gives rise to a frequency modulation of the rf signal. A single measurement cycle with a particular gradient applied thus gives a spatial Fourier transform of the excited nuclear resonance. A series of such cycles with the gradient varied in direction from cycle to cycle generates the full two- or three-dimensional Fourier transform of the nmr signal. Computer algorithms then select particular delay times (to get consistent intensities), apodize the data and perform the Fourier transforms that yield the sort of images shown on the cover and in figure 4.

When a proton scan is made sensitive only to density, a superbly uninteresting image results. All tissues have just about the same nmr-observable density of hydrogen. However, the decay times of the nmr signal turn out to depend strongly on physical and chemical factors that differ from tissue to tissue.[7] The biology of these differences is not as yet well understood.

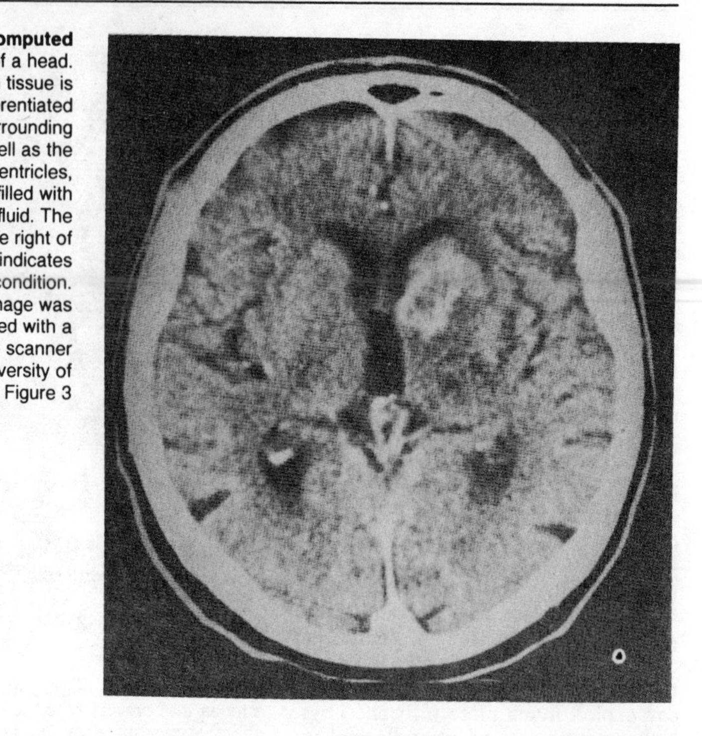

An x-ray computed tomogram of a head. The brain tissue is clearly differentiated from the surrounding bone as well as the (darker) ventricles, filled with cerebrospinal fluid. The light area to the right of the ventricle indicates a diseased condition. The image was obtained with a Siemens DR3 scanner at the University of Wisconsin. Figure 3

As we mentioned, there are several kinds of processes that affect the decay of the nmr signal:

▶ Processes that contribute to the establishment of thermal equilibrium of the spins with the rest of the body; one can think of these processes as analogous to the slow collapse of the spinning top, except that here, because the spin itself never decreases, the decay is to the aligned state. The characteristic time for these processes is called T_1.

▶ Processes that contribute to the loss of coherence among the various precessing spins. This, considerably shorter, relaxation time is generally called T_2.

By appropriately timing sequences of excitation pulses and selecting delay times for making the measurements, one can thus modulate the density-dependent nmr intensity by factors that depend on the local physical and chemical environment. Because the timing sequences as well as the field gradients can all be controlled, the operator not only acquires the data to produce an image, but in a profound sense also produces the specific subject to be imaged.

It even appears possible to observe directly the shifts in the Larmor frequency that arise from changes in the chemical environment, and thus obtain direct information about metabolic processes—for example, the relative levels and distribution of metabolic products incorporating phosphorus-31, say—in a living subject. The physician thus will be able to use nmr images to study normal and diseased tissues from a wide variety of viewpoints, anatomical, morphological, physiological or functional. Each of these uses of nmr data will of course be guided by—and contribute to—an already substantial knowledge of normal and pathological states. Thus, for example, while x-ray tomography yields highly detailed images of morphology and sensitivity to high-Z elements, an nmr scan can yield images that depend on physiology or functional properties. Most x-ray studies provide no distinction between a living subject and a cadaver; nmr scans can look totally different in the two cases.

Positron tomography

The information obtained from the imaging techniques we have discussed so far leads only indirectly to information about the chemical composition of the tissues imaged. Much disease, however, has a chemical origin; and changes in metabolism and biochemistry often come early, while changes in organ size, position or density generally occur late in the disease process. The use of radioactively labeled compounds to trace biochemical pathways has been a well-known research tool for quite some time. Applying the techniques of computed tomography allows one to map the distribution of the radionuclides.[8] To do this successfully, one must use radionuclides that meet two requirements: They must mimic the behavior of elements in the metabolic pathways and they must produce emanations whose paths one can determine. Positron emitters provide the natual choice for localization because the two 511-keV photons produced when the positron annihilates allow for easy reconstruction of the path. Metabolic equivalence then leads one to the short-lived positron emitters C^{11}, N^{13}, O^{15} and F^{18}.

Most positron tomography systems today look from the outside very much like x-ray tomography or nmr-imaging systems. Their discrete scintillators are arranged in rings; typically, there are about one hundred detectors per ring, with up to five rings in the gantry. A coincidence event between two detectors across the axis of the subject defines a line along which the positron annihilation must have occurred. The collection of millions of coincidence counts along the thousands of possible projection rays permits the reconstruction of the positron distribution through the now-standard backprojection techniques we described above. One can improve the resolution by spacing the detectors closer together. A group at UCLA, for example, is studying closely packed bismuth germanate detectors, with each detector indentified by cleverly adapted mercurous iodide photodiodes. Determining where along the ray the positron decayed can provide another improvement in resolution. This can be done with time-of-flight techniques and very fast counters.[9] Fast scintillators such as CsF and BaF_2 can produce time resolutions better than 300 picosec and can greatly help in suppressing image noise.

The amount of radioactive tracer material required to produce useful images is minuscule. One can use C^{11}-labeled carbon monoxide to trace blood flow to detect, for example, motion abnormalities of the cardiac walls by observing the heart contractions in a cine loop. A few millicuries of radiocarbon (its halflife is 20 minutes) will suffice for this purpose. This amount of radioactivity corresponds to 2×10^{11} molecules of carbon monoxide, about $\frac{1}{3}$ of a picomole. Carbon monoxide is a potent poison, but a factor of 10^{10} times as much as required to produce a detectable physiological response.

Nuclear medical procedures—that is, those that depend on introducing radioactive tracers—differ from other radiologic imaging techniques in that they are fundamentally dynamic. They are also sensitive to extremely small concentrations and can pinpoint highly specific receptors. The measurement of cardiac volume we mentioned above is not the best example of the sensitivity of positron emission tomography. The fluid volume of the heart is a simple mechanical property that can be obtained by many other imaging techniques; the receptors for carbon monoxide are simply hemoglobin, and are ubiquitous in the bloodstream.

An example of the sort of image unobtainable by other means is the use of rubidium-81 to trace the flow of blood to the heart muscle. In a normal heart this "myocardial perfusion" is about 50 ml/min for every 100 g of heart tissue. Rubidium chloride follows potassium pathways in the heart and is distributed by the blood flow. A dose of 16 mCi of Rb^{81}-labeled RbCl administered intravenously can provide a clear picture of the myocardial perfusion. Similarly, regional metabolic demands can be mapped through the kinetics of labeled fluorosugar analogs and fatty acids, probing the biochemical states of the heart and its fuel needs. Only nmr imaging can approach this type of information.

It is in the study of the brain, however, that positron tomography has made its greatest contributions. In the brain only oxygen and glucose serve as metabolic substrates. One can label glucose with carbon-11, but because the metabolic products of glucose (lactate and CO_2) are also mobile, the interpretation of an emission tomogram is difficult. However, a fluorinated analog of glucose, 2-fluoro-2-deoxyglucose can cross the blood–brain barrier but cannot be metabolized completely; it remains trapped[10] as the phosphate, 2-FDG-6-phosphate. Labeling 2-FDG with fluorine-18 allows one to make positron-emission maps of glucose uptake by the brain. Figure 1 shows such a set of maps for different kinds of activity. The differences in brain metabolism associated with differences in information processing are evident.

Oxygen does not have an analog capable of being trapped. One can, however, set up a steady state in which the subject comes into equilibrium with O^{15}-labeled oxygen inhaled over a period of 5–10 minutes. (The halflife of O^{15} is 2 min.) The regional input–output differences are balanced by decay and the activity thus reflects local oxygen needs. Regional blood flow can simply be mapped with O^{15}-labeled water. An even clearer picture can be obtained by observing the decay of O^{15} activity after the subject inhales a brief pulse of labeled oxygen. Such maps can show detail comparable to that obtained with labeled glucose analogs.

Diagnostic ultrasound

The last of the techniques we would like to discuss offers perhaps the least invasive method of probing the soft tissues of the body to produce images or detect disease states: the use of brief pulses of ultrahigh-frequency acoustic waves as a sort of sonar to map tissues within the body. As it travels through the body, an ultrasound pulse is reflected and scattered by changes in density and elasticity at tissue interfaces. Plotting the echo strength as a function of time allows one to visualize tissues as a function of distance within the body. For one-dimensional information one can simply plot the echo strength as a function of delay ("A-mode" display). For two-dimensional information, obtained by sweeping the source and detector across the patient, one generally just plots a point on a screen for echoes above a threshold, the distances along the line corresponding to delays

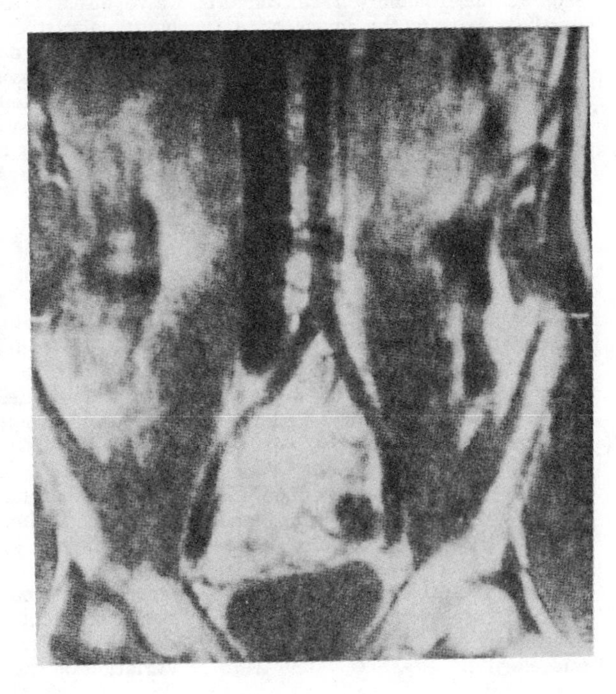

An nmr image of a coronal slice of an abdomen. Differences in brightness of the image correspond to differences in the decay times of the proton spin resonances and thus reflect different chemical environments—differentiating, for example, fatty tissue from muscle. (Photo courtesy Anne Deery, Siemens) Figure 4

Radiological imaging systems

	Digital subtraction angiography	Computed tomography	Nuclear magnetic resonance	Positron computed tomography	Ultrasound
What is detected	transmitted x rays	transmitted x rays	rf radiation from induced emfs	511 keV annihilation photons	acoustic echoes
What is imaged	electron density of object	electron density of object	induced nuclear magnetization	*in vivo* positron emitters	discontinuities in speed of sound
What causes structure	variation of electron density with composition and density	variation of electron density with composition and density	variation of strength and decay of magnetization with composition	differential uptake of labelled compounds	variation of density and elasticity of tissues
What is inferred	location of tissues filled with absorbing material	sizes and shapes of organs	physical and chemical variations among tissues	flow and metabolism of tracer material	sizes and shapes of organs; acoustical properties
Typical application	examination of arterial narrowing	detection of brain tumors	imaging of brain tumors	mapping of glucose metabolism in brain	fetal growth; tumor detection; cardiology
Signal source	x-ray tube	x-ray tube	precession of nuclei	ingested labelled compounds	piezoelectric transducer
Signal detector	image intensifier	x-ray detector	rf pickup coil	scintillation detectors	piezoelectric transducer
Image plane	longitudinal	transverse	any	transverse	any
Spatial resolution	½ mm	1 mm	2 mm	10 mm	2 mm
Temporal resolution	10^{-2} sec	1 sec	10^{-1} sec to 10^{2} sec	10^{1} sec to 10^{3} sec	10^{-2} sec
Typical radiation dose	2 rad (imaged field)	1 rad (imaged slice)	(not applicable)	10^{-2} rad (whole body)	(not applicable)
Cost (approximate)	$0.2 million	$1–2 million	$1–2 million	$1 million	$0.1 million
Chief use	clinical diagnosis	clinical diagnosis	physiological and clinical research	physiological research	clinical diagnosis

of the echo ("B-mode" display).

The ultrasound beam is produced and echoes are detected by a piezoelectric transducer. Manual-scanning instruments have the transducer attached to an electromechanical scan arm. The arm constrains the motion of the transducer to a plane. It also provides signals used to compute the position and orientation of the transducer for determining the path of the ultrasound beam, and thus the reflector locations. More commonly, the beam is swept automatically by a small hand-held scanning assembly; the operator positions and orients the scanner in the desired plane. Most instruments have an image memory or scan converter that presents the image on a tv monitor, and one can usually observe the image as it is produced. There are several techniques available for performing the automatic scans: mechanical motion of the transducer, rotation of a sound-deflecting mirror, electronic switching between elements in an array, or phased excitation of a multielement sectored array.

The rapid scanning speed and the "real-time" images of the hand-held automatic scanners offer several advantages to the clinician. By providing flexibility and more or less instantaneous feedback, the sonographer can move the scanner to find landmarks, such as blood vessels, that can help pinpoint the location of the tissue or organ under investigation. Such landmarks are especially important when there is only subtle acoustic contrast between the structure of interest and the background. Once the structure is found, the sonographer can position the scanner to provide optimum contrast and clarity.

The maximum possible scanning speed of acoustic imaging systems is limited by the speed of sound in tissue (about 1540 m/sec), the maximum depth of tissues of interest and the detail required for the image, which determines the number of lines per scan. Rapid scanners are very useful in providing the real-time images and the flexibility we have mentioned. They are also essential in viewing moving objects such as the heart. One can now obtain ultrasound images of the heart at a rate of more than 30 scans per second. The images are clear enough that one can diagnose cardiac problems such as mitral valve stenosis and other valve diseases as well as congential heart defects. Even faster scans are possible for examining superficial anatomic structures or for producing images with limited resolution.

One can add a further dimension to ultrasonic scans by coupling measurements of Doppler shifts with the detection of echoes so that one can, for example, measure blood flow at the same time that one is imaging the vessels through which the blood is flowing. Such "duplex" transducers could be very valuable in examinations of the heart, to detect abnormally high blood velocities or abnormal flow directions through defective valves.

A significant fraction of all ultrasound examinations performed involve studies of the abdominal organs (figure 5) or examinations of the fetus. In the abdomen, the images may help determine whether abnormal masses are likely to be malignant. Ultrasound provides a method of rapidly detecting stones and other obstructions in the gall bladder and related structures. Variations in the overall texture of the image of an organ are sometimes an indication of a disease condition. For example, early signs of rejection of kidney transplants may be detectable with ultrasound.

Fetal maturity studies—using measurements of head size or other anatomical dimensions from ultrasound scans to indicate the development of a fetus—are routine in most obstetrics departments and even in some physicians' offices. The resolution available with some devices allows one to distinguish fetal structures with sufficient detail permit diagnoses of many types of fetal abnormalities.[13] Most obstetricians now perform amniocentesis—in which a sample of the amniotic fluid is withdrawn for laboratory analysis, particularly to detect genetic problems—under ultrasonic guidance. By following the position of the probing needle on a screen, the obstetrician can avoid puncturing the fetus, placenta or umbilical cord.

The spatial resolution in an ultrasound image is related to the volume occupied by the sound pulse as it propagates through tissue. Ordinarily this volume can be divided into two components: one that depends on the lateral dimensions of the transducer beam and a second that depends on the time duration of the pulse, that is, its axial dimension. A typical instrument may use a 3.5 MHz transducer with a circular aperture 20 mm in diameter, focused at a depth of 12 cm. The diffraction-limited lateral resolution in the transducer's focal region is about 2.6 mm. The axial resolution for such a transducer is usually on the order of 1 mm. For a scanner whose focal length is fixed, the lateral resolution is, of course, less for reflectors located away

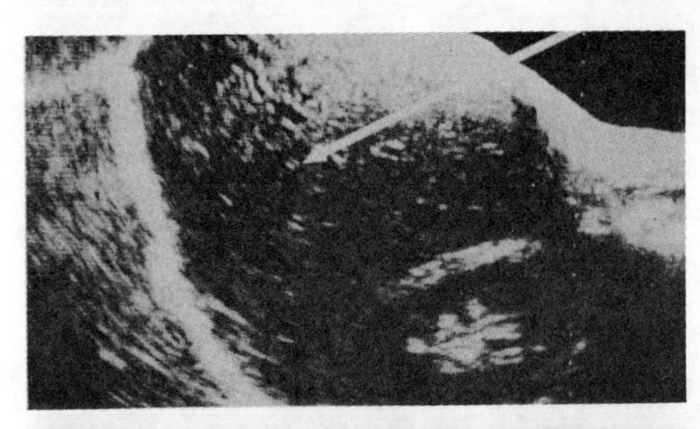

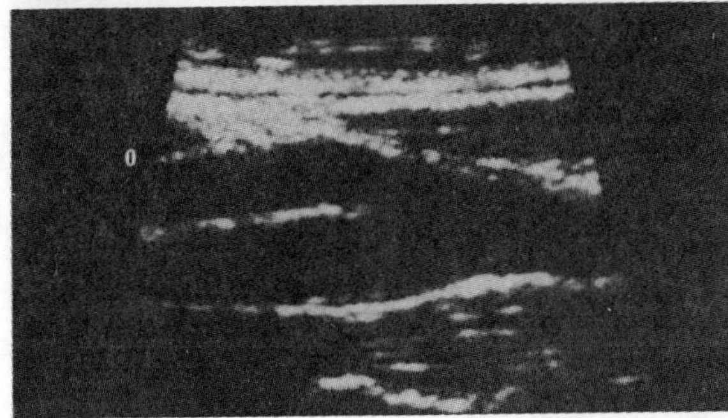

Ultrasound images. Above, an image of the abdomen of an adult; the bright band at upper right indicates the skin; the curved band at the left and bottom is the diaphagm; the liver is indicated by an arrow. Below, an image of the carotid artery, showing its bifurcation (into the external and internal carotid arteries) in the neck; such an image can be combined with measurements of the Doppler shift to give further information about blood flow through the artery. (Lower image courtesy of J. Buschell, Diasonics.) Figure 5

from the focal region.

One can improve the resolution by using higher frequencies. Unfortunately, however, higher frequencies are more strongly absorbed; the attenuation of ultrasound in soft tissues increases approximately linearly with frequency. The short-wavelength path to improved resolution can thus be used only for superficial structures, where the beam need not penetrate deeply into the body. One can, for example, use sound frequencies as high as 15 MHz to examine the eye for tumors or detachments of the retina. The lower part of figure 5 shows an image of the bifurcation of the carotid artery (in the neck) obtained with a scanner operating at 8 MHz; the image was produced in "real time." Such images can show atheromatous placque (which can tear loose and cause a stroke by blocking a smaller artery in the brain) early in its formation. Small calcifications in the arterial walls cast a recognizable acoustic shadow over deeper structures. Measurements of Doppler shifts, as we have mentioned, can provide additional information on the blood flow through the arteries, showing evidence of higher flow speeds through narrowed sections or turbulence associated with protrusions into the vessel.[14]

Recent technical improvements in diagnostic ultrasound equipment include incorporation of digital processing and digital image memories, improvements in the design of single-element and array transducers, fabrication techniques that increase the sensitivity and reduce off-axis radiation, and the development of special-purpose scanners for the breast, for superficial examinations and for other applications. In spite of all the advances that have been made, ultrasonography currently uses only part of the potential information that might be extracted from echo signals. The contrast and detail on B-mode images are based on the magnitude of echo signals reflected by soft tissues. In the future it may be possible to extract additional diagnostic information by applying more sophisticated signal processing to the returning echo signals, using, for example, waves scattered in directions other than backwards to construct the image, or using more information from the backscattered wave in computerized reconstructions and to give quantitative characterizations of the tissues being examined.[15]

To produce such more sophisticated images, we will have to understand better the interactions of high-frequency sound waves and soft tissues. The human body is acoustically a very complex structure, with many sources of scattering, reflection, and refraction. The bulk of our knowledge of the quantitative transmission characteristics of human tissue, including speeds of sound, ultrasonic attenuation and ultrasonic scattering, stems from measurements carried out on excised human and animal organs.[16] However, it is not always clear that the behavior of sound waves *in vivo* is always the same, and experimenters are now developing methods to make measurements of ultrasound characteristics in living subjects. It may become possible to observe specific tissues by their specific ultrasonic signatures. One useful tool[17] in such studies is an acoustic "phantom." Such phantoms are built to have precisely known acoustic properties at all points and to mimic to some extent the behavior seen in living organs; they are valuable in testing imaging and signal-processing algorithms and for checking[18] theories on how beams of ultrasound interact with soft tissues.

* * *

We would like to thank John R. Cameron for assistance in coordinating the contributions to this article and Charles A. Mistretta for help in preparing the discussion on digital subtraction angiography.

References

1. C. A. Mistretta, Opt. Eng. **13**, March/April 1974, page 134.
2. E. Rubenstein, E. B. Huges, *et al.*, Proc. Soc. Photo-Opt. Instrum. Eng. **314**, 42 (1981)
3. M. S. Van Lysel, J. T. Dobbins III, W. W. Peppler, *et al.*, Radiology, to be published; G. S. Keyes, S. J. Riederer, B. J. Belandger, W. R. Brody, Proc. Soc. Photo-Opt. Instrum, Eng. **347**, 34 (1982).
4. R. A. Kruger, J. A. Nelson *et al.*, presented at 68th Ann. Mtg. RSNA, Chicago, Ill, 28 November 1982.
5. G. N. Hounsfield, Br. J. Radiol. **46**, 1016 (1973); G. N. Hounsfield, Med. Phys. **7**, 283 (1980); A. M. Cormack, Med. Phys. **7**, 273 (1980).
6. P. C. Lauterbur, Nature **242**, 190 (1973); P. C. Lauterbur, Pure App. Chem. **40**, 149 (1974).
7. R. Damadian, Science **171**, 1151 (1971).
8. M. E. Phelps, E. J. Hoffman, N. A. Mullani, M. M. Ter-Pogossian, J. Nucl. Med. **16**, 210 (1975).
9. M. M. Per-Pogossian, N. A. Mullani, D. C. Ficke, J. Markham, D. L. Synder, J. Comput. Assist. Tomogr. **5**, 277 (1981).
10. B. M. Gallagher, J. S. Fowler, N. I. Gutterson, R. R. MacGregor, C.-N. Wan, A. P. Wolf, J. Nucl. Med. **19**, 1154 (1978).
11. T. Jones, D. A. Chesler, M. M. Ter-Pogossian, Br. J. Radiol, **49**, 339 (1976).
12. G. D. Hutchins, R. J. Nickles, Proc. XI Int. Symp. Cereb. Blood Flow and Metab. (in press).
13. R. C. Sanders, A. E. James, eds., *Ultrasonography in Obstertics and Gynecology*, 2nd ed., Appleton-Century-Crofts, New York (1980).
14. W. J. Zwiebel, ed. *Introduction to Vascular Ultrasonography*, Grune-Stratton, New York (1982).
15. M. Linzer, ed., *Ultrasonic Tissue Characterization II*, National Bureau of Standards Special Publication 525 (1979).
16. J. F. Greenleaf, S. K. Kenue, B. Rajagopolan, R. C. Bahn, *et al.*, in *Acoustic Imaging*, A. F. Metherell, ed., Plenum, New York (1980), vol. 8, page 599.
17. P. N. T. Wells, *AAPM Medical Physics Monograph 6*, G. Fullerton, J. Zagzebski, ed., AIP, New York (1980).
18. E. L. Madsen, J. A. Zagzebski, M. F. Insana, T. M. Burke, G. Frank, Medical Phys. **9**, 703 (1982); D. Nicholas, Ultrasound in Med. & Biol. **8**, 17 (1982). □

356 PROCEEDINGS OF THE IEEE, VOL. 71, NO. 3, MARCH 1983

Overview of Computerized Tomography with Emphasis on Future Developments

R.H.T. BATES, FELLOW, IEEE, KATHRYN L. GARDEN, AND TERENCE M. PETERS

Invited Paper

Abstract—The applications of computer-assisted, or comput(eriz)ed, tomography (CT) are reviewed. The major emphasis is on medical applications, but all relevant technical sciences are covered. A unified descriptive account of the underlying principles is presented (detailed reviews of algorithms and their mathematical backgrounds can be found elsewhere in this special issue). Deficiencies in existing hardware and software are identified and the possible means of remedying the more urgent of these are outlined. Promising approaches for future research and development into CT are suggested.

I. Introduction

THE TERMS "computer-assisted tomography" and "comput(eriz)ed tomography" CT for short—came into general use after the appearance of Godfrey Hounsfield's X-ray scanners, which as we all know have transformed diagnostic radiology. Now, a decade later, the two letters CT have come to refer to a host of combinations of imaging techniques and types of physical measurement.

Our present aim is to identify techniques which invoke CT principles that are likely to be worth pursuing during the next decade. To do this, we must first make clear what we think CT encompasses. Because so many measurement schemes routinely employed in engineering and applied science might by now be considered species of CT, there is a danger of the concept becoming so diffuse that it ceases to be useful. In order to be precise we introduce a few "portmanteau" terms:

a) The *experimenter* is anyone performing or wishing to perform any kind of CT (as defined following the introduction of these portmanteau terms).

b) The *body* represents any object whose interior the experimenter wishes to study.

c) The *emanations* represent any physical process (e.g., radiation, wave motion, static field, etc.) used to study the body from outside.

d) A *transducer* is any source or sink (e.g., transmitting or receiving antenna, probe, X-ray tube, scintillation counter, electrode, etc.) of the emanations (provided it is under the experimenter's control).

e) The *system* represents the combination of apparatus, measurement scheme, and emanations employed by the experimenter.

Manuscript received April 22, 1982; revised November 4, 1982. This work was supported by the New Zealand University Grants Committee and the Technicare Corporation, Cleveland, OH.

R.H.T. Bates and K. L. Garden are with the Electrical Engineering Department, University of Canterbury, Christchurch 1, New Zealand.

T. M. Peters is with the Institut Neurologique de Montréal, Montréal, P.Q., Canada H3A 2B4.

f) The *density* represents the spatial distribution of whatever material property of the body one wishes to reconstruct with the aid of the system.

g) A *neighborhood* is the smallest volume of a body, or smallest area of a cross section of a body, that can be resolved by the system.

h) A *clean image* of the density is one for which the value reconstructed at any spatial point is (ideally) uncontaminated by values of the true density outside the neighborhood of the point.

We hold the following definition:

> CT is the reconstruction of a clean image of the density from digital computational operations on measurements of emanations that have passed through the body.

It should be recognized immediately that a clean image cannot be formed with a conventional imaging instrument (e.g., telescope, microscope, etc.) because wave motions (e.g., light, ultrasound, electron beams, etc.) are continuous in space and time. Wave motion brought to a focus within a neighborhood of a particular point necessarily converges before and diverges after, thereby unavoidably contaminating the density reconstructed outside the neighborhood. So, the image of a body reconstructed with the aid of a conventional imaging system must be *contaminated*—the image of the surface of a body can be clean, of course, but that is irrelevant for CT because it is concerned with reconstructions of the *interiors* of bodies.

We maintain that the above definition captures the essence of what technical scientists now understand by the term CT. Table I lists a selection of emanations, their applications, the transducers that transmit and receive them and the densities that are sensed and probed.

Section II outlines the principles upon which all CT systems are based. The various systems are tabulated for convenient reference. We announce one of our own recent ideas in Section II-C. Consideration of the above mentioned principles together with our experience of several aspects of CT and our reading of the literature, convinces us that there are two general types of problem presently of importance for CT:

i) The overcoming of those technical difficulties which seriously hamper the further development of established CT applications.

ii) The recognition of those particular aspects of science (pure and applied), medicine, and engineering that have good chances of benefitting from the application of CT principles; and the devising of detailed means for realizing such benefits.

0018-9219/83/0300-0356$01.00 © 1983 IEEE

TABLE I
SPECIES OF EMANATIONS THAT ARE, OR COULD WITH ADVANTAGE BE, EMPLOYED IN PRACTICAL APPLICATIONS OF CT. ACTUAL AND PROMISING APPLICATIONS ARE LISTED, AS ARE THE TYPES OF TRANSDUCER WHICH ARE USED AND THE KINDS OF DENSITY WHICH ARE SENSED AND PROBED

Emanations	Densities	Transducers	Applications
X-rays	X-ray attenuation coefficients	X-ray sources; Scintillation detectors	Diagnostic Radiology (DR); Nondestructive Testing (NDT); Package and luggage surveillance
Gamma-rays from radio-isotope labeled ingested or injected substances	Concentration of radio-labeled substance	Scintillation counters	DR; Nuclear medicine [88]
Compton scattered X and γ rays	Electron density or distribution of atomic number	X-ray sources; Scintillation counters	DR; NDT; X-ray crystallography [40], [136]
Heavy particles; pions, alphas, protons, etc.	Scattering/absorbtion cross section	Linear accelerators; Stacked detectors	DR; NDT
Electron (wave) beams	Schrödinger potential distribution	Electron guns; Film; Photomultiplier/TV	Microscopy of "weak" specimens [119]
Ultrasound; also seismic or acoustic waves	Attenuation; refractive index; Changes in acoustic impedance	Any electro/mechanical transduction device	DR; NDT; Geological prospecting (GP)
Low-frequency electric currents	Electrical conductivity distribution	Electrodes	Crude imaging of blood vessels; GP; NDT
Magnetic fields	Distribution of nuclear spins; Blood flow	(Electro)magnets; RF coils; Magnetometers	NMR; MHD (see Table IV for descriptions of these techniques)
Radio frequency (RF) and microwave (μwave) fields	Electron spins; Permittivity and conductivity distributions	Capacitors; Coils; Loop, dipole, horn, etc., antennas	NMR; Mapping distributions of (RF and μwave) dielectric constant and conductivity
Spatially incoherent electro-magnetic radiation from μwave to soft X-rays	Volume temperature distributions; Celestial brightness distributions	Horn, etc., antennas; Radio, infrared, optical, ultraviolet, X-ray telescopes	Mapping temperature distributions inside human and other bodies; Astronomical imaging

Section III covers i), while ii) is dealt with in Sections IV and V. Our concern in Section IV is with some approaches to CT that have received little attention as yet but which we consider to be particularly promising. In Section V we state which CT developments we think are most worthwhile and we also suggest what seem to us to be profitable approaches to overall systems design.

II. BASIC PRINCIPLES

Any volume can be built up from a series of parallel cross sections, which idea is invoked in almost all medical diagnostic CT machines [27], [76]. What this means, of course, is that the principles of CT can be explained in two-dimensional terms. This is a welcome simplification because it is so much easier to visualize in two as opposed to three dimensions. Also, explanations of three-dimensional processes always seem to be much more complicated than their two-dimensional counterparts.

In this section, we accordingly restrict ourselves to discussing approaches to CT which reconstruct the density of an isolated cross section of a body. We comment on direct three-dimensional reconstruction schemes in Section IV.

It is convenient to introduce the *circumscribing circle* (represented by Γ_c in Fig. 1) which encloses the cross section being studied. The medium filling the space outside Γ_c is assumed to be *tenuous* (i.e., its density is negligible). What this means in practical terms is that the body is supported in such a way

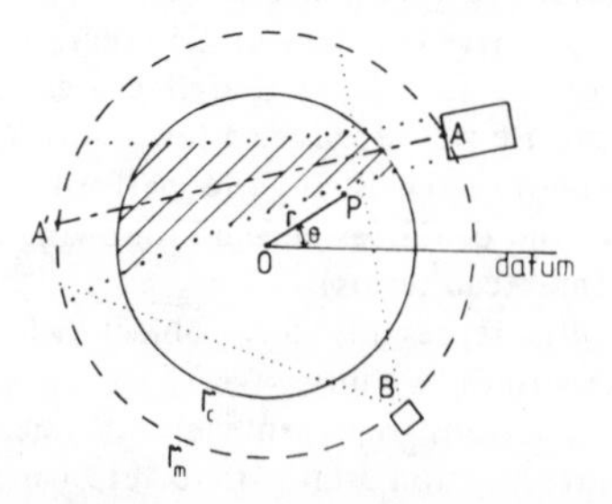

Fig. 1. Basic geometry for two-dimensional CT. Γ_m and Γ_c are, respectively, the measurement and circumscribing circles, centered at O. P is an arbitrary point, with polar coordinates r and θ, inside Γ_c. The large and small rectangles, at A and B, respectively, represent narrow-beam and wide-beam transducers, the "edges" of whose beams are indicated by the dotted lines. The cross-hatching depicts the intersection of the narrow beam with the interior of Γ_c.

that the emanations interact only with the body and the measurement system. In fact, if part of any supporting structure intercepts the emanations then it must be treated as part of the body.

There are two general classes of sources of the emanations, typified by the degree of control exerted over them by the experimenter. First, there are *interior sources* inside the body which are usually beyond the direct control of the experimenter—this relates to what we here call "remote-sensing CT." Second, there are *exterior sources* outside the circumscribing circle which are usually (ideally) completely under the experimenter's control–this relates to what we here call "remote-probing CT." In the following subsections we discuss measurement schemes that typify these two classes of CT separately and in combination. While the physical manifestations of these schemes are exceedingly diverse (refer to Table I), they are unified by common computational principles as we explain below.

The emanations are measured outside the circumscribing circle. Since the measured emanations are therefore entirely "outgoing" with respect to the body (i.e., they are flowing away from the body) it is clearly sufficient from an information-theoretic point of view to carry out the measurements on a single circle which encloses the circumscribing circle. Since noise and other imperfections are always present in practice, the accuracy of data gathering is obviously improved by increasing the number and range of measurements; but no new information can be inferred thereby. We choose a *measurement circle* (represented by Γ_m in Fig. 1) which is concentric with the circumscribing circle. Because we are only explaining principles in this section, we assume the measurements to be perfect. For remote-probing CT we take the exterior sources as well as the detectors of the emanations to lie on the measurement circle. All of this is illustrated in Fig. 1.

The resolution inherent in the reconstructed image is the main indicator of how useful a CT system is likely to be in any particular application. The sizes of the transducers limit this resolution because they determine the size of the neighborhood. Since the size of a transducer is usually related to its field of view, a vital transducer parameter is its beamwidth. It is useful to distinguish between *narrow-beam* and *wide-beam* transducers. Since the latter can in almost all cases be physically small, they seldom set a lower limit on the neighborhood's size–this is illustrated in Fig. 1 by representing the wide-beam transducer at B by a small square with the edges of its beam (defining the "edge" of a beam as where its amplitude is, say, 0.1 of its peak value) indicated by the two dotted lines diverging widely from B. However, the use of wide-beam transducers tends to lead to error-sensitive and mathematically awkward reconstruction algorithms. Typical of these are all viable schemes for imaging with low-frequency electric currents, either passively (as when attempting to deduce cardiac currents from ECG signals measured on the surface of the torso [22], [130]), or actively by injecting currents into the Earth during geophysical prospecting [96], [139] or the human body for imaging organs and blood vessels [75], [152]. (It is worth noting here that several attempts have been made to apply conventional CT concepts to electric current imaging, but they seem to rest on a misconception [15].) The transducers must be wide-beam because the frequency is always so low that the fields are effectively static.

The algorithm is almost always more stable, and the mathematics from which it is derived is usually much more straightforward, when a narrow-beam transducer can be used, because each measurement is sensitive only to that part of the density intercepted by the beam (e.g., the cross-hatched region of Fig. 1). On the other hand, the aperture dimension in wavelengths of a narrow-beam transducer is inversely proportional to the beamwidth, which does set a lower limit to the size of the neighborhood (because the width of the cross-hatched region is proportional to the aperture dimension) when the wavelength is appreciable (as it is, for instance, for ultrasound in medical applications). While the wavelengths of X-rays and electron beams are generally miniscule, the corresponding aperture dimensions are not always negligibly small because it is often necessary to use quite large apertures to obtain useful signal-to-noise ratios. An interesting point relevant here is that X-ray CT machines employing the now conventional "fan-beam geometry" [79], [133] are still primarily dependent (from the point of view of the reconstruction algorithm) on narrow-beam transducers–although the transmitting transducer is (comparatively) wide-beam, the receiving transducers are necessarily narrow-beam.

For ease of exposition in most of what follows, we assume that a wide-beam transducer radiates or receives with equal strength in all directions and that a narrow-beam transducer has an infinitesimal beamwidth. It should be remembered, however, that a wide beam usually exhibits significant variation over the range of angles subtended at the transducer by Γ_c, and that the width of a narrow beam is always appreciable in practice. The latter consideration usually sets the resolution limit, while allowing for the former tends to make a reconstruction algorithm more intricate. Nevertheless, the principles discussed below are essentially unaffected by these practical niceties.

It is convenient to introduce the symbol $\rho = \rho(r, \theta)$ to represent the density at an arbitrary point P inside Γ_c (refer to Fig. 1).

A. Remote-Sensing CT

In this subsection we consider how to perform CT when the sources of the emanations are inside Γ_c (and beyond the experimenter's direct control) and the system is entirely passive in the sense that it only receives emanations. (Systems in which extra emanations are transmitted are discussed in Section II-C below.) If the sources are embedded in a dense medium, it is, in general, impossible to disentangle the density of the sources from the density of the medium. A clean image of the source density, which is the quantity to be here represented by ρ, can only be reconstructed when the emanations are negligibly perturbed by the medium, which we then say is *tenuous* (as is assumed throughout this subsection).

Table II lists the measurement techniques and species of emanations that are of practical importance for remote-sensing CT. All of the interior sources that give rise to radiative emanations are *spatially incoherent*, which means in effect that there is no correlation between the source density at any two points separated by a distance of more than ξ, say; where ξ is very much less than the smallest conceivable linear dimension of the neighborhood. From a practical point of view, the most significant consequence of a source density being spatially incoherent is perhaps that its emanations can only produce interference patterns momentarily. The random fluctuations of the emanations cause the interference to be smoothed out when signals from receiving transducers are subjected to temporal averaging. Note that this smoothing process cannot occur for the electric currents that are carriers of ECG (often called EKG) and EEG signals, because they vary so slowly that they can be regarded with negligible error as if they were dc signals [22], [130].

Consider a wide-beam receiving transducer positioned at the point A on Γ_m (see Fig. 2). Since this transducer receives emanations from all sources inside Γ_c, there must exist a relation of the form

$$s(\phi) = \Lambda\{\rho(r, \theta); w\} \tag{1}$$

connecting the density and the received signal $s(\phi)$, where $\Lambda\{\cdot;\cdot\}$ is some integral operator spanning the interior of Γ_c

TABLE II
TECHNIQUES THAT HAVE BEEN, OR COULD BE, APPLIED FOR REMOTE-SENSING CT. THE PRINCIPLES ON WHICH THESE TECHNIQUES ARE BASED, THEIR INHERENT LIMITATIONS AND THEIR CURRENT PRACTICAL STATUS ARE ALSO LISTED

Technique	Basic Principle	Inherent Limitations	Status
Single Photon Emission (SPECT)	Radio-labeled substance introduced into body emits photons, detected by gamma cameras [92], [130]	Detected photon flux depends on attenuation through body, as well as on source density; Refer to possible improvement noted in Table IV	Routinely used for shadow imaging [97]
Positron Emission (PET), or Paired Photon Emission	Positronium-labeled substance introduced into body emits oppositely directed photons in pairs, detected by ring of counters and coincidence circuits	Same as above; Low photon flux probably inevitable	Developmental devices undergoing clinical trials [158], [160]
ECG (or EKG) and EEG	Current sources within body generate voltages, detected on body surface with electrodes	CT only possible for discrete point sources [107]; Specialized brain conduction may make EEG CT impossible [131]	Routinely used for mapping surface fields; No reported CT applications
Radiometric	Thermal radiation emitted because inside of body is hot	Reabsorbtion and reemission of radiation bound to make reconstruction algorithm very error sensitive	Routinely used for remote sensing of surfaces (e.g., Earth [138], [144], human [8], [19], [51]); No reported CT application
Super Synthesis (e.g., Earth rotation synthesis telescope)	Spatially incoherent surface sources in far field radiate, detected by coherent interferometer have movable elements on a moving platform	Resolution limited by maximum baseline and need for good final S/N; Response limited by need to separate received signals into narrow bands (widths inversely proportional to maximum baseline)	Routinely used for mapping the radio heavens [69], [87], [143]; Promising for future high-resolution optical synthesis telescopes [13], [50]

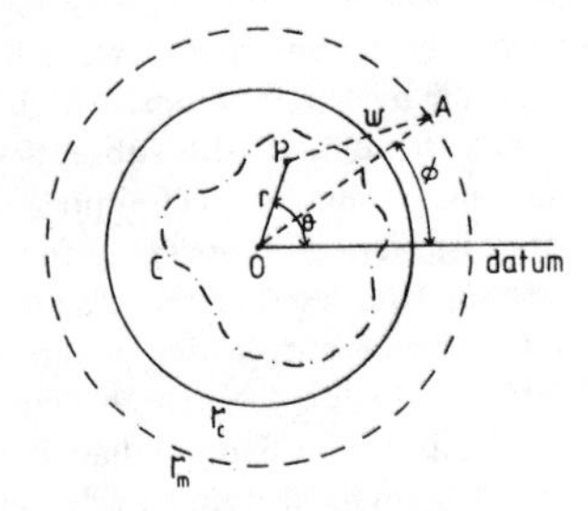

Fig. 2. Remote sensing of interior of Γ_c by wide-beam transducer at A, positioned at angle ϕ on Γ_m. The closed curve C is inside Γ_c.

and w is the distance from P to A. Note that $s(\phi)$ is "one-dimensional" in the sense that it depends only upon the angle the line OA makes with the *datum* (see Fig. 2). On the other hand, $\rho(r, \theta)$ is inherently "two-dimensional" in that it can vary arbitrarily with r and θ. In the special case of a cylindrically symmetrical density, for which $\rho(r, \theta)$ reduces to $\rho(r)$, the signal $s(\phi)$ is necessarily a constant, s say, independent of ϕ. Since $\Lambda\{\cdot\,;\cdot\}$ is an integral operator, there are infinitely many forms for $\rho(r)$ that give the same value of s. Similarly, when the density varies with θ as well as r, so that the signal depends upon ϕ, there is an infinity of functions $\rho(r, \theta)$ compatible with any given $s(\phi)$. We refer to this as the "dimensionality difficulty." It is related to the "ill-posedness" of inverse problems, which is now well appreciated in many sciences [7], [38].

If it is known *a priori* that the sources exist only at a finite number (N say) of discrete points inside Γ_c, measurement of $s(\phi)$ for N discrete values of ϕ is in principle sufficient to permit the form of $\rho(r, \theta)$ to be reconstructed. There is no longer a dimensionality difficulty because the discrete points at which $\rho(r, \theta)$ exists can be ordered as a one-dimensional array. When the value of N is given *a priori*, a stable computational algorithm can often be devised to estimate from $s(\phi)$ the positions and amplitudes of the point sources [107]. It is sometimes theoretically possible to estimate N from the measured data but it is likely to be as error-sensitive a procedure as, for instance, attempting to estimate the number of terms in data that are supposed to consist of a sum of decaying exponentials—this is a famous ill-conditioned numerical problem [1], [92].

Many fields whose space–time evolution is described by partial differential equations satisfy a "Huygens principle": if the sources are known to lie within some closed curve C which itself lies inside Γ_c (see Fig. 2) then the field outside C can be expressed exactly in terms of an equivalent source density on C [6]. Since this equivalent source density is "one-dimensional," because it is a function of a single coordinate (i.e., distance along C measured from any convenient point on C), it can be estimated unambiguously from $s(\phi)$. Note, however, that C must be specified. Note also that there is no unambiguous means for recovering the true source density from the equivalent source density.

It is worth emphasizing that the aforementioned "effectively dc nature" of ECG and EEG signals implies that their transducers are necessarily wide-beam, so that unique solutions to

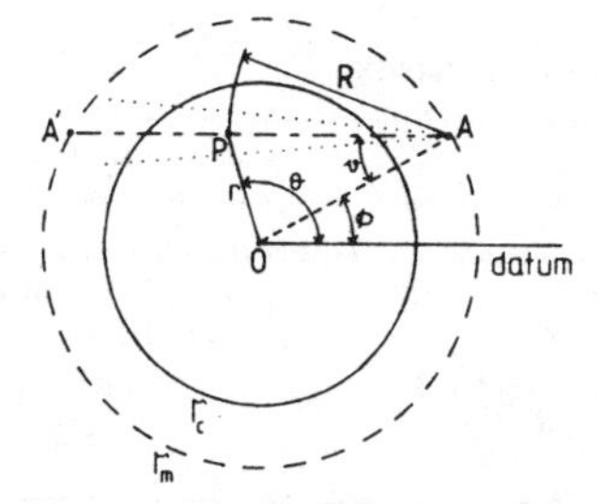

Fig. 3. Remote sensing/probing of interior of Γ_c by narrow-beam transducer at A, directed towards A' (the angles ϕ and ϑ define this direction).

the inverse problems of electrocardiography [22], [107] and electroencephalography [131] cannot be obtained unless the interior sources can be taken to be either discrete or confined to a given surface. Even then, the uniqueness may evaporate if the interior of the body is inhomogeneous [22] or anisotropic [131]. The other techniques listed in Table II are in a very different category because narrow-beam transducers can be designed and built for them. Furthermore, processing of the received signals is eased by the spatial incoherence of the sources, as is explained below.

Consider a narrow-beam receiving transducer at the point A of Fig. 3, directed towards the point A'. It is convenient at this juncture to recognize the finite beamwidth of the transducer (represented by the dotted lines). The power flow associated with the emanations from sources in the neighborhood of the point P is attenuated by a factor proportional to $1/R$ by the time it reaches A, because the power radiated by a point source in two dimensions must spread out cylindrically. However, the width of the beam at P is proportional to R. Consequently, within the limitation on the size of the neighborhood set by the width of the beam, the signal received at A is proportional to the integrated source intensity along the ray AA'. The signal received by the transducer can now be meaningfully written as $s(\phi, \vartheta)$ emphasizing that it is "two-dimensional," because it depends upon where A is on Γ_m and in which direction the transducer is pointing (refer to Fig. 3). Therefore,

$$s(\phi, \vartheta) = \int_{l(A)}^{l(A')} \rho(r, \theta)\, dl \tag{2}$$

where l is the coordinate along the ray, whose beginning and end points are denoted by $l(A)$ and $l(A')$, respectively. Note that $\rho(r, \theta)$ now represents the density of the time-averaged intensity of the sources inside Γ_c. The instantaneous source density of spatially incoherent sources is not usually of much interest because it is a very rapidly fluctuating quantity—the distribution throughout the interior of Γ_c of the time average of the square of the instantaneous source density is what matters in practice.

Note that (2) is free of any dimensionality difficulty. Both the measured quantity—i.e., $s(\phi, \vartheta)$—and the quantity that we wish to reconstruct—i.e., $\rho(r, \theta)$—are two-dimensional. So, given $s(\phi, \vartheta)$ for sufficient pairs of values of ϕ and ϑ to characterize $\rho(r, \theta)$ to the required resolution, a clean image of the latter can be reconstructed in a numerically stable manner. The formula (2) is, in fact, the basis of conventional CT. The ways in which it is handled in practice are discussed by Lewitt [104] and by Natterer and Louis [129] elsewhere in this issue.

B. Remote-Probing CT

The sources of the emanations are here taken to be located on Γ_m and are assumed to be completely under the experimenter's control. It is convenient to express the total emanations, which we write as $E = E(r, \theta, t)$ where t denotes time, as the sum of the *incident* and *secondary* emanations, $E_i = E_i(r, \theta, t)$ and $E_s = E_s(r, \theta, t)$, respectively,

$$E = E_i + E_s. \tag{3}$$

It is the density of the medium inside Γ_c that we wish to reconstruct. Table III lists the measurement techniques and species of emanations that are of practical importance for remote probing CT. Fig. 3 is relevant here.

Provided the two terms on the right-hand side (RHS) of (3) add up to the total emanations, it is of no account theoretically how we apportion the whole between the two parts. There are two conventional ways of doing this:

i) E_i is taken to be identical to what E would be if the medium inside Γ_c were tenuous; E_s is then what is needed to make up the difference between E and E_i. This is the "volume source" or "polarization source" approach to the interaction of fields with media [16]. While it is exact, in principle, it is in a sense "unphysical" because E_s cannot be interpreted straightforwardly in terms of progressive perturbations, due to the medium, of the emanations from the sources situated on Γ_m. The approach is what is often called a "self-consistent" one, in that the total perturbation is specified all at once.

ii) When the emanations are radiative or have the character of wave motion (as is true for the great majority of emanations having application to CT), it is intuitively appealing to think of "wavefronts" spreading out from the sources. Now, a wavefront can be distorted by reflection and/or refraction, thereby changing its shape and being attenuated. It can also suffer from dispersion, in that the relative amplitudes of the wavefront's frequency components can alter—e.g., if a pulse is emitted by the sources, the pulse shape changes as the wavefront progresses. However, wavefronts are easily detected in practice. Consider pulses of microwave energy or of ultrasound emitted into the kinds of media that figure in practical applications. We can always place small probes in the path of the emissions to sense the passing wavefronts, even if the media introduce appreciable attenuation, reflection, refraction, and dispersion. It also makes sense to think of an X-ray plate as a detector of a pulsed X-ray wavefront. It can be convenient, therefore, to take E_i to be the portion of E that follows (or, is inseparably associated with) the *initial wavefront*, which is the part of the emanations from the sources that first arrives at any point within Γ_c. The physical interpretation of E_s is then straightforward because it is comprised of all the scattered (and multiple-scattered) emanations from all points inside Γ_c.

When the medium is tenuous—i.e., $\rho(r, \theta)$ is negligible throughout the interior of Γ_c—the total emanations E reduce to E_i for both of the above approaches. There are then no secondary emanations in either case.

Whichever of the above two approaches is adopted, the secondary emanations must be given by some such formula as

$$E_s(\phi, t) = \Omega\{\rho(r, \theta); E(r, \theta, t); w\} \tag{4}$$

where $E_s(\phi, t)$ is the amplitude of $E_s(r, \theta, t)$ on Γ_m and $\Omega\{\cdot\,;\cdot\,;\cdot\}$ is some integral operator spanning the interior of Γ_c.

A digression is now helpful. There are two broad classes of problems in mathematical physics: *direct* and *inverse* problems. Invoking our notation, the archetypal direct problem is posed as: given the sources on Γ_m and the functional form of $\rho(r, \theta)$, calculate $E_s(\phi, t)$. It is often convenient to formulate this problem in terms of approach i) above, mainly because it is easy to compute the form of E_i throughout the interior of Γ_c.

The archetypal inverse problem—i.e., the "inverse scattering

TABLE III
TECHNIQUES THAT HAVE BEEN, OR COULD BE, APPLIED FOR REMOTE-PROBING CT

Technique	Basic Principle	Inherent Limitations	Status
Transmitted X-rays	Total X-ray attenuation measured along all rays through body; Image reconstructed from these data using standard CT algorithms [93], [104], [106], [151]	No serious limitations; Tricky problems: software for correcting for beam hardening; Design of collimators for increased rejection of scattered photons	This is the standard or conventional CT [27], [76] with which much of this special issue is concerned
Transmitted Heavy Particles	Same as above, in principle, with heavy particles substituted for X-rays	Same as above, in principle, but it is not yet clear whether the many practical difficulties will lead to any basic limitations	Interesting experiments in progress [18], [31]
Reflected Ultrasound	Echo-location measurements made at enough transducer positions to gather sufficient data for CT algorithms to be used	Not yet clear whether it is possible to allow for ray bending and variable propagation speed	This is the standard very successful B-scan [52], [53], [164]
Transmitted Ultrasound	Transmission measurements (of attenuation or pulse travel time) at enough transducer positions to allow CT algorithms to be used [93]	Same as above	Several promising experimental systems [12], [20], [35], [47], [63]–[65]; Some clinical trials [35], [63], [65]
Radar	Same as for Reflected Ultrasound except that μwave radiation is used	Very rarely any difficulty over ray bending and variable propagation speed, but usually impracticable to gather enough data to form a reasonable image using a CT algorithm	No recorded use of CT as yet; Proposed inversion of data from simple targets to reconstruct shapes [81]
Electrical Impedance	Voltages measured all round perimeter of cross section, for current source connected across enough pairs of points on perimeter to allow an image to be reconstructed [135], [152]	Analogies with conventional CT are illusory [15]; image reconstruction algorithms very error sensitive [152], but may be satisfactory for simple types of image	Promising experimental systems for medical [75], [135] and geophysical [48], [49] applications
Electron Microscopy	Specimen mounted on tilt stage; Micrographs recorded at enough inclinations of stage to allow CT algorithms to be used on basis of a micrograph being a "projection" [80], [119]	Specimen must be "weak" for micrograph to truly represent a projection; Operation of tilt stage is tricky; Specimen can be damaged by the measurement	Several useful results obtained in microbiology [119] specially with phages [159] and membranes [74]
Non-Destructive Testing (NDT)	Any of the other techniques listed in this Table could be used. The little that has so far been achieved [161] emphasizes the potential in this area		

problem" [36], [81], [117], [118] or "profile inversion problem" [38]–is posed as: given $E_s(\phi, t)$ and the sources on Γ_m, reconstruct the functional form of $\rho(r, \theta)$. Note that there is again a dimensionality difficulty because the given data $E_s(\phi, t)$ are two-dimensional, whereas the total emanations (whose form is *not* given *a priori* throughout Γ_c) are three-dimensional since they depend upon r, θ, and t. While this dimensionality difficulty, unlike the one noted in Section II-A, does not prevent one from obtaining unique solutions to the same extent as previously (cf. [7], [36], [38], [81], [117], [118], [152]), it tends to make inversion algorithms numerically unstable and very sensitive to any errors in the data. Furthermore, the theory of inverse problems is nowhere near as complete as that of direct problems. For instance, little headway has been made with approach i) above as a general formulation for the inversion problem. In fact, except in somewhat special cases, useful solutions to inverse problems can only be obtained by making rather drastic approximations (cf. [7], [36], [38], [81], [117], [118], [152]). One of the outstandingly challenging areas of mathematical physics/engineering is the development of new, applicable (in practice) approximations to inverse theory.

The crudest and most obvious "inverse approximation," and the one that is perhaps most frequently invoked, is based on the approach i). This is the Rayleigh–Gans [91] or Born [148] approximation, in which E_s is assumed to be very small compared with E_i throughout the interior of Γ_c. This applies very accurately in nonrelativistic quantum mechanics and X-ray crystallography, for instance, because the effective refractive indices of the relevant media are extremely close to unity. In many situations of practical interest involving macroscopic wave motion, however, this approximation is useless because it fails when, after passing through the interior of Γ_c, the E_i of approach ii) is delayed with respect to the E_i of approach i) by more than a quarter period of the highest significant frequency component of the emanations. Consequently, either the linear dimensions of the medium must be small or the emanations must be of low frequency. It follows that approximations based on approach i) are of practical use in ECG and EEG studies and in electrical prospecting [15], [22], [75], [96], [107], [130], [131], [139]. They can also be applied in elementary echo-location contexts, typified by (comparatively) small scatterers embedded in an effectively tenuous medium.

When the medium exhibits significant variations in its refractive index, as is usual for instance for medical ultrasonic B-scan, it is obviously preferable to invoke approach ii) for analyzing echo location. However, no generally useful means of formulating the inverse problem in terms of this approach has yet been found.

Most species of emanations, other than very low frequency ones, can usually be usefully approximated in practice as bundles of rays. In fact, the replacement of full-wave theory by ray-optical techniques has enormously simplified the design of antennas, optical systems, electron lenses, microwave and acoustic waveguides, etc., and it has made it quite a straightforward matter to calculate the interaction of wave motion with complicated structures [72], [84], [134]. There is no essential difficulty in taking into account the weakening of E_i, as defined for approach ii), by reflection and attenuation on its passage through the interior of Γ_c. Diffraction is of no importance for CT applications involving gamma-rays and X-rays, and it can be neglected in some instances for electron beams and ultrasound. Furthermore, when diffraction effects have to be accounted for (as virtually always with microwaves, for instance, and often for acoustics and ultrasound) they can usually be adequately handled by the geometrical theory of diffraction [84].

Despite the successes of ray techniques enumerated in the previous paragraph, which relates, by the way, entirely to direct problems, it is proving very difficult to incorporate them into a general "inverse" formulation.

There is one situation of great practical importance for which approach ii) can be combined with ray techniques to produce an inverse formulation that is virtually perfect. This is when the rays travel in straight lines, which applies most notably to conventional CT. It is convenient to discuss it with reference to Fig. 3. We postulate that narrow-beam transmitting and receiving transducers are positioned at A and A', respectively. The density $\rho(r, \theta)$ corresponds to either incremental attenuation or incremental delay of the ray in the neighborhood of P. The signal picked up at A' is thus expressible as

$$L \exp\left(-K \int_{l(A)}^{l(A')} \rho(r, \theta)\, dl\right)$$

where L and K are constants and the integral is the same as the one on RHS of (2). It is straightforward, therefore, to recover from the received signals estimates of the values assumed by this integral for all pairs of values of ϕ and ϑ. So, the processed signals are conveniently written as $s(\phi, \vartheta)$, which still satisfies (2).

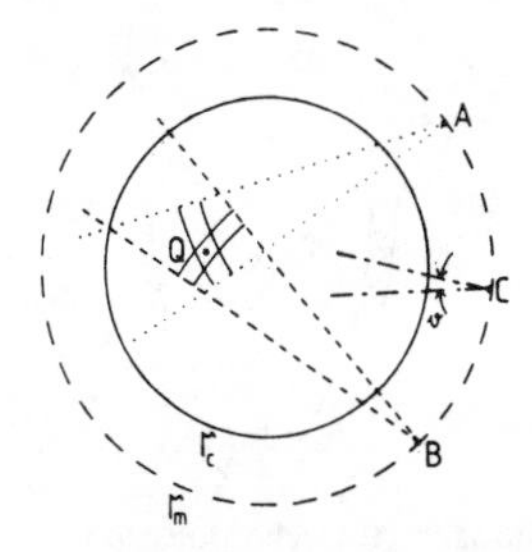

Fig. 4. Echo location of targets inside Γ_c. Narrow-beam radar transceivers are at A and B—the pairs of parallel circular arcs indicate their respective "range resolutions" in the neighborhood of Q. Narrow-beam ultrasonic B-scan transceiver is at C.

The principle of conventional CT has been extended to ultrasonic transmission imaging [12], [20], [35], [47], [63], [64], [93]. Useful results have only been obtained, however, for media that exhibit comparatively small spatial variations of refractive index, such as the female breast and various excized organs of animals. Furthermore, the images have been reconstructed using conventional CT algorithms, predicated upon the rays traveling in straight lines (i.e., no bending of the rays, due to spatial variation of the refractive index, is taken into account). Compensation for ray curvature has been attempted with several iterative schemes [14], [90], [163] and, most intriguingly, with the recently introduced "back-propagation" technique [44], [45] which is presently being extended to vector wave motion [21]. The data for all of these methods are measured characteristics of "wavefronts"—as defined in ii) above—so that it is not clear whether they can overcome a particular limitation [114] that is inherent for any algorithm predicated upon a single ray path from A to A' (see Figs. 1 and 3).

We think it useful to fit the principles of echo location into our unified approach to CT. In conventional radar, the position of a target is determined by the delay of the echo and the direction in which the antenna is pointing. The resolution in range is usually much better than the angular resolution. Suppose that there are narrow-beam transducers at A and B in Fig. 4 transmitting short pulses towards a point target at Q and receiving the reflections. The beams necessarily diverge with increasing range, while the pulsewidth is, of course, independent of range. The intersection of the two "range resolution cells" in Fig. 4 immediately suggests how the overall resolution might be improved by "triangulation." Sophisticated radar systems employ this routinely, and it is successful when (as is usual) the targets are sufficiently isolated to prevent ambiguities arising.

Fig. 4 also relates to standard medical ultrasonic B-scan, in which a transmit/receive transducer (at C say) probes the interior of Γ_c. The scanning is effected either by moving C along Γ_m or by changing the direction (represented by the angle ϑ) of the beam or (as is often the case) by a combination of both scanning motions [53], [164].

There is a generalized approach to echo-location CT that can be outlined with the aid of Fig. 5. Suppose there are wide-beam transducers at A and A', the former transmitting and the latter receiving. The two quasi-parallel curved lines labeled α–α are ellipses with A and A' as foci. The separation of these ellipses indicates the resolution of the system. Reflections from all targets within these ellipses arrive at A' at the same instant. The total reflection is proportional to the target density integrated over the area between the ellipses. Recognizing

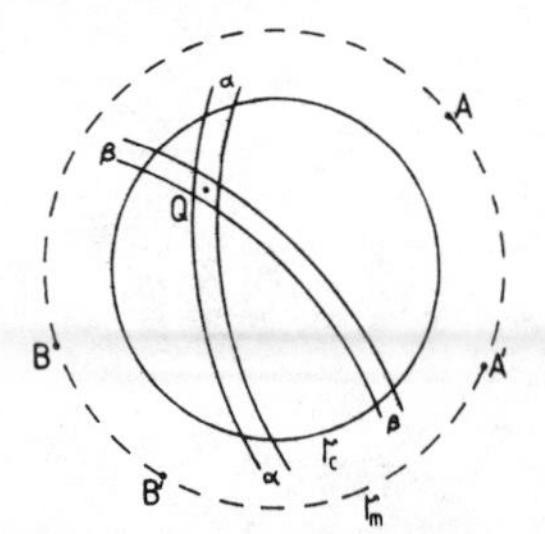

Fig. 5. Generalized approach to echo location. Transmitting and receiving transducers at A and B and at A' and B', respectively. The pairs of parallel elliptical arcs, labeled α–α and β–β, respectively, identify the respective resolution capabilities of the transducer pairs AA' and BB'. The neighborhood of Q is the envelope of all pairs of elliptical arcs corresponding to all pairs of transducers belonging to the CT system.

that the separation of the ellipses equals the diameter of the neighborhood, it makes sense to integrate the density along the mean ellipse (identified below by the symbol α). Consider targets within the neighborhood of the point Q. They can be isolated by changing the positions of the transmitting and receiving transducers; one such pair of positions is B and B', for which the corresponding pair of ellipses are the curved lines labeled β–β. By taking sufficient pairs of transducer positions, a clean image of the density can be reconstructed. Note that the signal, which is conveniently written as $s(A, A'; \alpha)$, picked up at A' from the targets between the ellipses labeled α–α can be usefully approximated by the integral of the density along the mean ellipse. In conventional radar and ultrasonic/B-scan situations, A' and A are coincident so that $s(A, A'; \alpha)$ reduces to a two-dimensional quantity, $s(A; \alpha)$ say, which when appropriately normalized can be expressed as

$$s(A;\alpha) = \int_{\alpha} \rho(r, \theta)\, dl \tag{5}$$

where dl is now arc length along the mean ellipse, which reduces, of course, to a circle because the two foci A and A' coincide. As with (2), there is no dimensionality difficulty. While a reconstruction algorithm based on (5) cannot be as convenient numerically as an algorithm based on (2), because of the nonrectilinear path of integration in (5), it is certainly theoretically tractable, as Cormack showed nearly twenty years ago [39] and has recently been reconfirmed [46].

Radar systems that employ the synthetic-aperture principle—in aircraft for terrain mapping [28], [153] or glaciology [66], [67], and in earth satellites for remote sensing of land masses and oceans [138], [144]—come closest to being capable of taking advantage of the approach to image reconstruction outlined in the previous paragraph. It is difficult to conceive of a physical situation in which it would be both feasible and worthwhile to construct a radar system in which the antenna(s) could be moved right round the targets. However, there are many practical applications of ultrasonic imaging (in nondestructive testing and medical diagnosis, for instance) in which the transducers can be scanned all around a body. Although some resolution improvement, by approaches equivalent to that outlined in the previous paragraph, is being claimed [141], [149] it is limited by the appreciable spatial variations in propagation speed exhibited by most of the media that are important in practice. These variations restrict the accuracy with which distances can be estimated by echo location. Interesting parallels between synthetic-aperture radar and CT have been reported recently [2], [124].

C. Combined Probing-Sensing CT

There are situations which call for the combined application of remote-sensing and remote-probing techniques.

We first consider interior sources embedded in a dense medium. This is the situation in nuclear medicine when the distribution of a radio-labeled substance within the body is imaged with a gamma camera (refer to Table IV). There are two densities to be taken into account: say ρ_1 of the sources and ρ_2 of the medium. Gullberg [68] has studied iterative schemes for correcting for the attenuation of the emanations from ρ_1 due to the presence of ρ_2. While the convergence of the iterations cannot be assured in general, it is numerically manifest in examples reported by Gullberg. Perhaps because the iterations tend to be computationally protracted, the only correction commonly applied to gamma cameras is predicated upon a constant attenuation coefficient (its value is chosen on the basis of experience)—the effectiveness of such correction is materially increased if the actual boundary of the body cross section is incorporated into the attenuation-compensation algorithm. The crudest correction procedure, which sometimes improves images noticably, is to average projections measured from diametrically opposite sides of the body [100].

Another form of combined probing-sensing situation arises when interior sources are induced by exterior sources, but the emanations from the induced sources can be distinguished from those that do the inducing. Such a situation occurs with three of the schemes (i.e., Nuclear Magnetic Resonance (NMR), Magneto-Hydro-Dynamic (MHD), and Compton scatter) listed in Table IV. Once the interior sources have been induced, the reconstruction proceeds as outlined in Section II-A above.

We think it appropriate here to suggest a new CT technique which we call corrected Compton scatter (see Table IV). Although it relies on the inducing of interior sources by emanations transmitted from outside, and requires remote-probing and remote-sensing CT measurements to be carried out independently, a clean image of the Compton-scatter density is reconstructed very simply by a neat trick. McKinnon [113] has independently, and more or less simultaneously, come up with the same sort or idea for quantitative calibration of ultrasonic scattering from biological tissue.

An X-ray transducer at A transmits a narrow beam towards A' (refer to Fig. 6). A narrow-beam receiving transducer at Q is set up to sense the emanations scattered from the neighborhood of P. The signal picked up at Q depends on the attenuation from A to P, the scattering at P, and the attenuation from P to Q. So, the signal can be represented as the penultimate paragraph of the previous subsection, but with the integral there replaced by the sum of two integrals plus a scattering exponent. Consequently, on taking the negative of the logarithm of this signal and subtracting out any extraneous constants, we get a quantity, written here as $s(Q, P, A)$, which can be expressed as

$$s(Q, P, A) = \int_{l(A)}^{l(P)} \rho(r, \theta)\, dl + \beta(P, \psi) + \int_{l(P)}^{l(Q)} \rho(r, \theta)\, dl \tag{6}$$

where the integrals are to be interpreted similarly to the integral discussed in the penultimate paragraph of the previous subsection and where $\exp(-\beta(P, \psi))$ is the Compton-scatter coefficient at angle ψ from the neighborhood of P. The transmitting and receiving transducers are now moved to A' and Q', respectively, with the former pointing at A and the latter

TABLE IV
TECHNIQUES THAT HAVE BEEN, OR COULD BE, APPLIED TO COMBINED PROBING-SENSING CT

Technique	Basic Principle	Inherent Limitations	Status
Nuclear Magnetic Resonance (NMR)	Static or low-frequency magnetic fields induce sources (aligned "spins" processing around field lines); Pulsed RF fields amplify the precession at points in the body where the RF frequency corresponds to the magnetic field strength; RF field relaxation is measured [78]	Image resolution depends mainly on accuracy to which magnetic field gradients can be maintained (it is too soon to assess practical consequences); Not yet known how tissue character related in detail to spin relaxation times [122]	Commercial systems now available [78], [94], [95], [165]
Corrected Emission	Internal sources imaged conventionally, after attenuation of emanations through body is estimated, taking body outline into account	Attenuation compensation based on constant attenuation coefficient throughout body; Not known whether iterative algorithms can correct varying attenuation	Constant attenuation algorithms incorporated into working systems [32]
Uncorrected Compton Scatter	Induced emanations are the photons scattered from electrons in the body	This technique is simple provided attenuation of incident and scattered X-rays is negligible	Interesting experimental results [17], [71]
Corrected Compton Scatter	Same as above except that attenuation of incident and scattered X-rays found from preliminary Transmitted X-ray CT measurement	Main limitation likely to be accuracy with which X-ray beams can be collimated on transmission and reception	Only exists as an idea as yet
Magneto-Hydro-Dynamic (MHD)	Low-frequency magnetic field applied to body having conducting fluids flowing across cross section; Induced voltages measured at perimeter of cross section	Main limitation is the huge magnetic field needed to generate sensible voltages	Interesting experiments [146] give results showimage reconstruction possible

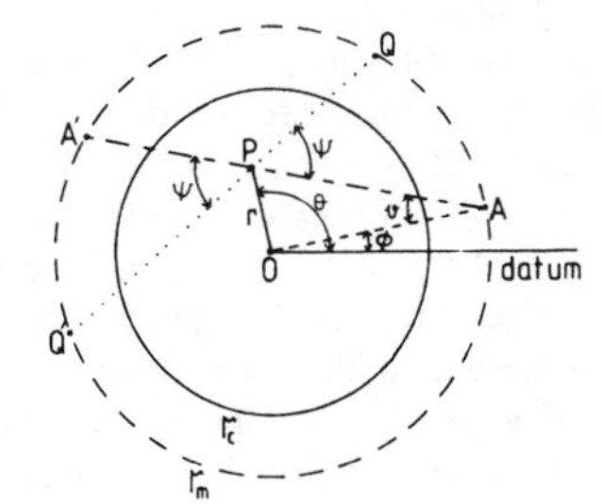

Fig. 6. Scatter CT (Compton or ultrasonic). Narrow-beam transmitting and receiving transducers are at A and A' and at Q and Q', respectively. All are directed at P, so that the angles ϕ, ϑ, and ψ characterize the measurements. When A (or A') transmits then Q (or Q') receives.

pointing at Q. The processed signal picked up at Q' is

$$s(Q', P, A') = \int_{l(P)}^{l(A')} \rho(r, \theta)\, dl + \beta(P, \psi) + \int_{l(Q')}^{l(P)} \rho(r, \theta)\, dl. \tag{7}$$

Adding the two processed signals together gives

$$s(Q, P, A) + s(Q', P, A') = \int_{l(A)}^{l(A')} \rho(r, \theta)\, dl + \int_{l(Q')}^{l(Q)} \rho(r, \theta)\, dl + 2\beta(P, \psi). \tag{8}$$

The uncorrected-Compton-scatter scheme (see Table IV) is predicated upon the attenuation throughout the interior of Γ_c being either negligible or effectively independent of A, A', Q, and Q'. The integrals in (6)–(8) are, therefore, neglected, so that it is assumed $\beta(P, \psi)$ is given by either $s(Q, P, A)$ or $s(Q', P, A')$. Inspection of (8) reveals that it is a simple matter, in principle, to correct the measurements by subtracting out the integrals on RHS of (8). All that is necessary is to make conventional remote-probing CT measurements first, because each of the integrals represents one such measurement. This corrected-Compton-scatter scheme only requires simple additions and subtractions to produce a clean image, which is the neat trick referred to above. A significant source of error is *beam hardening* (the Compton-scatter coefficient varies with the photon energy such that a beam generally contains a greater proportion of higher energy photons after it has traversed a body), but this can be avoided by using a monoenergetic source of photons (e.g., an intense isotope source). As mentioned above, McKinnon has suggested that the same sequence of measurements be applied to ultrasonic imaging. While it is easy enough to avoid effects akin to beam hardening with ultrasonics (e.g., variation of attenuation with frequency), by, for instance, using a sufficiently narrow band of frequencies when making the measurements, we would expect the Compton-scatter application to be inherently more accurate because it does not suffer from the diffraction effects and ray curvature that are bound to figure significantly in the ultrasonic application. On the other hand, the Compton-scatter measurements are likely to suffer from a poor signal-to-noise ratio because it is usually either difficult or dangerous to generate a sufficiently intense incident beam to produce scattered beams intense enough to be effectively free of noise (characterized mainly by photon-counting statistics [83]) [17].

D. Measurements in Fourier Space

It is convenient to give the name *image space* to the particular region (e.g., the interior of Γ_c) of the body we are examining. While we are always interested in forming a clean image in that space, we may be forced by practical constraints to gather data in a different space. The character of the latter is determined by the relation to the density (as this term is defined in Section I) of the quantity that is actually measured. When this quantity is the Fourier transform of the density, we say that the data are gathered in *Fourier space*.

All data gathered by X-ray crystallographers [136] and those radio astronomers who use the giant synthesis telescopes [109] are gathered in Fourier space. Electron micrographs are often exposed in the back focal plane (equivalent to Fourier space). Even in optical astronomy, data [9] and image [25] processing are being based increasingly on Fourier techniques, mainly because of the successful development of speckle interferometric imaging methods [9].

From a theoretical point of view, it makes no difference whether CT data are gathered in image space or Fourier space. There are definite conceptual advantages in combining both spaces when developing algorithms [105]. Some interpolation procedure is a central part of most CT algorithms [4], [70], [73], although we emphasize that NMR measurements can be made to correspond to a rectangular grid of points in Fourier space thereby implying not only that interpolation is unnecessary but also that images can be formed by more or less direct Fourier transformation of measured data [78]. When interpolation is needed, the quality of reconstructed images depends critically upon the space in which the interpolation is carried out [106]. Details of algorithms are thus highly dependent upon whether the data are gathered in image space or Fourier space [11], [104].

III. Problem Areas for Existing CT Systems

For CT to be useful, it must provide, with considerable spatial and temporal resolution, images of meaningful material properties of bodies. Keeping this in mind permits us to isolate several serious technical difficulties, which are discussed in the following subsections. The gaps in our understanding of the physical processes involved in the interactions of the emanations with the body is limiting the development of certain types of CT, as we argue in Section III-A. As is indicated in Section III-B, some CT applications are less powerful than one might hope because it is not yet possible to devise accurate, stable computational algorithms from the basic physical theories (even when they are expressible in concise mathematical forms). Section III-C deals with limitations imposed by deficiencies in certain hardware, while in Section III-D we attempt to identify those clinical considerations that seem most urgent.

A. Areas Requiring Improved Basic Understanding

It is probably safe to state that the behavior of X-rays is adequately understood, as far as their employment for CT is concerned. They travel in straight lines, are not subject to diffraction effects, and are attenuated by a combination of scattering and absorption [86], [88], in accordance with relatively simple laws. It is possible to distinguish many pathological conditions from healthy tissue by inspection of X-ray images [166]. Some other kinds of emanations are less well understood. Ultrasound, which seems to be harmless at the intensity levels ordinarily employed in clinical practice [103], [155], does not produce properly quantifiable images, mainly because of our ignorance of fine details of the interaction of the wave motion with biological tissue. Ultrasonic B-scan machines have reached a high state of perfection as far as their image-display capabilities are concerned [99], [132], but it seems that potentially interesting information is being thrown away because we do not know how to make use of it.

It has been suggested that improved understanding of ultrasonic CT images might be obtained by employing perturbation techniques to construct inverse solutions to the Helmholtz equation—by carrying out the reconstruction in the spatial frequency domain, the algorithm can be made computationally efficient [123]. However, we do not see how such an approach can overcome the difficulties, noted in Section II-B, associated with full-wave approaches to inverse scattering.

NMR (see Tables I and IV and refer to the paper by Hinshaw and Lent [78] in this issue) is showing great promise as a clinical imaging modality. Much remains to be done, however, in the way of quantitative relation of the parameters measured by this technique (proton density or hydrogen content, and spin relaxation times) to the properties of biological tissue.

Despite current ingenious developments [110], the imaging time for a single body cross section seems certain to remain rather long (e.g., 4 to 8 min) if the image quality is to be good—but by careful selection of pulse sequences, many separated cross sections can be measured simultaneously. In fact, it may soon prove possible to gather data from the whole of a patient's body in less than a quarter of an hour [95].

One must not lose sight of the diagnostic function of all CT systems. If the images produced by a particular system are repeatable and free from artifacts which tend to obscure diagnostic information, and if clinicians can learn to relate features of images to normal and abnormal anatomy, it could be argued that a lack of basic physical understanding is of little consequence. For example, CT images of the female breast formed from ultrasonic time-of-flight measurements, and reconstructed using the conventional "straight-ray" algorithms, are providing useful diagnostic information [63], [65]. However, if no effort is made to understand the tissue/ultrasound interaction better then there is a danger that improvements in hardware and software, while permitting the previous information to be displayed to finer resolution perhaps, will not uncover any new kinds of information.

B. Algorithms that Need Improvement

There are several areas in medical X-ray CT which could benefit from small improvements. These include compensation for hardening of the X-ray beam, correcting for the effects of scattered radiation, and eliminating the artifacts produced by partial volume effects of small dense structures smaller than the width of the X-ray beam.

Iterative approaches to the beam-hardening problem have been reported [26], [125]. Since the density is unknown initially, the only feasible way to estimate its distribution throughout the interior of Γ_c is to invoke a conventional CT algorithm (cf. [106]). Once this estimate has been formed, beam hardening due to objects that appear to be inside the body can be predicted from the known behavior of X-rays [88]. The measured data can then be "corrected" so that a new image can be reconstructed from which the beam hardening can be repredicted. This iterative process is continued until the predicted beam hardening settles down. Apart from being computationally protracted and therefore expensive, there is unfortunately no guarantee that any such scheme will converge. It is worth noting that the degree of beam hardening can be crudely, though usefully, estimated directly from the observed attenuation of the X-ray beam—such crude correction is employed in commercial scanners [26].

Another approach to the beam-hardening problem requires that two scans be made at different X-ray energies [55], [56]. By ingenious hardware design these scans may be made simultaneously by utilizing a split filter in the beam of a third- or fourth-generation X-ray CT scanner. Alternate detectors must be provided with different X-ray beam filtration, or special detectors yielding high- and low-energy signals must be constructed.

A significant problem for fourth-generation scanners (i.e., fixed-detector ring geometry) is the proportion of direct-to-scattered radiation seen by the detectors. This can lead to "cupping" artifacts which distort the reconstructed CT numbers [121]. Techniques for recognizing such artifacts and ameliorating them are needed.

When imaging structures close to the base of the skull (*posterior fossa*) one often encounters rapid density changes perpendicular to the cross section being imaged (i.e., from bone to soft tissue). This is known to cause severe streaking artifacts which degrade the diagnostic usefulness of the reconstructed images. It is worth noting that such artifacts seem to be absent from NMR images of the *posterior fossa* [33].

The only current method of correcting for the streaking artifacts noted in the previous paragraph is to materially reduce the width of the cross section being examined. However, this either increases the noise level or greatly increases the dose to the patient. Processing techniques for compensating for this and other nonlinear effects are urgently needed.

Theoretically valid and computationally stable (and efficient) image reconstruction algorithms are more urgently needed for emission CT than for conventional X-ray CT. Corrections are required for spatially varying resolution and for attenuation of the emanations on their passage through the body.

CT measurement data are sometimes sparse (e.g., when attempting to form images from "low-dosage" measurements, or when performing ECG-gated CT [145]). Efficient interpolation methods are then required. While useful accuracies have been demonstrated [4], [70], [73], there is need for improvements in efficiency.

Conventional CT algorithms can produce useful images from measured data of either the attenuation or the time of flight of ultrasonic pulses transmitted through bodies—this is Ultrasound Transmission CT [63] (see Table III). The propagation is thought of in terms of bundles of rays, which is usually justifiable (refer to Section II-B). The rays must in general follow curved paths. The only thing known for certain about the convergence of the several schemes [14], [90], [163] proposed for correcting for ray curvature is that they are unlikely to converge [114]. Perhaps Devaney's new approach [44], [45] will point the way to sure means for compensating for the bending of ultrasonic rays.

Because professional mathematicians have increasingly retired from the real world during the past half century, physicists and engineers have tended to devise their own mathematical techniques, and with considerable success, because it has been largely physical intuition that has been needed. The beautiful simplicity of the modified backprojection algorithms used in conventional CT image reconstruction is a good example. In many branches of technical science, however, very severe computational problems have arisen which require for their solution a degree of specialization that only the professional numerical analyst can provide. Regularization techniques [43] have already produced striking results in several fields [126]. We suggest that CT theorists might do well to immerse themselves in these rather new numerical methods. It is possible that the "ray-bending" problem could yield to sufficiently sophisticated regularization approaches. We feel that Electrical Impedance CT (see Tables I and III) is almost entirely a matter of numerical analysis because such large systems of simultaneous linearized equations have to be solved and in a stable manner. Conventional CT algorithms are inappropriate for Electrical Impedance CT [15], as is emphasized by the uncertain results obtained with early intuitively based approaches [135], [150].

In the longer term there is the general problem of how to organize and display three-dimensional data. The main difficulty is how to do it tomographically, because it seems impossible to form a clean three-dimensional image (in the sense defined in Section I).

It is impossible to look through a surface without having one's view of internal structures at least partly obscured by other closer objects. One way out of this difficulty might be to use a vari-focal mirror to display an image in three dimensions with the ability to interactively select any arbitrarily defined plane for separate two-dimensional display.

In another approach, due to Herman [77], the surface of an organ with a clearly defined boundary is rapidly computed and projected onto a two-dimensional surface. The impression of an isolated three-dimensional structure is enhanced by rotating the image.

It is not clear at present whether any of these ideas are potentially superior to the simple conventional technique of selecting images in sagittal and/or coronal planes, rapidly computed from series of stored transverse tomographic images.

Real-time imaging of rapidly moving structures (e.g., the human heart which beats roughly once a second, whereas the fastest commercial X-ray scanner needs at least 2 s to complete a scan) is of current interest. Techniques at present routinely employed for the study of such organs are palpably invasive (sometimes dangerously so, as on occasion with cardiac angiography [162]) but they are simple, useful, inexpensive, and well understood. However, even accepting the inevitably greater expense involved, safe noninvasive alternatives to traditional methods are obviously worth investigating. It is not yet clear whether true real-time imaging should be attempted, as with the dynamic spatial reconstructor (DSR) [140] or by cardiovascular computerized tomography (CVCT) [24], or whether a stroboscopic approach, like ECG-gated CT [115], [127], might be more appropriate. The DSR and CVCT approaches are potentially much more powerful but they are bound to be significantly more expensive than ECG-gated CT, if only because new apparatus must be developed, purchased, and installed.

C. Hardware Limitations

To obtain significantly better resolution from many CT systems, narrower X-ray beams are needed. The main difficulty here is to maintain the photon flux without subjecting the X-ray tube to excessive heating stresses. Boyd's CVCT scanner [24], which utilizes an anode effectively surrounding the patient, thus having a high heat capacity because of its large mass, can be regarded as a partial solution to this problem. There is the possibility of materially reducing the 99-percent photon loss, caused by collimation and shielding, suffered by conventional X-ray CT systems.

The timing resolution of Emission CT detectors (particularly for positron-emission CT) needs to be improved without sacrificing sensitivity. The chief requirement for NMR systems is magnetic fields with greater spatial and temporal uniformity—a further need is for the optimization of wide-band radio-frequency probe design. For ultrasonic CT, the main need is for efficient wide-band transducers. For nonmedical CT applications, existing hardware does not constitute a limiting factor, because improved performance can only come from increases in the size and power of whole systems.

In some medical applications, parts of the interiors of bodies can move comparatively quickly. It is important therefore to be able to speed up CT measurements. This has already been discussed in the final paragraph of Section III-B above and it is further touched on in Section IV-A. Of all medical CT techniques, perhaps Emission CT is the one most in need of improved hardware. It is difficult to design collimators and detectors to reject random events and unwanted scatter. The radioactive substances commonly used as sources of the gamma-ray emissions are weak, in the sense that the signal-to-noise ratio is generally limited by photon-counting statistics. Gamma cameras suffer from comparatively poor resolution and from variations in spatial sensitivity, together with calibration difficulties stemming from the point-spread functions of collimators being dependent upon the distances of events from the surfaces of the collimators [85].

While the main difficulties associated with Electrical Impedance CT are algorithmic (refer to Section III-B above), the reliable injection of currents into bodies (and the reliable measurement of voltages on the surfaces of bodies) is far from trivial in practice [41]. Standard designs do not exist for arrays of electrodes, all of which can be connected effectively identically to humans or animals, or even geological or archaeological specimens. The electrical prospecting techniques routinely employed in the field [96] are in reality very crude when regarded as imaging methods.

Two techniques in their infancy are Heavy-Particle CT [18], [31] and Radiometric CT [8], [19], [51]. Problems associated with image reconstruction algorithms for these two techniques have been so little studied that we cannot comment on them usefully here. Neither can we meaningfully assess the hardware problems likely to be experienced with Heavy-Particle CT. Even if adequate algorithms can be developed for Radiometric CT, the resolutions of the processed images are bound to be adversely effected, at microwave frequencies at least, by the physical sizes of the transducers (antennas in this case) [82]—this difficulty would be eased by sensing infrared radiation, provided it can be shown in both theory and practice that interesting information is implicit in feasible sets of measurements.

At present, the most useful CT technique for medical applications is undoubtedly that which uses X-rays. Ultrasonic CT and NMR are beginning to represent some sort of challenge to X-ray CT. The chief problem area for NMR is in the design of the associated magnets. The spatial and temporal stability of the magnetic fields is crucial. The future of NMR, as a CT technique, depends upon how conveniently and economically such fields can be generated and calibrated—of course, it could be argued that the machines already on the market demonstrate that these problems have been solved [55], [78], [94].

As intimated in Section III-B, the main problems for Ultrasonic CT are algorithmic. However, there are also significant problems associated with transducer hardware. Whenever ultrasonic (or acoustic, for that matter) radiation is used for remote probing (e.g., in medical diagnosis, nondestructive testing, atmospheric sounding, or by animals such as bats or porpoises) it seems that the wavelength is not much smaller than the detail which it is hoped to resolve. This means that the achievable resolution tends to be limited by the sizes of the apertures of narrow-beam transducers. Consequently, there is a definite need to devise systems that improve resolution by appropriate combinations of ingenious algorithms, clever signal processing, and optimized transducer design [57], [89].

D. Clinical Considerations

Special considerations arise when the body subjected to CT measurements is a human one, and the purpose is to aid medical diagnosis.

While CT is certainly noninvasive by comparison with surgery or with insertion of catheters, it must not be forgotten that something has to invade a body if information concerning its interior is to be gathered. If this "something" is a beam of ionizing radiation (e.g., X-rays) then tissue can be damaged with a consequent risk to the patient's subsequent health [112]. It is, therefore, mandatory to keep X-ray dosage to a minimum (this is discussed further in Section IV).

It seems generally agreed that ultrasound is harmless to humans at the intensity levels routinely employed in commercially available medical apparatus [103], [155]. Although there is as yet no appreciable body of clinical experience with NMR machines, it is very unlikely that their emanations could represent any sort of risk to patients. The few studies carried out so far record no sensible harmful effects for magnetic-field strengths up to about 1 T [23], [30]. It is difficult to believe that much stronger fields could be generated for routine clinical use.

Perhaps the most important clinical consideration is what has come to be known as *tissue characterization* [37]. The purpose of any medical diagnostic measurement is to determine whether the patient is or is not healthy, according to some clinical criterion. The difficulty here is that the state of any kind of tissue can vary enormously between individual patients and yet still be healthy—there is usually a vast range of "normals," as they say in the medical business. It is much easier, of course, to gauge how a particular patient's health has varied between measurements made at well separated times. However, we often want to make a definite diagnosis at the first attempt. Consequently, there is frequently a real need for major improvements in the precision with which tissues can be characterized by inspection of CT images.

IV. Some Promising Approaches

It is useful to distinguish between what we here call *sequential* and *instantaneous* CT systems. The former term implies that the complete data set needed for a CT measurement consists of subsets that are measured one after the other—it is further implied that the time needed to complete the measurements is longer than the shortest interval within which the density can, under extreme conditions, exhibit sensible temporal variations—it is also of course implied that, under many circumstances of practical interest, the density does not exhibit significant changes during the time needed to complete a set of sequential CT measurements. An instantaneous system performs a complete set of measurements either (effectively) simultaneously or so rapidly that they are gathered within an interval shorter than that required by any interior part of the body to exhibit a sensible change. There is considerable current interest in instantaneous medical CT systems because of the wish to image the beating heart—refer to the paper by Robb *et al.* [140] in this issue.

There are several CT systems which, while they are certainly instantaneous by the above definition, all suffer from the particular disadvantage discussed later in Section IV-A—because of this it is not obvious that these systems offer a clear advantage over possible modifications of more conventional and much cheaper systems. Section IV-B is concerned with applications of CT in prospecting. We indicate in Section IV-C how recently developed solutions to the multidimensional Fourier phase problem might prove to be of major significance in astronomy and microscopy.

The approaches mentioned in the preceding paragraph are by no means the only ones that can be considered promising. For instance, we certainly classify Ultrasonic CT and NMR among the promising approaches, but there is no need for us to state more than that in this section because, for the purposes of this paper, we say all that is necessary about these two techniques in Sections II, III, and V. There does not seem to have been much attempt to apply CT principles to nondestructive testing and to the surveillance of packages and luggage [161]. This surprises us because we feel that these are two applications of CT that seem to be both useful and full of commercial promise.

A. Comments on Potentially Instantaneous X-Ray Systems

The conventional X-ray CT scanners now installed in so many hospitals and clinics round the world are sequential. Patients being examined with the aid of these machines must lie still. A few seconds are needed by the fastest of these machines to complete a set of CT measurements, implying that a conventional image of the beating heart is unavoidably blurred.

With the primary aim of imaging the beating heart, three ingenious types of machine have been devised for performing instantaneous CT. One of these due to Mayden, Shepp, and Cho [111] relies on extremely rapid rotation of an electron gun. The other two, the DSR [140] and the CVCT machine [24], require no mechanical motion to generate an instantaneous image. It should be noted that the image-processing power of the DSR is greatly enhanced by continuous rotation of its X-ray sources and detector systems [3].

All of the devices referred to in the previous paragraph have the potential for forming images within intervals short compared with the duration of a single heart beat, and the DSR has already done so [140]. However, these devices form "cone-beam" projections rather than the familiar "fan-beam" projections. Furthermore, the projections in the DSR span an angular range of only 162°, as opposed to the 180° range in commercial CT machines. Consequently, care and ingenuity has to be exercised to prevent significant image degradation in practice. It should be emphasized nevertheless that the DSR gathers data over a volume (as opposed to a single cross section) more efficiently than conventional CT scanners.

While commenting on CT systems that are potentially instantaneous, it is worth mentioning Paired Photon Emission CT (see Table II) or PET as it is often called. The active agent in the sources used for PET is a positron-emitting isotope. When an electron annihilates with a positron, two gamma-ray photons are emitted in opposite directions (necessary to conserve momentum and energy). A ring of detectors surrounds the cross section to be imaged—i.e., detectors are placed side by side all round Γ_m (refer to any of the figures). Coincidence circuits interconnect all pairs of detectors, implying that annihilations from any point inside Γ_c can be sensed at any time. So, PET is instantaneous in principle. Unfortunately, however, practical flux levels of paired photons are so low that the data gathering rates of PET systems are appreciably slower than those of conventional X-ray CT apparatus.

B. Prospecting CT

The injection of low-frequency electric currents into the Earth, together with measurement at the Earth's surface of the resulting voltages, has been used for several decades as a means of prospecting for minerals [96]. Although this technique is useful, it is so crude from an imaging viewpoint (as already noted in Section III-C) that it hardly rates classification as a CT system.

There have recently been several fairly extensive investigations of the application of CT principles to geological prospecting [48], [49]. Because of the large physical scale involved it is, of course, impractical to gather data sets that are complete, as this term might be understood in, for instance, conventional medical applications of CT. Nevertheless, worthwhile results have been achieved and we feel that further developments should be pursued.

The CT measurements are usually made by lowering appropriate transducers into boreholes in the ground. The emanations that have been employed are acoustic (seismic) waves [49], low-frequency electric currents [48], and radio waves of frequencies up to 100 MHz [102]. The reconstruction algorithms have been based on those developed for X-ray CT [62], [156], with attempts to correct for ray curvature [101], [109] and the special nature of low-frequency electric currents [108]. We suggest that these attempts at correction are doomed to failure [15], [114], but we also feel that the conventional algorithms will be sufficiently applicable often enough to provide useful results.

C. Fourier Space and the Phase Problem

We point out in Section II-D that certain measurement systems which gather their data in Fourier space can usefully be regarded as CT systems. The image reconstruction capabilities of some of these are limited mainly by their inability to measure phase as accurately as intensity. It is highly significant, therefore, that recent studies [10], [29], [54], [58], [59] seem to show that phase distributions can almost always be computed from given intensity distributions, provided the measurements are made in Fourier spaces of more than one dimension. It must be emphasized that these results cannot be applied directly to crystallography because Fourier space must be oversampled (as compared with the requirement set by the usual sampling theorem) by a factor of at least 2 in each direction in Fourier space. It is worth remembering the general lack of unique solutions to one-dimensional Fourier-phase problems [157]. These new results may be of use in Very Long Baseline Radio Astronomical Interferometry, even though the so-called closure phase procedure has gone a long way towards solving the phase problems that arise there [137]. There may well be applications to electron microscopy [60], [119], [147]. However, the greatest benefit is likely to be to optical astronomy [9], [34], [42], [98], [142].

V. Conclusions and Prognostications

Sections III and IV are already replete with specific conclusions, which it would be merely repetitive to summarize here. Our purpose in this section is to outline what we feel are likely to be profitable future approaches to the development of CT. This is the main reason why we have attempted a comparative review of the many kinds of system that employ CT principles. It must, in our opinion, be constructive to identify the underlying similarities between what may initially appear as very different methods of forming clean images by computational operations on measured data.

Except for examining the lower abdomen (especially in pregnant women) for which ultrasonic B-scan (i.e., Ultrasonic Reflection CT) is perhaps the ideal diagnostic imaging technique, we feel that X-ray CT is by far the most effective of the established modalities. On the other hand, NMR is rapidly catching up with X-ray CT and is already in many ways superior to it. NMR has the potential for sensing many different tissue characteristics [95], [165]. A particularly exciting possibility is that it may eventually prove feasible to estimate the state of metabolic function by appropriately organized NMR measurements. However, it is worth emphasizing that it is very unlikely that it will ever be possible to form a high-quality image of a single cross section anywhere near as rapidly with NMR as with X-ray CT.

Some responsible medical opinion advocates mass screening for detecting various prevalent diseased conditions [116]. The basic difficulty with this is not so much the expense, which can easily be justified if the screening process is successful by and large, but the fact that people do not like being ordered around by "experts" (the same consideration often deters people from having regular checkups), especially if the procedure is lengthly or embarrassing or awkward in any way at all. The worldwide success of the programs for mass X-ray screening for tuberculosis may have been due to the simplicity and impersonality of the procedure.

There is a premium then upon reliable "one-shot" diagnoses. We feel that this provides the main justification for research into new and even very expensive medical applications of CT. Note, however, that this reasoning goes beyond the obvious rationale of studying a technique, such as Ultrasonic Transmission CT or NMR, to see if it is suitable for mass screening for cancer in the female breast without being in danger (as some people think X-ray techniques may be [5], [128]) of inducing carcinomas as well as detecting them.

There seems to be growing evidence that different kinds of CT can characterize tissues in significantly different ways. So, if it becomes feasible to form a variety of CT images of a particular part of a particular patient during a single visit to a clinic, the chances of making an accurate initial diagnosis must be improved.

A worthy challenge to engineers is to design systems incorporating all the major medical CT techniques in such a way that they function reliably, without demanding continuous multifaceted servicing, and without requiring increased numbers of support personnel. Technology by itself is not contributing all that much to the increasing cost of health care. The trouble is that every time a fresh invention takes the fancy of our medical colleagues, a new horde of salaried acolytes housed in new buildings is deemed necessary to minister to it. Our profession should be capable of reversing this trend. Something else engineers should be concerned with is simplifying the routine operation of CT systems, which will certainly be vitally important if several types of CT are ever incorporated into a single overall system.

Finally, we emphasize that one is likely to miss profitable opportunities by attempting to restrict the concept of CT to a single area, such as medical diagnostics. As we show in Section II, the principles of CT apply widely. As in all science nowadays, advances are much more likely to come from crossing interdisciplinary boundaries than from unswerving allegiance to a narrow field.

References

[1] F. S. Acton, *Numerical Methods that Work.* New York: Harper & Row, 1970, pp. 245–257.

[2] M. F. Adams and A. P. Anderson, "Synthetic aperture tomographic (SAT) imaging for microwave diagnostics," *Proc. IEE,* pt. H., vol. 129, pp. 83–88, 1982.

[3] M. D. Altschuler, Y. Censor, P.P.B. Eggermont, G. T. Herman, Y. H. Kuo, R. M. Lewitt, M. McKay, H. K. Tuy, J. K. Udupu, and M. M. Yau, "Demonstration of a software package for the reconstruction of the dynamically changing structure of the human heart from cone beam X-ray projections," *J. Med. Syst.*, vol. 4, pp. 289–304, Apr. 1980.

[4] N. Baba and K. Murata, "Image reconstruction from limited-angle projections," *Optik*, vol. 60, pp. 327–332, Mar. 1982.

[5] J. C. Bailar, III, "Screening for early breast cancer, pros and cons," *Cancer*, vol. 34, pp. 2783–2795, June 1977.

[6] B. B. Baker and E. T. Copson, *Mathematical Theory of Huygens' Principle*, 2nd ed. Oxford, England: Clarendon Press, 1953.

[7] H. P. Baltes, Ed., *Inverse Source Problems in Optics.* Berlin, Germany: Springer, 1978.

[8] A. H. Barret and P. C. Myers, "Subcutaneous temperatures: A method of non-invasive sensing," *Science*, vol. 190, pp. 669–671, Nov. 1975.

[9] R.H.T. Bates, "Astronomical speckle imaging," *Phys. Rep.*, vol. 9, pp. 203–297, Oct. 1982.

[10] —, "Fourier phase problems are uniquely solvable in more than one dimension: I: Underlying theory," *Optik*, vol. 61, pp. 247–262, June 1982.

[11] —, "Phase recovery, speckle and image reconstruction," in *Transformations in Optical Signal Processing*, W. T. Rhodes, Ed. Bellingham, WA: Soc. Photo-Opt. Instrum. Eng., 1982.

[12] R.H.T. Bates and G. R. Dunlop, "Inverse scattering and tomography," in *Ultrasonics Int. 1977 Conf. Proc.* Guildford, England: IPC Sci. Technol. Press, 1977, pp. 104–110.

[13] R.H.T. Bates and W. R. Fright, "Towards imaging with a speckle-interferometric optical synthesis telescope," *Monthly Notices Royal Astronom. Soc.*, vol. 198, pp. 1017–1031, 1982.

[14] R.H.T. Bates and G. C. McKinnon, "Towards improving images in ultrasonic transmission tomography," *Australasian Phys. Sci. Med.*, vol. 2, pp. 134–140, Mar./Apr. 1979.

[15] R.H.T. Bates, G. C. McKinnon, and A. D. Seagar, "A limitation on systems for imaging electrical conductivity distributions," *IEEE Trans. Biomed. Eng.*, vol. BME-27, pp. 418–420, July 1980.

[16] R.H.T. Bates and F. L. Ng, "Polarization-source formulation of electromagnetism and dielectric-loaded waveguides," *Proc. Inst. Elec. Eng.*, vol. 119, pp. 1568–1574, Nov. 1972.

[17] J. J. Battista and M. J. Bronskill, "Compton-scatter tissue densitometry: Calculation of single and multiple scatter photon fluences," and "Compton scatter imaging of transverse sections: an overall appraisal and evaluation for radiotherapy planning," *Phys. Med. Biol.*, vol. 23, pp. 1–23, Jan. 1978, and vol. 26, pp. 81–99, Jan. 1981.

[18] E. V. Benton, M. R. Cruty, R. P. Henke, and V. A. Tobias, "Heavy particle radiography—Potential for three-dimensional reconstruction," in *Image Processing for 2-D and 3-D Reconstruction from Projections: Theory and Practice in Medicine and the Physical Sciences* (Dig. Tech. Papers), Stanford, CA, Stanford Univ., p. PD.1, Aug. 1975.

[19] J. Bigu del Blanco, C. Romero-Sierra, and J. A. Tanner, "Some theory and preliminary experiments on microwave radiometry of biological systems," in *IEEE Symp. Rec.*, IEEE Publ. 74CH 0838-3MTT (IEEE/GMTT Int. Microwave Symp., Atlanta, GA, June 1974), pp. 41–43.

[20] B. Blyth, G. Cavayne, M. Cleary, R. Fleming, R. Koch, L. Meara, J. McCaffrey, and J. Whiting, "An ultrasonic computer-assisted tomographic scanner," *Australasian Phys. Sci. Med.*, vol. 2, pp. 141–153, Mar./Apr. 1979.

[21] W. M. Boerner, V.K.S. Mirmira, A. C. Manson, C. W. Yang, and A. K. Buti, "Electromagnetic reconstruction techniques," presented at the 26th Int. Symp. and Instrument Display of SPIE, Aug. 25, 1982, to be published in *Proc. SPIE.*
[22] P. J. Bones and R.H.T. Bates, "Computational assessment of electrocardiographic measurement sensitivity," *Med. Biol. Eng. Comput.*, vol. 19, pp. 717-724, Nov. 1981.
[23] P. A. Bottomley and E. R. Andrew, "RF magnetic field penetration, phase shift and power dissipation in biological tissues: Implications for NMR imaging," *Phys. Med. Biol.*, vol. 23, pp. 630-643, July 1978.
[24] D. P. Boyd, "Future technologies, Transmission CT," in *Radiology of the Skull and Brain, Technical Aspects of Computed Tomography*, T. H. Newton and D. G. Potts, Eds. St. Louis, MO: C. V. Mosby Co., 1981, ch. 130.
[25] R. N. Bracewell, "Computer image processing," *Annu. Rev. Astron. Astrophys.*, vol. 17, pp. 113-114, 1979.
[26] R. A. Brooks and G. Di Chiro, "Beam hardening in X-ray reconstructive tomography," in *Image Processing for 2-D and 3-D Reconstruction from Projections: Theory and Practice in Medicine and the Physical Sciences* (Dig. Tech. papers), Stanford, CA, Stanford Univ., p. PD.3, Aug. 1975.
[27] ——, "Principles of computer assisted tomography (CAT) in radiographic and radioisotopic imaging," *Phys. Med. Biol.*, vol. 21, pp. 689-732, 1976.
[28] W. M. Brown and L. J. Porcello, "An introduction to synthetic-aperture radar," *IEEE Spectrum*, vol. 6, pp. 52-62, Sept. 1969.
[29] Y. M. Bruck and L. G. Sodin, "On ambiguity of the image reconstruction problem," *Opt. Commun.*, vol. 30, pp. 304-308, Sept. 1979.
[30] T. F. Budinger, "Thresholds for physiological effects due to RF and magnetic fields used in NMR imaging," *IEEE Trans. Nucl. Sci.*, vol. NS-26, pp. 2821-2825, Apr. 1979.
[31] T. F. Budinger, K. M. Crowe, J. L. Cahoon, V. P. Elischer, K. H. Huesman, and L. L. Kanstein, "Transverse section imaging with heavy charged particles: Theory and applications," in *Image Processing for 2-D and 3-D Reconstruction from Projections: Theory and Practice in Medicine and the Physical Sciences* (Dig. Tech. Papers), Stanford, CA, Stanford Univ., pp. MAI.1-MAI.4, Aug. 1975.
[32] T. F. Budinger, G. T. Gullberg, and R. H. Heusman, "Emission computed tomography," in *Topics in Applied Physics, Image Reconstruction from Projections*, G. T. Herman, Ed. Berlin, Germany: Springer, 1979, ch. 5, pp. 207-218.
[33] G. M. Bydder, R. E. Steiner, I. R. Young, A. S. Hall, D. J. Thomas, J. Marshall, C. A. Pallis, and N. J. Legg, "Clinical NMR imaging of the brain: 140 cases," *Amer. J. Roentgenol.*, vol. 139, pp. 215-236, Aug. 1982.
[34] F. M. Cady and R.H.T. Bates, "Speckle processing gives diffraction-limited true images from severely aberrated instruments," *Opt. Lett.*, vol. 5, pp. 438-440, Oct. 1980.
[35] P. L. Carson, D. E. Dick, G. A. Thieme, M. L. Dick, E. J. Bayly, T. V. Oughton, G. L. Dubuque, and H. P. Bay, "Initial investigation of computed tomography for breast imaging with focused ultrasound beams," in *Ultrasound in Medicine*, D. N. White and E. A. Lyons, Eds. New York: Plenum, 1978, pp. 319-322.
[36] K. Chadan and P. C. Sabatier, *Inverse Problems in Quantum Scattering Theory.* Berlin, Germany: Springer, 1977.
[37] R. C. Chivers, "Tissue characterization," *Ultrasound Med. Biol.*, vol. 7, pp. 1-20, Jan. 1981.
[38] L. Colin, Ed., "Mathematics of profile inversion," NASA Tech. Memor. TM X-62, NASA Ames Research Center, Moffet Field, CA, Aug. 1972.
[39] A. M. Cormack, "Representation of a function by its line integrals, with some radiological applications, I & II," *J. Appl. Phys.*, vol. 34, pp. 2722-2728, Oct. 1963, and vol. 35, pp. 2908-2912, Sept. 1964.
[40] J. M. Cowley, *Crystal Structure Determination by Electron Diffraction.* Oxford, England: Pergamon, 1968.
[41] L. F. Cromwell, F. S. Weibell, E. A. Pfeiffer, and L. B. Usselman, *Biomedical Instrumentation and Measurements.* Englewood-Cliffs, NJ: Prentice-Hall, 1973.
[42] J. Davis and W. J. Tango, Eds., *Higher Angular Resolution Stellar Interferometry* (Proc. Int. Astronom. Union Colloquium no. 50), School of Physics, Univ. Sydney, NSW 2006, Australia.
[43] G. A. Deschamps and H. S. Cabayan, "Antenna synthesis and solution of inverse problems by regularization methods," *IEEE Trans. Antennas Propagat.*, vol. AP-20, pp. 268-274, May 1972.
[44] A. J. Devaney, "A computer simulation study of diffraction tomography," submitted to *IEEE Trans. Biomed. Eng.*
[45] A. J. Devaney, "A filtered backprojection algorithm for diffraction tomography," *Ultrason. Imag.*, vol. 4, pp. 336-350, Oct. 1982.
[46] D. R. Dietz, S. I. Parks, and M. Linzer, "Expanding-aperture annular array," *Ultrason. Imag.*, vol. 1, pp. 56-75, Jan. 1979.
[47] K. A. Dines and A. C. Kak, "Ultrasonic attenuation tomography of soft tissues," *Ultrason. Imag.*, vol. 1, pp. 16-33, Jan. 1979.
[48] K. A. Dines and R. J. Lytle, "Analysis of electrical conductivity imaging," *Geophysics*, vol. 46, pp. 1025-1036, July 1981.
[49] ——, "Computerized geophysical tomography," *Proc. IEEE*, vol. 67, pp. 1065-1073, July 1979.
[50] R. D. Ekers, "Convergence of optical and radio techniques," in *Optical Telescopes of the Future*, F. Pacini, W. Richter, and R. N. Wilson, Eds. Geneva, Switzerland: Feb. 1978, pp. 387-389.
[51] B. Enander and G. Larson, "Microwave radiometric measurements of the temperature inside a body," *Electron. Lett.*, vol. 10, p. 317, July 1974.
[52] M. Fatemi and A. C. Kak, "Ultrasonic B-scan imaging: Theory of image formation and a technique for restoration," *Ultrason. Imag.*, vol. 2, pp. 1-47, Jan. 1980.
[53] H. Feigenbaum, *Echocardiography.* Philadelphia, PA: Lea and Febiger, 1976.
[54] J. R. Feinup, "Phase retrieval algorithms: A comparison," *Appl. Opt.*, vol. 21, pp. 2758-2769, Aug. 1982.
[55] A. Fenster, "Split xenon detector for tomochemistry in computed tomography," *J. Comp. Assist. Tomography*, vol. 2, pp. 243-252, July 1978.
[56] A. Fenster, D. Drost, and B. Rutt, "Correction of spectral artifacts and determination of electron density and atomic number from CT scans," in *Applications of Optical Instrumentation in Medicine VII* (Proc. 7th Meet. on Applications of Optical Instrumentation in Medicine), Toronto, Ont., Canada, Mar. 1979, also in *Proc. SPIE*, vol. 173, pp. 333-340, 1979.
[57] F. S. Foster, M. S. Patterson, M. Arditi, and J. W. Hunt, "The conical scanner: A two transducer ultrasound scatter imaging technique," *Ultrason. Imag.*, vol. 3, pp. 62-82, Jan. 1981.
[58] W. R. Fright and R.H.T. Bates, "Fourier phase problems are uniquely solvable in more than one dimension: III: Computational examples for two dimensions," *Optik*, vol. 62, pp. 219-230, 1982.
[59] K. L. Garden and R.H.T. Bates, "Fourier phase problems are uniquely solvable in more than one dimension: II: One-dimensional considerations," *Optik*, vol. 62, pp. 131-142, 1982.
[60] R. W. Gerchberg and W. O. Saxton, "A practical algorithm for the determination of phase from image and diffraction plane pictures," *Optik*, vol. 35, pp. 237-246, Apr. 1972.
[61] G. H. Glover and N. J. Pelc, "Nonlinear partial volume artifacts in X-ray computed tomography," *Med. Phys.*, vol. 7, pp. 238-248, May/June 1980.
[62] R. Gordon, "A tutorial on ART/Alegbraic Reconstruction Techniques," *IEEE Trans. Nucl. Sci.*, vol. NS-21, pp. 78-93, June 1974.
[63] J. F. Greenleaf, "Computerized tomography with ultrasound," this issue, pp. 330-337.
[64] J. F. Greenleaf, S. A. Johnson, and A. Lent, "Measurement of spatial distribution of refractive index in tissues by ultrasonic computer assisted tomography," *Ultrasound Med. Biol.*, vol. 3, pp. 327-339, Oct. 1978.
[65] J. F. Greenleaf, S. K. Kenue, B. Rajagoplan, R. C. Bahn, and S. A. Johnson, "Breast imaging by ultrasonic computer-assisted tomography," *Acoust. Imag.* (A. F. Metherell, Ed.), vol. 8, pp. 599-614, 1980.
[66] P. Gudmansen, "Layer echoes in polar ice sheets," *J. Glaciology*, vol. 15, pp. 95-101, 1975.
[67] ——, "Remote sensing of sea ice—A Danish project," *ESA J.*, vol. 1, pp. 89-102, 1980.
[68] G. T. Gullberg, "The attenuation radon transform: Theory and application in medicine and biology," Ph.D. dissertation, Lawrence Berkeley Lab., Univ. Calif., Donner Lab., Berkeley, CA, 1979.
[69] R. Hanbury-Brown, *Intensity Interferometer: Its Application to Astronomy.* London, England: Taylor and Francis, 1974.
[70] K. M. Hanson, "CT reconstruction from limited projection angles," to be published in *Proc. Appl. Opt. Instrum. Medicine X* (New Orleans, LA, May 13-19, 1982), also in *Proc. SPIE*, vol. 347, 1982.
[71] G. Harding, "X-ray imaging with scattered radiation," submitted to *J. Physics E: J. Sci. Instrum.*
[72] P. W. Hawkes, *Electron Optics and Electron Microscopy.* New York: Academic Press, 1978.
[73] P. B. Heffernan and R.H.T. Bates, "Image reconstruction from projections VI: Comparison of interpolation methods," *Optik*, vol. 60, pp. 129-142, Mar. 1982.
[74] R. Henderson and P. T. Unwin, "Three-dimensional model of purple membrane obtained by electron microscopy," *Nature*, vol. 257, pp. 28-32, Sept. 1975.
[75] R. P. Henderson and J. G. Webster, "An impedance camera for spatially specific measurements of the thorax," *IEEE Trans. Biomed. Eng.*, vol. BME-25, pp. 250-254, May 1978.
[76] G. T. Herman, "Introduction to the foundations of X-ray computed tomography," in *Computer Aided Tomography and Ultrasonics in Medicine*, J. Raviv, J. F. Greenleaf, and G. T. Herman, Eds. Amsterdam, The Netherlands: North-Holland,

1979, pp. 3-6.

[77] G. T. Herman and H. K. Liu, "Three-dimensional display of organs from computed tomograms," *Comput. Graph. Image Proces.*, vol. 9, pp. 1-21, Jan. 1979.

[78] W. Hinshaw and A. Lent, "Nuclear magnetic resonance reconstruction," this issue, pp. 338-350.

[79] B.K.P. Horn, "Fan-beam reconstruction methods," *Proc. IEEE*, vol. 67, pp. 1616-1623, Dec. 1979.

[80] H. E. Huxley and A. Klug, Eds., "A discussion on new developments in electron microscopy," *Philosophical Trans. Royal Soc. London*, vol. B261, pp. 1-230, May 1971.

[81] *IEEE Trans. Antennas Propagat.* (Special Issue on Inverse Methods in Electromagnetics), vol. AP-29, pp. 185-417, Mar. 1981.

[82] J. H. Jacobi, L. E. Larsen, and C. T. Hast, "Water-immersed microwave antennas and their application to microwave interrogation of biological targets," *IEEE Trans. Microwave Theory Tech.*, vol. MTT-27, pp. 70-78, Jan. 1979.

[83] E. Jakeman and P. N. Pusey, "Photon-counting statistics of optical scintillation," in *Inverse Scattering Problems in Optics*, H. P. Baltes, Ed. Berlin, Germany: Springer, 1980, pp. 73-116.

[84] G. L. James, *Geometrical Theory of Diffraction.* Stevenage, England: Peter Peregrinus, 1976.

[85] R. J. Jaszczak, R. E. Coleman, and F. R. Whitehead, "Physical factors affecting quantitative measurements using camera-based single photon emission computed tomography (SPECT)," *IEEE Trans. Nucl. Sci.*, vol. NS-28, pp. 69-80, Feb. 1981.

[86] F. Jaundrell-Thompson, *X-Ray Physics and Equipment.* Oxford, England: Blackwell Sci. Publ., 1970.

[87] R. C. Jennison, *Radio Astronomy.* London, England: Newnes, 1966.

[88] H. E. Johns and J. R. Cunningham, *Physics of Radiology*, 3rd ed. Springfield, IL: Thomas, 1971.

[89] S. A. Johson, J. F. Greenleaf, B. Rajagopalan, R. C. Bahn, B. Baxter, and D. Christensen, "High spatial resolution ultrasonic measurement techniques for characterization of static and moving tissues," in *International Symposium: Ultrasonic Tissue Characterization II*, M. Linzer, Ed. Washington, DC: Nat. Bur. Stand., 1979.

[90] S. A. Johnson, J. F. Greenleaf, W. F. Sanayoa, F. A. Duck, and J. D. Sjostrand, "Reconstruction of three-dimensional velocity fields and other parameters by acoustic ray tracing," in *1975 Ultrason. Symp. Proc.*, IEEE Cat. 75CH0994-4SU, pp. 46-51, 1975.

[91] D. S. Jones, *Theory of Electromagnetism.* Oxford, England: Pergamon, 1964, pp. 335-337.

[92] R. S. Julius, "The sensitivity of exponentials and other curves to their parameters," *Comput. Biomed. Res.*, vol. 5, pp. 473-478, Oct. 1972.

[93] A. C. Kak, "Computerized tomography with X-ray, emission and ultrasound sources," *Proc. IEEE*, vol. 67, pp. 1245-1271, Sept. 1979.

[94] G. X. Kambic, private communication, Technicare Corp., Solon, OH, Feb. 1982.

[95] L. Kaufman, L. E. Crooks, and A. R. Margulis, *Nuclear Magnetic Resonance Imaging in Medicine.* New York: Igaku-Shoin, 1982.

[96] G. V. Keller and F. C. Frischknecht, *Electrical Methods in Geophysical Prospecting.* Oxford, England: Pergamon, 1966.

[97] G. Knoll, "Single-photon emission computerized tomography," this issue, pp. 320-329.

[98] K. T. Knox and B. J. Thompson, "Recovery of images from atmospherically degraded short-exposure photographs," *Astrophys. J.*, vol. 193, pp. L45-L48, Oct. 1974.

[99] G. Kossof, "Display techniques in ultrasound pulse echo investigations: A review," *J. Clin. Ultrasound*, vol. 2, pp. 61-72, Mar. 1974.

[100] D. E. Kuhl and R. Q. Edwards, "Reorganizing data from transverse section scans of the brain using digital processing," *Radiology*, vol. 91, pp. 975-983, Nov. 1978.

[101] M. La Porte, J. Lakshmanan, M. Lavergne, and C. Wilson, "Mesures sismiques par transmission: Application au genie civil," *Geophys. Prosp.*, vol. 21, pp. 146-158, Mar. 1973.

[102] E. F. Laine, "Detection of water-filled and air-filled underground cavities," Rep. DACW 39-80-M-3385, U.S. Dept. Interior (Bureau of Mines Project H0202007) and U.S. Army Corps of Engineers, Waterways Exper. Sta., Vicksburgh, MI, Dec. 1980.

[103] P. P. Lele, "Safety and potential hazards in the current applications of ultrasound in obstetrics and gynecology," *Ultrasound Med. Biol.*, vol. 5, pp. 307-320, Oct. 1979.

[104] R. M. Lewitt, "Reconstruction algorithms: Transform methods," this issue, pp. 390-408.

[105] R. M. Lewitt and R.H.T. Bates, "Image reconstruction from projection I: General theoretical considerations," *Optik*, vol. 50, pp. 19-33, Feb. 1978.

[106] R. M. Lewitt, R.H.T. Bates, and T. M. Peters, "Image reconstruction from projections III: Modified back-projection methods," *Optik*, vol. 50, pp. 85-109, Mar. 1978.

[107] M. S. Lynn, A.C.L. Barnard, J. H. Holt, and L. T. Sheffield, "A proposed method for the inverse problem in electrocardiography," *Biophys. J.*, vol. 7, pp. 925-945, Nov. 1967.

[108] R. J. Lytle and K. A. Dines, "An impedance camera: A system for determining the spatial variation of electrical conductivity," Rep. UCRL-52413, Lawrence Livermore Lab., Univ. Calif., Jan. 1978.

[109] —, "Iterative ray tracing between boreholes for underground image reconstruction," *IEEE Trans. Geosci. Remote Sensing*, vol. GE-18, pp. 234-240, July 1980.

[110] P. Mansfield and P. G. Morris, "NMR imaging in biomedicine," in *Advances in Magnetic Resonance.* New York: Academic Press, 1982, suppl. 2.

[111] D. Maydan, L. A. Schepp, and Z. H. Cho, "A new design for high speed computerized tomography," *IEEE Trans. Nucl. Sci.*, vol. NS-26, pp. 2870-2873, Apr. 1979.

[112] E. C. McCullough and J. T. Payne, "Patient dosage in computed tomography," *Radiology*, vol. 129, pp. 457-463, Nov. 1978.

[113] G. C. McKinnon, private communication, Forschungslaboratorium, Philips GmbH, Hamburg, West Germany, Nov. 1981.

[114] G. C. McKinnon and R.H.T. Bates, "A limitation on ultrasonic transmission tomography," *Ultrason. Imag.*, vol. 2, pp. 48-54, Jan. 1980.

[115] —, "Towards imaging the beating heart usefully with a conventional CT scanner," *IEEE Trans. Biomed. Eng.*, vol. BME-28, pp. 123-127, Feb. 1981.

[116] V. Menges, "Mammography, the most reliable examination method for the early detection of breast cancer (parts 1 and 2)," *Electromedical (Siemens Rev.)*, vols. 2 and 3, pp. 42-49 and 98-104, 1979.

[117] R. P. Millane, "Inverse methods and modelling," Ph.D. dissertation, Engineering Library, Univ. Canterbury, Christchurch, New Zealand, 1981.

[118] R. P. Millane and R.H.T. Bates, "Inverse methods for branched ducts and transmission lines," *Proc. Inst. Elec. Eng.*, vol. 129, pp. 45-51, Feb. 1981.

[119] D. L. Misell, "The phase problem in electron microscopy," in *Advances in Optical and Electron Microscopy*, R. Barer and V. E. Cosslett, Eds., vol. 17, pp. 185-279, 1978.

[120] H. Mizutani, "A new approach to the fan beam geometry image reconstruction for a CT scanner," *IEEE Trans. Nucl. Sci.*, vol. NS-27, pp. 1312-1320, Aug. 1980.

[121] P. R. Moran, W.S.Y. Kwa, and S.A.G. Chennery, "The influence of scattered radiation in the CT numbers of bone on a scanner with a fixed detector array," submitted to *Med. Phys.*, 1982.

[122] P. G. Morris, P. Mansfield, I. L. Pykett, R. J. Oldridge, and R. E. Coupland, "Human whole body line scan imaging by nuclear magnetic resonance," *IEEE Trans. Nucl. Sci.*, vol. NS-26, pp. 2817-2820, Apr. 1979.

[123] R. K. Mueller, M. Kaveh, and R. D. Iverson, "A new approach acoustic tomography using diffraction techniques," *Acoust. Imag.*, A. F. Metherell, Ed., vol. 8, pp. 615-628, 1980.

[124] D. C. Munson and W. K. Jenkins, "A common framework for spotlight mode synthetic aperture radar and computer-aided tomography," in *Proc. 15th Asilomar Conf. Circuits, Systems and Computers* (Pacific Grove, CA, Nov. 9-11, 1981), IEEE Cat. B1CH1742-6, pp. 217-221.

[125] O. Nalcioglu and R. Y. Lou, "Post-reconstruction method for beam-hardening in computerized tomography," *Phys. Med. Biol.*, vol. 24, pp. 330-340, Mar. 1979.

[126] M. Z. Nashed, Ed., *Generalized Inverses and Applications.* New York: Academic Press, 1975.

[127] N. Nassi, W. R. Brody, P. R. Cipriano, and A. Macovski, "A method for stop-action imaging of the heart using gated computed tomography," *IEEE Trans. Biomed. Eng.*, vol. BME-28, pp. 116-122, Feb. 1981.

[128] National Institutes of Health, U.S. Dept. of Health, Education and Welfare, *NIH Consensus Development Conf. Summaries*, p. 8, 1977/78.

[129] A. Louis and F. Natterer, "Mathematical problems of computerized tomography," this issue, pp. 379-389.

[130] C. V. Nelson and D. B. Geselowitz, Eds., *Theoretical Basis of Electrocardiography.* Oxford, England: Clarendon Press, 1976.

[131] P. L. Nunez, "A study of origins of the time dependencies of scalp EEG: I, Theoretical basis," *IEEE Trans. Biomed. Eng.*, vol. BME-28, pp. 271-280, Mar. 1981.

[132] J. Ophir and N. F. Maklad, "Digital scan converters in diagnostic ultrasound imaging," *Proc. IEEE*, vol. 67, pp. 654-663, Apr. 1979.

[133] T. M. Peters and R. M. Lewitt, "Computed tomography with fan beam geometry," *J. Comput. Assist. Tomography*, vol. 1, pp. 429-436, 1977.

[134] L. B. Phelsen and N. Marcuvitz, *Radiation and Scattering of Waves.* Englewood-Cliffs, NJ: Prentice-Hall, 1973.

[135] L. R. Price, "Electrical impedance computed tomography (ICT): A new CT imaging technique," *IEEE Trans. Nucl. Sci.*,

vol. NS-26, pp. 2736-2739, Apr. 1979.

[136] G. N. Ramachandran and R. Srinivasan, *Fourier Methods in Crystallography*. New York: Wiley-Interscience, 1970.

[137] A.C.S. Readhead, R. C. Walker, T. J. Pearson, and M. H. Cohen, "Mapping radio sources with uncalibrated visibility data," *Nature*, vol. 285, pp. 137-140, May 1980.

[138] R. G. Reeves, Ed., *Manual of Remote Sensing, I–Theory, Instruments and Techniques*. Falls Church, VA: Amer. Soc. Photogrammetry, 1975, pp. 454-475.

[139] L. Rijo, W. H. Pelton, E. C. Feitosa, and S. H. Ward, "Interpretation of apparent resistivity data from Apodi Valley, Rio Grando do Norte, Brazil," *Geophysics*, vol. 42, pp. 811-822, June 1977.

[140] R. Robb, E. A. Hoffman, L. J. Sinak, L. D. Harris, and E. L. Ritman, "High-speed three-dimensional X-ray computed tomography: The dynamic spatial reconstructor," this issue, pp. 308-319.

[141] D. E. Robinson and P. C. Knight, "Computer reconstruction techniques in compound scan pulse-echo imaging," *Ultrason. Imag.*, vol. 3, pp. 217-234, July 1981.

[142] C. Roddier and F. Roddier, "Imaging with a coherence interferometer in optical astronomy," in *Image Formation from Coherence Functions in Astronomy*, C. van Schooneveld, Ed. Dordrecht, The Netherlands: Reidel, 1979, pp. 175-185.

[143] M. Ryle, "Radio telescopes of large resolving power," *Rev. Mod. Phys.*, vol. 47, pp. 557-566, July 1975.

[144] F. F. Sabins, Jr., *Remote Sensing Principles and Interpretation*. San Francisco, CA: Freeman, 1978, pp. 187-190.

[145] S. S. Sagel, E. S. Weiss, R. G. Gillard, G. N. Hounsfield, R.G.T. Jost, R. J. Stanley, and M. M. Ter-Pogossian, "Gated computed tomography of the human heart," *Investigat. Radiol.*, vol. 12, pp. 563-566, Nov./Dec. 1977.

[146] S. X. Salles-Cunha, J. H. Battocletti, and A. Sances, Jr., "Steady magnetic fields in electromagnetic flowmetry," *Proc. IEEE*, vol. 68, pp. 149-155, Jan. 1980.

[147] W. Saxton, *Computer Techniques for Image Processing in Electron Microscopy*. New York: Academic Press, 1978.

[148] L. I. Schiff, *Quantum Mechanics*. New York: McGraw-Hill, 2nd ed., 1955, pp. 161-171.

[149] H. Schomberg, "An improved approach to reconstructive ultrasound tomography," *J. Phys. D: Appl. Phys.*, vol. 11, pp. 1181-1185, Dec. 1978.

[150] —, "Reconstruction of spatial resistivity distribution of conducting objects from external resistance measurements," in *Proc. Annual Scientific Conf. of Gesellschaft fur Angewandte Mathematik und Mechanik (GAMM)* (Brussels, Belgium, Mar. 1973).

[151] H. J. Scudder, "Introduction to computer aided tomography," *Proc. IEEE*, vol. 66, pp. 628-637, June 1978.

[152] A. D. Seager, "Probing with electric currents," Ph.D. dissertation, Engineering Library, Univ. Canterbury, Christchurch, New Zealand, 1982.

[153] M. I. Skolnik, *Introduction to Radar Systems*, 2nd ed. Tokyo, Japan: McGraw-Hill Kogakusha, 1980, pp. 517-529.

[154] P. R. Smith, U. Aebi, R. Josephs, and M. Kessel, "Studies of the structure of the T4 bacteriophase tail sheath: I. The recovery of three-dimensional structural information from the extended sheath," *J. Mole. Biol.*, vol. 106, pp. 243-275, 1976.

[155] A. Sokollu, "Irreversible effects of high frequency ultrasound on animal tissue and related threshold intensities," *Acoust. Holography*, vol. 3, pp. 97-107, 1971.

[156] K. Tanabe, "Projection method for solving a singular system of linear equations and its applications," *Numerische Mathematik*, vol. 17, pp. 203-214, 1971.

[157] L. S. Taylor, "The phase retrieval problem," *IEEE Trans. Antennas Propagat.*, vol AP-29, pp. 386-391, Mar. 1981.

[158] M. Ter-Pogossian, "Positron pair emission computerized tomography," to be published.

[159] A. R. Thompson, B. G. Clark, C. M. Wade, and P. J. Napier, "The very large array," *Astrophys. J. Suppl. Ser.*, vol. 44, pp. 151-167, Oct. 1980.

[160] C. J. Thompson, "Data handling in positron emission tomography," presented at the MARIA Workshop on Positron Emission Tomography, Edmonton, Alta., Apr. 22, 1982.

[161] J. C. Urbach, B. Brenden, and R. Aprahamian, Eds., *Imaging Techniques for Testing and Inspection* (Seminar in Depth), Bellingham, WA, Society of Photo-Optical Instrumentation Engineers (Conf. Proc., vol. 29), 1972.

[162] D. Verel and R. G. Grainger, *Cardiac Catheterization and Angiocardiography*. London, England: Churchhill-Livingstone, 1973.

[163] G. Wade, R. K. Mueller, and M. Kaveh, "A survey of techniques for ultrasonic tomography," in *Computer Aided Tomography and Ultrasonics in Medicine*, J. Raviv, J. F. Greenleaf, and G. T. Herman, Eds. Amsterdam, The Netherlands: North-Holland, 1979, pp. 165-214.

[164] P.N.T. Wells, *Biomedical Ultrasonics*. London, England: Academic Press, 1977.

[165] R. L. Witcofski, N. Karstaedt, and C. L. Partain, *NMR Imaging* (Proc. Intnl. Symp.), Bowman Gray School of Medicine, Wake Forest Univ., Winston-Salem, NC, 1982.

[166] J. Wittenberg, H. V. Fireberg, J. T. Ferrucci, Jr., J. F. Simeone, P. R. Mueller, E. van Sonnenberg, and R. H. Kirkpatrick, "Clinical efficacy of computed body tomography, II," *Amer. J. Roentgenology*, vol. 134, pp. 1111-1120, June 1980.

Reprinted from *Physics in Medicine and Biology*, **29**, (2) pp. 115–120, by kind permission of The Institute of Physics (UK). ©1984

Review of 40 years of Medical Physics by Founder Members of the Hospital Physicists' Association

40 years of development in radiotherapy

W J Meredith OBE DSc FInstP, Founder Member
Formerly of Christie Hospital and Holt Radium Institute, Manchester

A Rip Van Winkle hospital physicist who fell asleep in 1943 and did not awaken until 1983 would find the face of radiotherapy greatly changed from that which he knew. He would be astonished and impressed by the size and range of the equipment and resources that are available but he would also soon realise that, though the face had changed, the heart had not. The underlying principles and the basic methods in use would be familiar to him, for they have altered little in forty years.

The development of radiotherapy

In the 1930s radiotherapy moved from being mainly a palliative procedure to being a major curative agent in the treatment of cancer. At the beginning of that decade it was the province of a very few talented and experienced specialists who varied the parameters of any treatment (dose, field size and position, overall treatment time) according to experience, whim or fancy and the patient's condition day by day. Their knowledge and experience could only be effectively passed on to immediate assistants or apprentices so that the dissemination of the best techniques was slow. Throughout the decade, therefore, some of the new generation of radiotherapists were focussing attention on making treatments more quantitative and more easily specified so that they were more reproducible and could be carried out in any centre by those with adequate, but not necessarily lengthy, experience.

By 1943, thanks mainly to the work of British radiotherapists like Ralston Paterson and Frank Ellis and British hospital physicists like L H Gray, W V Mayneord and H M Parker, the principles of modern radiotherapy were established and were becoming widely practised. These were (i) that it was necessary to establish the size, shape and position of the volume of tissue to be treated; (ii) that this volume should be raised, as uniformly as possible to a known (and pre-determined) dose, in a laid-down time-scale pattern; (iii) that the treatment be designed to achieve minimal dosage outside the 'treatment' volume, and (iv) to use consistent treatments for each site and disease type for considerable numbers of patients in order to accumulate adequate information on which '. . . to determine the exact lethal dose for each type of tumour' (Paterson 1933).

As further evidence of the revolutionary ideas being put forward, and on which the above principles were based, three more statements by Paterson (1935, 1939) may be quoted. On the subject of dosage (1939): 'Put shortly, tumour dose is to be thought of first, skin dose next and last of all the technique by which the dose will be delivered'. On the importance of the time pattern in any treatment scheme: 'These considerations lead to one important conclusion, namely that very much more attention should be paid clinically to this factor of time' (1935), and then (in 1939): 'In clinical work statement of quantity (dose) alone is inadequate. The duration of the treatment has a profound influence on the effects produced. Thus any consideration of the correct dose to be given to produce a particular effect requires that a statement be made of the overall treatment time'.

But not only were the principles upon which radiotherapy is still based established by 1943, but the basic methods for their achievement were also available. The roentgen, re-defined in 1937, was accepted as the unit of radiation quantity, even though its validity as a unit for radium gamma rays was questioned by some. For gamma ray therapy the 'Manchester' dosage system (see, for example, Meredith (1967)) provided a means for using radium (or radon) sources to give known and 'uniform' doses in a wide variety of situations and, by 1943, had already been in successful clinical use over a number of years.

Limiting the irradiation outside the desired treatment zone in x-ray therapy implies the use of beams just adequate to 'cover' that zone—'small' fields which can only be successful if accurately applied at each and every session of the treatment. For this purpose 'beam direction' devices such as the 'front and back pointer' and the 'pin and arc' (basically the forerunner of the isocentric mounting of modern equipment) had been developed (Dobbie 1939) and again had already been in clinical use for a number of years. Also by 1943 rotation therapy was being used in a number of departments, and wedge filters (Ellis and Miller 1944), and beam flattening filters (Meredith and Chester 1945) had been developed, though it must be confessed that the full and great value of each could not fully be exploited until they were used with megavoltage radiations.

The coming of megavoltage radiations

Between 1943 and 1983 then, the principles and, with one exception, the basic methods have changed little. The exception is in the field of determining the size, shape and position of the tumour and hence of the treatment zone, the ability to do which has been transformed by the development of the CAT scanner. Otherwise our Rip Van Winkle would be struck by the markedly changed balance of the treatment methods in use; by their general simplicity compared with those he knew previously and by the overall improvement in the results achieved, changes which stem mainly from the two factors, namely the coming of megavoltage radiation and advances in dosimetry.

Today there can be few, if any, radiotherapy departments in this country which do not have megavoltage radiation beam equipment of one sort or another. In 1943, with the exception of the pioneer one million volt machine at St Bartholomew's Hospital, London, and a couple of 500 kV machines elsewhere, radiotherapy in Great Britain was carried out with 200–250 kV machines. Consequently, not only was gamma-ray brachytherapy used in intracavitary gynaecological treatments, as it is today, but also for those sites (chiefly in the head and neck) where bone was present in the treatment field. This was to avoid the greater bone necrosis risk and shielding effects inherent in the kilovoltage radiations.

In 1943 some 40% of the 4000 odd radiotherapy treatments carried out at the Holt Radium Institute, Manchester, were by gamma-ray brachytherapy. In 1982 the total number of radiotherapy patients had increased a little and, as the table shows, the total number of gynaecological treatments remained practically unchanged. It will be noted, however, that nowadays a considerable fraction is treated by 'afterloading'. This technique, which affords much greater radiation protection for all the staff con-

Table. Changed pattern of radiotherapy techniques 1943–1982. Approximately same total number (about 4000) of radiotherapy treatments in each year.

Technique	No of cases treated 1943	1982
Moulds		
Single	156 (Radium)	45 (Co-60)
Double	55 (Radium)	2 (Co-60)
Single	288 (Radon seed)	81 (Au-198 seed)
Double	35 (Radon seed)	16 (Au-198 seed)
Implants		
Needle	261 (Radium)	19 (Radium)
Seed	283 (Radon)	40 (Au-198)
Intracavitary gynaecological		
Direct radium insertion	443	262
'After-loading'	—	161 (Cs-137)

cerned with the treatment as well as the opportunity for the adjustment of applicator positions before the radioactive sources are inserted should radiography show that this is desirable, was not available in 1943. On the other hand 'mould' and implant treatments have been largely replaced by megavoltage beam treatments exploiting wedge filters or, to some extent, by electron therapy. Thus, this form of brachytherapy is now used for about 5% of all treatments whereas it was about 30% forty years ago: a change advantageous to the patient and certainly for the physicist. Dosage control for brachytherapy treatments represented a considerable fraction of his daily work. Calculations, either from radiographs or reconstructions, of the dose-rates for ten implants per week and the checking, where possible, by measurements (using Sievert condenser chambers or much more temperamental home-made versions, since we seldom could afford the genuine article) of a like number of 'moulds' were a drain on departmental time and a chore of which physicists are happy to be relieved.

More important, of course, are the advantages to the patient. Our returned physicist would quickly be aware of the almost total absence of skin reactions, which were often painful and unpleasant for the patient as well as being potential pathways for infection: he would certainly miss the bedaubings with Gentian Violet which made patients look like latter-day Ancient Britons! Due to increased radiation 'output' treatments are shorter and, therefore, less wearisome for the patient and, because of greater percentage depth doses and reduced scattering, fewer and better defined fields are used, resulting in a very desirable reduction of the dose to other than the treatment zone. Like skin reactions, radiation sickness is far less a feature of radiotherapy than it used to be. And finally, especially through the use of wedge filters, the desired uniform irradiation patterns can be achieved both on paper and (because, for megavoltage radiation there is reduced, and easily compensated, disturbance by bones and cavities) in the patient. Megavoltage-beam therapy is markedly better than kilovoltage therapy and the cure rates seem to show it.

Dosimetric developments

In the forty years since the HPA was born there have been great advances in the sensitivity, accuracy and availability of radiation dosemeters, as Professor Boag will describe. There has also been an enormous increase in the range and amount of dosage information available.

Even by 1940 it was not unusual for a radiotherapy department to lack central depth dose data for many of its treatment fields, though the situation was soon greatly improved by the publication by Mayneord and Lamerton (1941) of tables of central axis percentage depth dose data for a range of field sizes, focus–skin distances and radiation qualities. Isodose charts were, of course, even rarer since their production, with the dosemeters available to most departments, was extremely laborious, time-consuming and possibly not very accurate. However, the Mayneord and Lamerton tables stimulated the development of calculation methods, such as those of Clarkson (1941) and of Meredith and Neary (1944). Such methods made the production of isodose charts generally more accurate and quicker and meant that it no longer had to be done in evenings or weekends which often were the only times that x-ray equipment was available to physicists for their measurements.

Aided by the Diagrams and Data scheme set up by the HPA isodose charts could, by the late 1940s, be available for the main planes of all beams in use with kilovoltage radiation. Planning and charting (albeit in one plane only) of the dose pattern for treatments thus became possible (assuming that the patient was a tank of water!) and was quite widely undertaken even though it was time consuming and laborious. By the time megavoltage radiation was coming into clinical use, more sensitive dosemeters, which were more suitable for depth dose measurements, had been developed and this, along with the larger number of physicists to undertake the work, meant that each department was quickly able to produce its own set of central axis data and isodose charts or obtain one from another department. This exchange of data was much easier with megavoltage radiation than with kilovoltage because of its simpler physical characteristics and because of the smaller range of radation energy, focus–skin distance and defining systems.

Today all the data can be stored in a computer and complete isodose patterns for multi-field treatments can be produced almost in the twinkling of an eye (once the appropriate programmes have been rather more laboriously devised and written) and not only for the plane of the central axis but of any selected plane thus possibly revealing previously unsuspected weaknesses in a plan. Furthermore the result of modifications of field position, size or dose can be revealed as quickly resulting in treatment improvements that could only have been achieved, in 1943, after hours of work, if at all. Finally, allowance (previously never attempted) for the presence of bones and body cavities can be made.

Our re-awakening physicist would find that, instead of the 'dose' unit with which he had been familiar, the roentgen, there was a new quantity (absorbed dose) in use, its unit being the rad. He would quickly recognise that the change had been for the good, even though the roentgen had served radiotherapy well. The new unit represented the quantity that really mattered—the energy absorbed in the tissue of interest—rather than, indirectly, the energy absorbed in air. Further it had the serendipitous feature that the numbers involved in most dosage statements remained practically unchanged since the 'roentgens to rads' conversion factor for muscle tissue (the only tissue dose really in use in clinical practice) was within a few percent of unity.

Of course in a small, but increasing number of departments he would find quite a different unit being introduced—the gray. He would certainly welcome the honour to one who would almost certainly have been his friend, one of the founder members of the HPA and one of the greatest hospital physicists. He would also realise that it satisfies the pedants but he might wonder whether there was any real gain to offset the undoubted upset and the possibilities of error that are being introduced. (In just the same way as he might wonder what real advantage had accrued (except to some traders) in the change from pounds, shillings and pence to pounds and decimal pence!) At least he can take comfort that the magnitude of the gray is not so inappropriate as that of the new unit of (radio) activity!

New lamps for old—how much light?

The changes in radiotherapy between 1943 and 1983 have then been small as far as fundamentals are concerned—does this mean that radiotherapy has itself been something of a Rip Van Winkle with no new ideas coming forward, no innovations being tried, no attempts being made to exploit new agencies?

What has been tried; with what success?

The answer, of course, is the familiar one of many hopes raised and almost as many dashed. Soundly based ideas that did not fulfil their promise; outlandish ideas that did not sustain their hopes. Who now, for example, remembers the lithium salt of pontamine sky-blue 6B, the lithium salt of a dye claimed to be selectively absorbed by malignant cells. It was hoped that this property would lead to higher doses in tumours than elsewhere due to capture in the lithium when the tumour was irradiated with neutrons. Unfortunately the promise of enhanced uptake proved false.

Hopes more securely based, but no less completely dashed, were placed in unsealed sources of radioactive isotopes which became available in therapeutically useful quantities from the late 1940s. It was thought, for example, that more radioactive iodine might be taken up by the more rapidly growing thyroid cancers than by the normal thyroid tissues and that tumours would, thereby, 'commit suicide'. Unfortunately almost the reverse is true. In practice, though radioactive isotopes have great value as sealed sources for brachy- and beam-therapy and enormous use in diagnostic studies; they have very little application as unsealed sources in cancer therapy.

Electrons, unlike photons have a definite range and produce most of their ionisation in a narrow 'peak' close to the end of their track (the 'Bragg peak'). This property encouraged the hope that electrons of appropriate energy might be able to produce a limited high dose zone around a tumour well inside the body. As ever higher electron energies became available their direct use in treatment, rather than for the generation of X-rays, gained support. In spite of the fact that the LET of the individual primary electrons would be very similar to that of electrons generated by X-rays of similar energy, it was argued that the very high instantaneous ionisation densities that could be produced by electron beams would ape the high LET (and hence the hoped for biological advantages) of neutrons. In practice the dose patterns at tumour depths for electron beams of therapeutically useful dimensions are quite unlike those from single particles or narrow beams and in general are less therapeutically acceptable than those that can be produced by photons. No biological advantage has been shown. The definite range, however, can be exploited in a limited number, mainly superficial, of clinical situations. For example, for the treatment of cancer of the lip at least one hospital now prefers electron therapy to its well-established, effective but uncomfortable radium 'double-mould'. Thus, though the more extravagant hopes and claims for electron therapy have not been fulfilled or substantiated, the facility to be able to extract electron beams from a linear accelerator primarily used for X-ray production can be valuable.

Of all the new methods that have been proposed in the last forty years those with the best radiobiological 'pedigree' are, undoubtedly, the use of 'hyperbaric' oxygen and the use of neutrons. In a classic series of experiments carried out round about 1940, using bean roots as their test material, Gray and Read (1942 for example) showed that the biological effect of X or gamma rays (low LET radiation) depended on the degree of oxygenation of the irradiated material whereas the effect of neutrons (high LET) was much less dependent. These results suggested that if normally poorly oxygenated tumours could have their oxygenation increased, or could be treated with neutrons, improved results might be obtained. Treatments of patients breathing oxygen at increased pressure (hyperbaric oxygen treatments) and treatments with neutron beams produced by cyclotrons or the D–T reaction in specially designed equipment have had fairly intensive trials on which the probable verdict at the present time should be 'Not proven'. Once again high hopes have not been realised but in these cases (and especially in the case of neutrons) the clinical results should not discourage persistence with further trials since, for example, it is by no means certain that optimum dose levels and the best fractionation regimes have been established.

One lesson that might be learned from this experience is that *Homo sapiens* is a much more complex organism than *Vicia faba* and that results obtained with one are not necessarily directly applicable to the other. Oxygenation is far from the whole story as far as cancer radiotherapy is concerned; but equally it would be foolish to discount the results from radiobiology or, for example, different fractionation regimes or the possible role of sensitisers. Radiotherapy cannot afford to ignore any leads that it may be given and, in fact, has never done so.

The potential of pi-mesons and high energy protons has yet to be fully tested but past experience suggests that 'classical' radiotherapy, with the improved techniques and tools of today will continue to be a main agent in the treatment of cancer for a long time to come. It has changed much since 1943 and yet it has changed little: one aspect that would encourage our Rip Van Winkle is that in spite of computers and other automation, radiotherapy needs physicists as much as ever. And physicists still need radiotherapy, still the largest, continuous and continuing user of his services.

References

Clarkson J R 1941 *Br. J. Radiol.* **14** 265
Dobbie J L 1939 *Br. J. Radiol.* **12** 121
Ellis F and Miller H 1944 *Br. J. Radiol.* **17** 90
Gray L H and Read J 1942 *Br. J. Radiol.* **15** 11, 39, 72 and 320
Mayneord W V and Lamerton L F 1941 *Br. J. Radiol* **14** 255
Meredith W J 1967 *Radium Dosage: the Manchester System* (Edinburgh: Livingstone)
Meredith W J and Chester A E 1945 *Br. J. Radiol.* **18** 382
Meredith W J and Neary G J 1944 *Br. J. Radiol.* **17** 75, 126
Paterson R 1933 *Br. J. Radiol.* **6** 218
—— 1935 *Br. J. Radiol.* **8** 155
—— 1939 *Radiology* **32** 221

IEEE TRANSACTIONS ON BIOMEDICAL ENGINEERING, VOL. BME-31, NO. 1, JANUARY 1984 3

Hyperthermia for the Engineer: A Short Biological Primer

GEORGE M. HAHN

***Abstract*—The current concepts of cancer treatment center around the need for preferential killing of malignant cells. Hyperthermia may offer a means of doing so, not because of the inherent heat sensitivity of cancer cells, but because of the physical environment in which many cells of solid tumors find themselves. Nutritional deprivation, low pH, and chronic hypoxia characterize the interior of many tumors and are conditions that also render cells heat sensitive.**

Heat also enhances the effectiveness of X-irradiation, and magnifies the cytotoxicity of many anticancer drugs. For this reason, the combination of hyperthermia and traditional anticancer treatments is receiving much attention. Studies utilizing mouse tumors as experimental systems have shown that, indeed, combined treatments result in superior tumor responses that can be obtained with either modality alone.

Heating of solid tumors must take into account the cooling action of blood flow. Nonuniform heating is almost always associated with variable blood flow characteristics. Nonuniform heating may also result from the nonuniform absorption of energy, either because of applicator characteristics or due to tissue absorption variability.

I. Introduction

THE idea that elevated temperatures should have healing powers is hardly a novel thought. Roman and Greek physicians, and probably others before them, were struck by the association of fever not so much with disease but with the healing process. Not unnaturally, many of them attributed beneficial powers to fevers. In the days when medicine was not a science, not even an empirical one, this certainly was an appealing hypothesis. Hard data showing that fevers really contribute toward overcoming specific illnesses, however, are even today difficult to come by. Some bacteria, adapted as they are to normal body temperatures (approximately 37°C), find it difficult to proliferate at temperatures only a few degrees higher. Fever temperatures may also inhibit, or at least slow down, multiplication of some viruses. But we simply are not sure whether or not any of these findings are really relevant to the treatment of bacterial or viral disease. On the other hand, we are quite sure that very high fevers (>42°C) can have dire consequences if allowed to reach excessive temperatures. Convulsions, permanent brain damage, and even death can result from uncontrolled fevers. If we know so little about possible beneficial effects of fevers, but are so aware of the dangers, what gives us the idea that hyperthermia can be of use in the treatment of cancer?

Actually, attempts to treat malignancies with heat date back to the end of the last century. These attempts, described at times with great enthusiasm by the practitioners involved, did not lead to any systematic efforts to investigate the behavior of cells and organized tissue at mildly elevated temperatures. Only in the last decade, in laboratories both in the U.S. and elsewhere, have attempts been made to investigate cytotoxic (i.e., cell killing) effects of heat, using either heat by itself or in conjunction with X-rays and anticancer drugs. These experiments were performed on tissue culture cells, tumor-bearing rodents, large animals, and, as preliminary clinical trials, on humans. The results obtained have been sufficiently encouraging to warrant enlarged clinical trials. Obviously they have also provided the impetus for the current engineering effort to develop equipment to heat arbitrary tissue volumes within the human body and to measure the temperature distributions achieved. These are not simple tasks, and to accomplish them requires a cooperative effort of engineers and biologists (or physicians), who are able to talk each other's language and who have a reasonable understanding of the problems faced by both groups. Among other things, this means that engineers engaged in hyperthermia work need to have at least some level of understanding of thermal biology. In turn, biologists will have to perform experiments that are designed to develop the kind of information needed by the engineers.

In my paper I will attempt to provide an introduction to thermal biology that concentrates on those areas of experimental finding that relate most directly to equipment design considerations. This means concentrating on empirical descriptions of thermal effects, without devoting much space to mechanistic considerations. For those of you interested in a more complete presentation, there are now several books published that present much of what we currently know about the ways hyperthermia affects cells, tissues, and people [1]-[4].

II. Cell Survival at Elevated Temperatures

A. The Importance of Cell Studies

Current cancer treatments all concentrate on eliminating as many malignant cells as possible without excessively impairing vital normal tissue. Surgery removes cancerous cells by cutting out affected tissue volumes; the limitation of surgery is that no vital parts of the anatomy may be removed. Radiation therapists kill malignant cells by exposing them to lethal doses of X-rays. Here treatment is limited by the response of normal tissue within the treatment volume. Anticancer drugs also act by killing individual malignant cells. Because of the systemic nature of drug treatment, the limiting tissue is the cell system most sensitive to the particular drug

Manuscript received April 14, 1983; revised September 27, 1983.
The author is with the Department of Radiology, Stanford University School of Medicine, Stanford, CA 94305.

0018-9294/84/0100-0003$01.00 © 1984 IEEE

used. Hyperthermia also kills cells; and for localized heating (in contrast to whole body heating) it is the response of the normal tissue within the heated volume that decides what heat "dose" can be applied. In most situations it is the relative killing of malignant and specific normal cells that determines success or failure of a treatment protocol.

B. Heat Dose

Fundamental to any quantitation of a biological phenomenon is the dose response curve. For example, in the case of X-irradiation, the dose response curve relates the amount of absorbed energy to cell survival. For hyperthermia it is found that two quantities are of equal importance in determining the probability of survival of any cell. These two are temperature and time at that temperature. The relationships between cell survival and these parameters are nonlinear (Fig. 1), and therefore no single physical variable or linear combination of variables can be identified that could define a heat dose. Rather, a complete record of time and temperature needs to be provided. The need for a definition of heat dose is one of the most urgent requirements in clinical hyperthermia. The dose response picture is further complicated by the finding that most cells behave differently at temperatures above 43°C than at lower temperatures, as is indicated in Fig. 1.

C. Parametric Studies of Variables that Influence Cell Survival

Most survival curves of cells in tissue culture are performed under what might be called standard conditions: cells are sufficiently supplied with nutrients and oxygen, growth temperature is at 37°C, and pH is that of normal tissue (i.e., 7.4). But except for the growth temperature, these are not the conditions prevailing in the interior of many tumors. There nutrients and oxygen exist only in limited supply, and the pH is frequently well below that of normal tissue. Experiments have shown that nutritional deprivation, chronic hypoxia, and low pH all tend to make cells heat sensitive [5]-[7]. These findings constitute perhaps the major reasons which suggest that at least some tumors should be more heat sensitive than normal tissue.

Many studies purport to show that malignant cells are inherently more sensitive than are normal cells of similar origin, even when the experimental conditions are identical. This would, of course, be a very fortunate condition indeed. If it were true, we would only need to heat patients for appropriate lengths of time and all malignant cells would be eliminated—the patient would be cured. Indeed, the rationale for whole body hyperthermia rests on early experimental evidence that suggested such a finding. Unfortunately, a thorough analysis of more recent data in the literature does not warrant such a rosy view. Many tumor cells are fully as resistant to heat as their normal counterparts, and sometimes are even more resistant [8].

D. Multiple Heat Doses: A Paradox?

In the clinic, localized hyperthermia is not given as one dose, but the treatment is fractionated, much as radiation therapists fractionate a course of X-irradiation. Furthermore, heating is not instantaneous; before a tumor reaches the desired treatment temperature, say 44°C, much time may have to be spent at lower temperatures. Therefore, the response of cells to multiple heat treatments is of importance. Here a curious finding has been made: if cells are treated at temperatures of 43°C or higher for short treatment times, the surviving cells are, for a few fours at least, much more sensitive to subsequent treatments at lower temperatures ("step-down heating") [9]-[11]. If, however, the initial treatment is below 43°C, then cells are much more resistant to heating at any temperature. Finally, if cells are treated at 43°C or higher, then returned to the incubator at 37°C for times up to as much as 100 h, they are much more resistant than are cells not previously heated. In these last two situations, cells are said to be thermotolerant. Much work is currently being expended in working out the exact kinetics of the induction and decay of both thermotolerance and step-down heating. Thermotolerance has now been demonstrated to occur in tumors *in vivo* and in normal tissues. Even whole animals can be protected against heat by preexposures to mild heating.

The biology of these phenomena is very fascinating and is currently the subject of much research [12]. But for our purposes, these findings are mainly of interest because they must be taken into account when considering design of equipment and its use in the clinic. Excessive heating at the beginning of treatment, even if only for a few minutes, can lead to step-down heating, and therefore to unexpected cell killing and perhaps associated toxicity. More importantly, slow heating, say during patient setup, can lead to the induction of thermotolerance, and hence inadequate cytotoxicity. Thus, it is of importance that heating be rapid, and, whenever possible, that the temperature distributions in tumors and in normal tissue to controlled within well-defined limits. Thermotolerance must also be considered when physicians design protocols for fractionated treatments.

E. Hyperthermia and the Effectiveness of X-rays and of Certain Drugs

While heat treatments by themselves are of use against tumors, perhaps the greatest attraction that hyperthermia has to offer is that it enhances the effectiveness of many other anticancer agents. Several examples of this are shown in Fig. 2. Perhaps of greatest potential, and certainly of greatest current use, is the synergism between heat and X-rays [13]. Radiation therapy is the second most widely used anticancer weapon, and any method of making it more effective will therefore surely be translated into improved cure rates. As is shown in panel (a) of Fig. 2, the cytotoxicity of X-rays becomes appreciably increased if the cells are exposed to elevated temperatures either before or after irradiation. After 1000 rads of X-rays, 1 in 100 unheated cells survives, but only 1 in 10 000 of heated cells survives. Another way of looking at the data is to notice that it takes approximately 30 percent less dose to kill to similar survival levels (say 10^{-3}) when heat is added to the radiation. The usual terminology is to say that heat has effected a dose modification factor of 1.3. Hypoxic cells are known to be highly resistant to X-irradiation. Most human tumors probably contain at least some hypoxic cells,

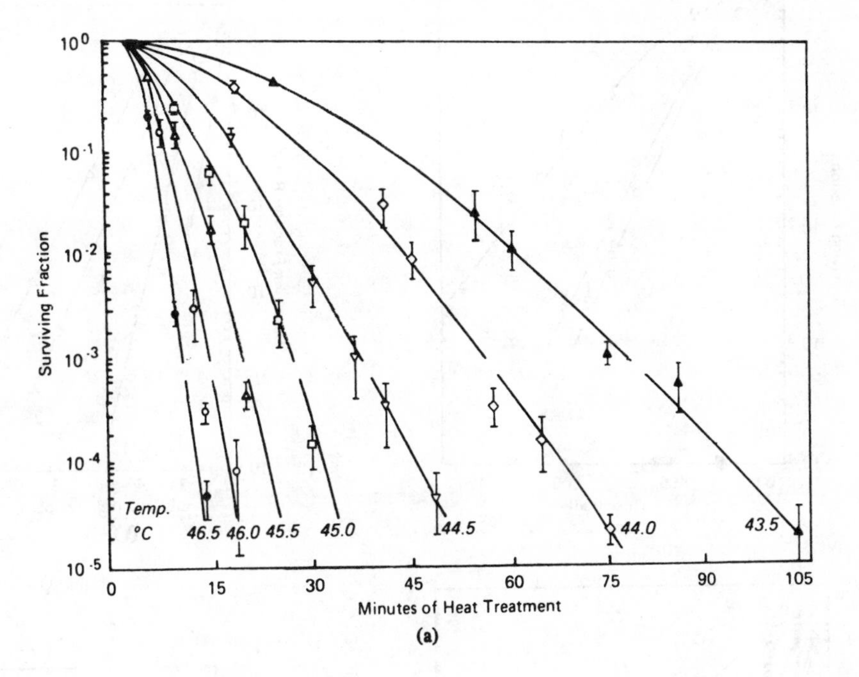

(a)

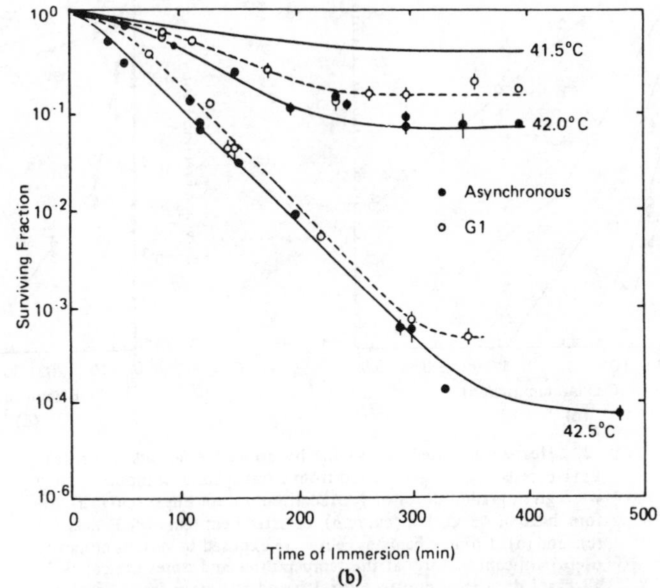

(b)

Fig. 1. Heat response of mammalian cells. Chinese hamster cells (CHO) were exposed to appropriate temperatures for the indicated times and then the ability of individual cells to multiply indefinitely ("colony formation") was assayed. Two major features of the resulting survival curves should be noted: first, the nonlinearity of the survival data with respect to temperature, and second, the differences between the shape of the survival curves obtained at temperatures above 43°C [panel (a)] and below 43°C [panel (b)]. Shown in that last curve are data for cells growing randomly as well as for "synchronized" cells, i.e., cells treated in such a way as to find themselves in a specific part of the cell cycle. The importance of the fact that these two sets of curves are so similar is the demonstration that the "tail" of the curve very likely does not reflect the response of cells engaged in a specific portion of the cycle during which they might be unusually heat resistant. Data are redrawn from [13] and [18].

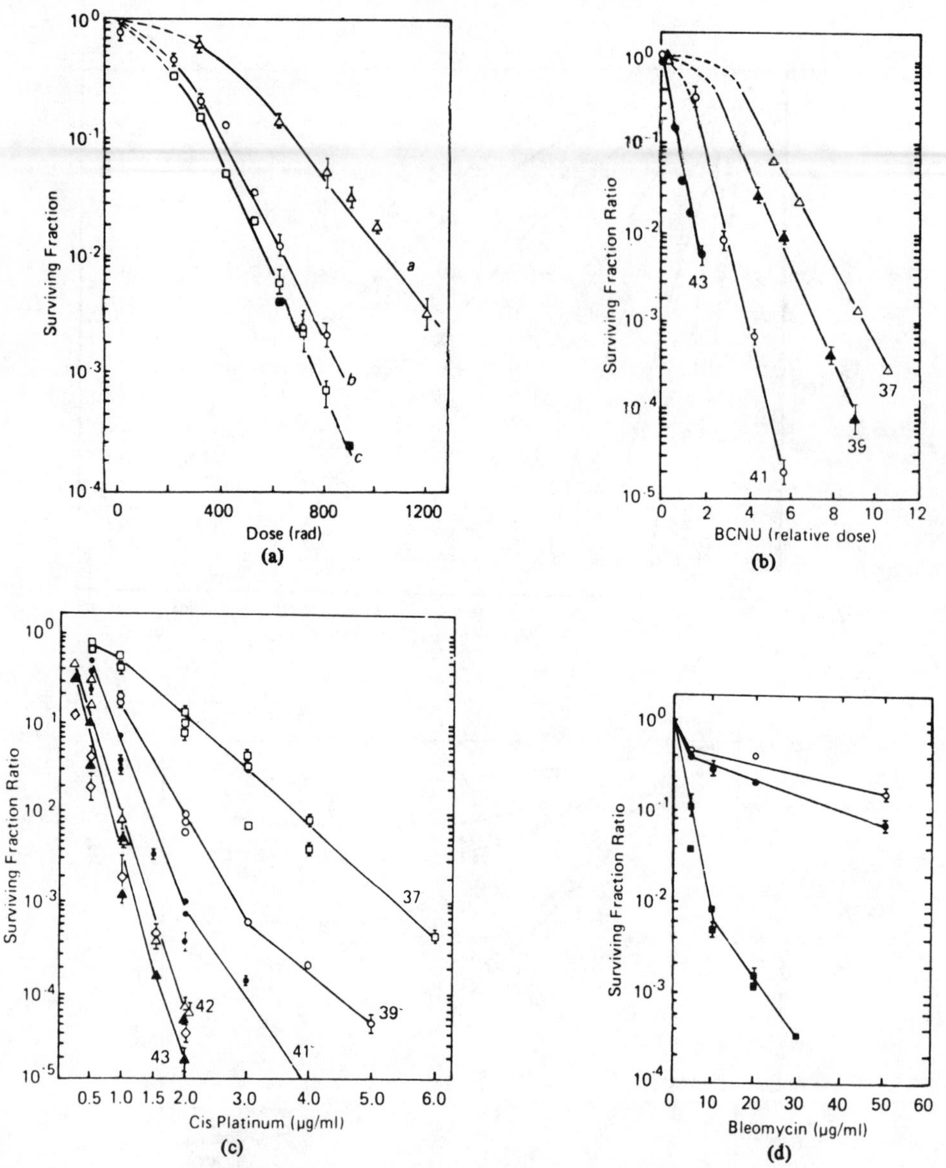

Fig. 2. Heat-accentuated cell killing by anticancer agents. Panel (a): EMT-6 cells (a cell line derived from a transplantable mouse tumor) were given graded doses of X-irradiation without heat (curve *a*), before heat of 43°C, 1 h (curve *b*), or after heat (curve *c*). Panels (b) (c), and (d): Chinese hamster cells were exposed to various clinically useful anticancer drugs at the temperatures and doses indicated. In all cases drug exposure was for 1 h and the assay for survival was colony formation. In panel (d), open circles denote exposure at 37°C, closed circles at 41°C, and closed squares at 43°C. Data are redrawn from [2].

and these are thought to be responsible for many failures of radiotherapy. Heat kills hypoxic cells at least as efficiently as it does well oxygenated cells. Of course, the therapist can only take advantage of these increased benefits if damage to normal tissue is not equally accentuated. The easiest way that this can be accomplished, conceptually at least, is via preferential heating of tumor volume; so we are back to the task of the equipment designer.

Heat can also increase the cell killing ability of many drugs [Fig. 2, panels (b), (c), and (d)]. Chemotherapy is used primarily against those cancers that have spread throughout the body. Thus, the benefits of local treatments are not as obvious. Nevertheless, the use of hyperthermia has considerable potential. A treatment situation frequently encountered is that one or more large perhaps painful or even life threatening lesions do not respond to the chemotherapy. Then the combination of heat treatment of those tumors and systemic drug therapy may induce regressions in these lesions. The combination appears to be beneficial also in chemotherapeutic perfusions of tissue [14]. Whether or not wholebody hyperthermia coupled with chemotherapy offers any benefits remains to be seen.

III. Responses of Tumors *In Vivo*

In this section I will limit myself to a discussion of some of the factors that influence the response of tumors in experimental systems, specifically transplantable tumors in mice. The reason for this restriction is that only on these have enough studies been performed so that I can talk with any assurance about response rates and possible relationships to causative factors.

A. Responses of Murine Tumors

Some mouse tumors are exquisitely sensitive to heat, while others show considerable resistance. For example, in my laboratory we have examined four different transplantable tumors [15]. Treatment was under conditions that were standardized with respect to site of implantation, size of the lesions, and method of heating. Yet one of these tumors could be cured with high efficiency by exposures to 44°C for as short a time as 30 min, while such a heat "dose" did not even appreciably slow the growth of the most resistant tumor type. The other tumors had responses that fell between these two extremes. While to some extent the tumor responses may have been the results of inherent differences in cellular heat sensitivities, other factors were at least equally as important.

The combination of heat and X-rays is more effective against mouse tumors than either treatment is by itself, thus reflecting the results obtained with cells in tissue culture. There is much controversy in the literature as to the most effective sequence of heat and X-irradiation; i.e., should heat precede or follow X-rays and what is the optimum time interval between the two treatments? The discussions are of interest to equipment designers primarily because it may be necessary for the two modalities to follow each other within a few minutes or at most an hour. This means that setup times must be held to reasonable limits.

Some drugs are clearly more efficient against mouse tumors when the treatment is combined with simultaneous, localized heat exposures. Yet there are exceptions, and these are very likely due to different pharmaco-kinetics and to variations in the detailed temperature distributions within the tumors [1]-[4], [16].

B. Blood Flow

Blood flow plays an extremely important role in determining the response of tumors to localized hyperthermia. First of all, blood flow can rather dramatically affect temperature distributions. This is not surprising because the removal of heat from the tumor volume is primarily carried out by blood flow, and therefore nonuniform blood flow invariably results in nonuniform temperature distributions. Furthermore, blood flow does not remain constant during heat treatment. In normal tissue, the body increases flow within the heated volume in an attempt to effect cooling. Some tumors, although not all, behave similarly. In other tumors, this homeostatic control response is either reduced or missing altogether. After extended periods of heating (>30 min), and particularly at temperatures of 43°C or higher, blood flow in some murine tumors stops almost entirely. For these reasons, temperature monitoring during heating is essential. Finally, blood flow to a large extent determines the availability to tumor cells of oxygen, nutrients, as well as drugs during chemotherapy. All of these can influence the response of cells within tumors to treatments. Blood flow and its role in hyperthermia are discussed in greater detail in the paper by Song *et al.* in this issue [17].

C. Immune Response

One of the major factors that determines how a tumor responds to any therapy is how well the host is able to recognize the foreign nature of the malignant cell. Hyperthermia is no exception; antigenic tumors are more easily eradicated than are those whose cells are not perceived by the immune system as not belonging to the host. But, in spite of frequent assertions in the literature to the contrary, there is no real evidence that heated cells evoke any enhanced form of immune response. The same comments about the role of antigenicity apply equally well to X-ray therapy or to surgery.

D. Uniformity of Heating

I have pointed out already that current forms of cancer therapy are concerned with eradication of cancer cells. Some cancers are so virulent that the survival of one or at most a few of these cells can lead to regrowth of the tumor and hence treatment failure. Obviously, then, if a heated tumor contains even small volumes of inadequately heated sections, its eradication by hyperthermia becomes highly unlikely. (One gram of tissue contains upwards of 10^9 cells!). What constitutes inadequate heating? I have already drawn your attention to the fact that cell killing is highly nonlinear with respect to temperature. In Fig. 1 the survival curves indicate that even changes of 1°C can cause appreciable differences in cell killing rates. This means that when heat alone is relied upon for treatment, either the treatment temperature is so high that sufficient cell killing occurs even in the cold spots, or that temperature uniformity within the heated tumor is maintained to within about 1°C. Either of these approaches entails great technical difficulties.

How do cold spots come about? I have already discussed the problems that blood flow can cause. In those tumors where such flow is highly nonuniform, avoidance of cold spots becomes a particularly difficult problem. If any major blood vessels traverse the lesion, then uniform heating is probably impossible. Blood flow in those vessels is so great that in their vicinity some degree of cooling is inevitable. In addition to blood flow, the mode of heating can create nonuniform temperature distributions. If ultrasound is used to deposit energy within the tumor volume, shielding by bony structures may prevent heating in some volumes. Bones also reflect ultrasound, thereby also contributing to nonuniformities. Bones are reasonably good conductors of heat, and therefore near bones, areas of low temperatures may be encountered. Nonuniformity of either electric field strength or of current density can also result in the occurrence of cold spots when these heating techniques are used. These result from nonuniform power deposition patterns, or from nonuniform absorption characteristics of the tissue traversed. Examples of histograms of tumor temperatures of murine tumors implanted in the thighs of mice are shown in Fig. 3.

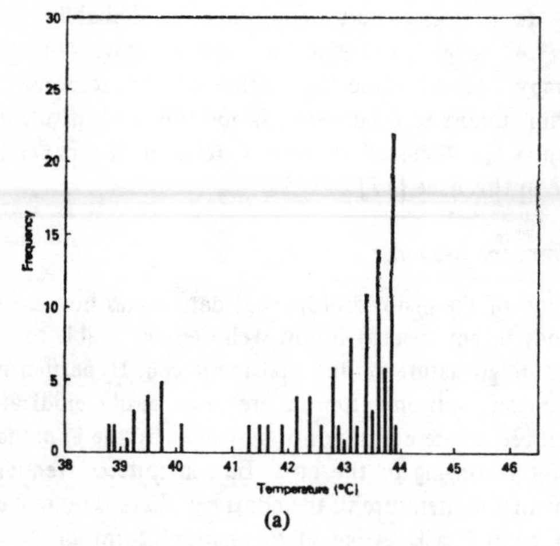

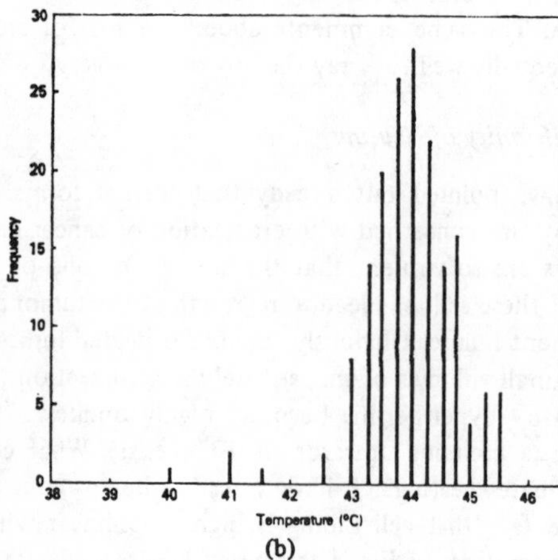

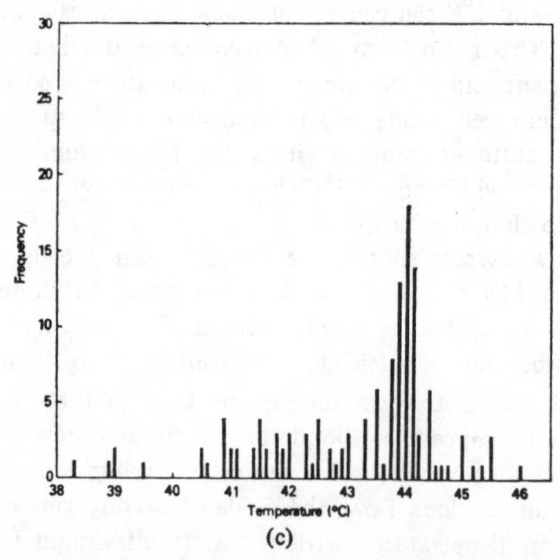

Fig. 3. Heat distributions in tumors implanted into the thighs of mice and heated with a variety of techniques. Panel (a): hot water baths. Panel (b): capacitively coupled VHF (13.56 MHz). Panel (c): ultrasound (2 MHz). Each histogram represents results from at least 20 tumors and each tumor was sampled at least 3 times at different locations. The very low temperature measurements were obtained from regions where the tumor began to infiltrate bone, and they probably reflect heat conductivity of bone. For details see [16].

IV. Conclusion

I have presented some of the biological evidence that indicates why human tumors should respond to heat exposures, particularly when these are combined with more traditional modes of cancer treatment. At the same time, I have tried to show the complexity of the biology involved, and to hint at the difficult problems facing equipment designers. I hope that I have not been too discouraging. While there are difficult problems to tackle, these are not insurmountable. Surely in Roentgen's days no one could have foreseen the development of linear accelerators for radiotherapy; yet today we regard these as necessary components of a therapy clinic. I have little doubt that within the next few years rapid progress will be made in the development of their hyperthermia equivalents.

References

[1] L. A. Dethlefsen and W. C. Dewey, Eds., "Cancer therapy by hyperthermia, drugs and radiation," U.S. Dept. HHS, NCI Monograph 61, 1982.

[2] G. M. Hahn, *Hyperthermia and Cancer*. New York: Plenum, 1982.

[3] F. K. Storm, Ed., *Hyperthermia in Cancer Therapy*. Boston, MA: G. K. Hall, 1983.

[4] G. Nussbaum, Ed., "Physical aspects of hyperthermia," Amer. Inst. Phys., New York, AAPM Monograph 8, 1983.

[5] G. M. Hahn, "Metabolic aspects of the role of hyperthermia in mammalian cell inactivation and their possible relevance to cancer treatments," *Cancer Res.*, vol. 34, pp. 3117–3123, 1974.

[6] L. E. Gerweck, "Modification of cell lethality at elevated temperatures: The pH effect," *Radiat. Res.*, vol. 70, pp. 224–235, 1977.

[7] W. C. Dewey, M. L. Freeman, G. P. Raaphorst, E. P. Clark, R. S. Wong, D. P. Highfield, J. S. Spiro, S. P. Tomasovic, D. L. Denman, and R. A. Coss, "Cell biology of hyperthermia and radiation," in *Radiation Biology in Cancer Research*, R. E. Meyn and H. R. Withers, Eds. New York: Raven, 1980, pp. 589–623.

[8] G. M. Hahn, "Comparison of the malignant potential of 10T1/2 cells and transformants with their survival responses to hyperthermia and to amphotericin B," *Cancer Res.*, vol. 40, pp. 3763–3767, 1980.

[9] E. W. Gerner and M. J. Schneider, "Induced thermal resistance in Hela cells," *Nature*, vol. 256, pp. 500–502, 1976.

[10] K. J. Henle and D. B. Leeper, "Interaction of hyperthermia and radiation in CHO cells: Recovery kinetics," *Radiat. Res.*, vol. 66, pp. 505–518, 1976.

[11] K. J. Henle, "Sensitization to hyperthermia below 43°C induced in Chinese hamster ovary cells by step-down heating," *J. Nat. Cancer Inst.*, vol. 64, pp. 1479–1483, 1980.

[12] M. J. Schlesinger, M. Asburner, and A. Tissieres, Eds., "Heat shock; from bacteria to man," Cold Spring Harbor Lab., 1982.

[13] A. Westra and W. C. Dewey, "Heat shock during the cell cycle of Chinese hamster cells *in vitro*," *Int. J. Radiat. Biol.*, vol. 19, pp. 467–477.

[14] J. S. Stehlin, Jr., B. C. Giovanella, P. D. de Ipoli, and R. F. Anderson, "Eleven years experience with hyperthermic perfusion for melanoma of the extremities," *World J. Surg.*, vol. 3, pp. 305–307, 1979.

[15] G. M. Hahn, G. C. Li, J. B. Marmor, and D. W. Pounds, "Thermal and nonthermal effects of ultrasound," in *Radiation Biology in Cancer Research*, R. E. Meyn and H. R. Withers, Eds. New York: Raven, pp. 623–637, 1980.

[16] G. M. Hahn, "Does the mode of heat induction modify drug anti-tumor effects?," *Br. J. Cancer*, vol. 45, suppl. V, pp. 238–242, 1982.

[17] C. W. Song, J. G. Rhee, and S. H. Levitt, "Implication of blood flow in hyperthermia treatment of tumors," *IEEE Trans. Biomed. Eng.*, this issue, pp. 9–16.

[18] S. A. Sapareto, L. E. Hopwood, W. C. Dewey, M. R. Raju, and J. W. Gray, "Effect of hyperthermia on survival and progression of Chinese hamster ovary cells," *Cancer Res.*, vol. 38, pp. 393–400, 1978.

George M. Hahn, for a photograph and biography, see this issue, p. 2.

Reprinted from *Physics in Medicine and Biology*, **29**, (2) pp. 157–161, by kind permission of The Institute of Physics (UK). ©1984

40 years of development in radioisotope non-imaging techniques

H Miller OBE PhD DSc FInstP

Formerly of The University Department of Medical Physics, Sheffield

In 1911 a young man in digs in Manchester suspected that his landlady was serving up for meals left-overs from previous meals. Surreptitiously he spiked some meat with a radioactive material and was able to prove that her claims to provide fresh food daily were false. He was told promptly to seek fresh lodgings. This little story is intended to remind us that radioactive tracer techniques had quite long history before the birth of the HPA. The young man of the story was George Hevesy, then aged 26. Rutherford had set him to attempt a separation of RaD from lead. Hevesy did not succeed but he realised that RaD could be used as a marker for lead and the concept of isotopic tracers was born. In 1923, for example, Hevesy used RaD to study the metabolism of lead in plants and in 1934 he used deuterium to estimate the elimination of water from the human body.

The use of artificially produced radioactive elements as tracers was developed rapidly during the mid and late 1930s. February 1934 was the date when the Joliot–Curies announced their production of artificial radioactivity and shortly afterwards Ernest Lawrence produced radioactive isotopes in his cyclotron. Hevesy himself used phosphorus-32 to study dynamic processes in the bones of rats, also in 1934, and John Lawrence and his colleagues used it as a therapeutic agent for leukaemia as early as 1936. Cyclotron poduced iodine-131 was used both for the diagnosis and therapy of hyperthyroidism by Hamilton, Hertz and others in the early 1940s. Though it was a good deal more difficult to obtain these isotopes then than it is now, the stage was set for the subsequent 40 years of development.

An event that made this development possible took place on the morning of December 2nd 1942, a little before the HPA was born. This was the morning when, under Fermi's guidance, the first experimental nuclear reactor became critical. It was built under the stadium at a base-ball ground in Chicago. It did however take four more years for the first trickle of pile produced isotopes to appear, but the trickle soon became a substantial flow. In August 1946 there was a public announcement in the States that reactor generated radionuclides were available for civilian applications, though for another year the McMahon Act prohibited the export of these isotopes from the USA. The first iodine-131 samples, carrier free and pile produced, came over from Oak Ridge at the end of 1947 and about that time sodium-24 became available from the experimental pile at Harwell. It was called GLEEP—the Graphite Low Energy Experimental Pile. There had been a paper in December of the previous year in the *British Journal of Radiology*, by Professor Joseph Mitchell in which he summarised the availability and possible applications of radioisotopes in medicine. He gave some details of both pile-produced and cyclotron-produced materials and considered their use in diagnosis and investigation, and also for therapy. The first paper published in the UK describing clinical work with isotopes seems to be one by Ansell and Rotblat in the BJR in 1948 entitled, 'Radioactive iodine as a diagnostic aid for intrathoracic goitre'. They used an end-window Geiger counter and did some point by point activity distribution plotting.

In the next few years the clinical applications of tracer techniques developed rapidly as other isotopes became available. Many early members of the HPA will recall the enthusiastic rush to apply the new materials in clinical tests of all kinds. not a few strange and unexpected things happened in the early days. I recall measuring in 1947 the decay of activity in the urine from a patient undergoing our first test with sodium-24 for estimating total exchangeable sodium. The urine activity after a short time decayed with a half-life of 14 days. This was a puzzle at first but the phosphorus-32 contaminant in the active sodium sample was eliminated by a change in the target material irradiated in the pile.

I need give only a few illustrations of the rapidity with which the tracer technique was exploited. Hevesy published a massive volume *Radioactive Indicators* in 1948. Its 550 pages contained accounts of work done with a wide variety of radioisotopes in animal physiology, pathology and biochemistry and it even included a long catalogue and price list of isotopes available from the US Atomic Energy Commission and most of those of importance in biological investigations were included.

In July 1951 a very successful Isotope Techniques Conference sponsored by AERE, Harwell, was held in Oxford with some 75 papers covering therapeutic and diagnostic applications and biochemical and metabolic studies. By this time a larger pile had been commissioned at Harwell for isotope production (BEPO) and the Radiochemical Centre had been set up, then as an out-station of the Harwell Isotope Division. By the end of the 1950s more than 150 radioisotopes could be obtained commercially and hundreds of labelled compounds. Hospitals had by then developed a wide variety of clinical tests and were using the active products in a large numer of clinical research projects.

Two publications in this decade were important to those working in the field. In 1950 Professor Mayneord produced Supplement No 2 of the BJR. *Some Applications of Nuclear Physics to Medicine* and in 1958 appeared Veall and Vetter's *Radioisotope Techniques in Clinical Research and Diagnosis.* These two books gave magnificent scientific and technical support to those of us embarking on the new techniques.

The growth of the isotope services in UK hospitals continued. The radiochemical Centre reported a seven-fold increase in diagnostic shipments in four years in the mid 1960s. Nevertheless, my impression is that the general pattern of the clinical non-imaging work was well established by the early 60s and though important developments have taken place since, the most exciting and non-routine field was captured from then on by imaging techniques which Norman Veall will describe in the next session.

Let me pick out some of the more obvious developments in the non-imaging field: firstly the equipment. The early work was done by Geiger counters. These were commercially available in 1947 so you did not need to make your own. A wide variety could be obtained; end-window counters such as the ones used by Ansell and Rotblat, or the glass walled tubes like the M6 for liquids, or the cylindrical gamma counters or the gas flow tubes for assay of tritium or carbon-14.

During the period 1946–1955 annual meetings were set up by AERE called 'Counter Meetings' at which representatives of industry, University and Hospital departments met to consider the development of GM counters and their standardisation.

Towards the end of the 40s however the possibilities of scintillation counting in solid fluors transparent to their own radiation were being investigated. The advantages were obvious. Sensitivities were much greater than the 1% of the GM tube and pulse times of less than one microsecond compared favourably with the 100 μs of the GM tube.

In 1947 Kallman described the counting of particles and quanta using light flashes in naphthalene, and in the following year thallium activated sodium-iodide was developed by Hofstadter. No superior scintillator seems to have been developed since for general use. In 1950 Kallman developed the method of the liquid scintillation counter using a suitable scintillator in an organic solvent into which the radioactive materials could be dissolved.

These detector changes meant changes in the ancillary equipment. In 1947 we first used equipment from AERE, scalers, probe units, power packs, ratemeters etc., and home-made bits too. Dr Osborn has said he once used a bunch of keys on a calculated length of string as a timing device in his first measurements of venous blood flow rates in the leg in 1948. He was not satisfied with the time constant of the ratemeters then available. In the early 1950s electronic developments in medical physics departments often consisted of building scalers and power packs. But by the end of the decade commercial electronic equipment was excellent, the home-made stuff tended to become museum pieces, and the electronics work of a department was often devoted to keeping equipment going to cope with the pressures from routine clinical tests, meantime hoping that money for new equipment would appear soon. At this time GM tubes for thyroid measurements for example gave way to scintillation counters, at first home made. After 1964 the thyroid uptake counter of the standard IAEA design was being generally adopted. This development was no doubt stimulated by the work of Dr Marshal Brucer of Oak Ridge who looked at the need for standardisation in these measurements. In 1957 he toured the world, or some of it, with two wax female models with radioactive thyroids. 'Mock Iodine' was used in the gland—an isotope mixture giving a gamma-ray spectrum close to that of iodine-131 and centres were invited to measure the uptake in the thyroid. An extraordinary range of values was revealed. Obviously some counting techniques left much to be desired and some standardisation was very necessary.

A significant step in the development of diagnostic procedures was the introduction of new isotopes of short half-life with low internal energy deposition, and in particular their production in generator columns so that they could be readily available in local hospitals using simple preparation methods satisfacory from a pharmaceutical point of view. The first such columns were the tellurium-132–iodine-132 ones available from about 1956. The most notable one commercially produced, first about 1965, is the one which gives technetium-99m. This isotope introduced a new era in radioactive diagnostic work since it rapidly became perhaps the most used isotope in clinical routine.

In the 1950s there was an interest in the measurement of the natural gamma-ray activity of the body and in the exposure of the general population arising from fall-out from nuclear weapon tests. Apparatus using high pressure ionisation chambers in well shielded rooms was developed by Sievert in 1951 and Spiers in 1953, and a few years later liquid or solid scintillation detectors were adopted in Leeds and elsewhere. These were very expensive pieces of equipment but the possibility of using the great sensitivity in studies of body composition and isotope turnover studies was attractive. A variety of somewhat less expensive experimental facilities for whole-body counting were devised. Indeed in 1955 a simple whole-body counter using a few Geiger tubes was described, and some were built for turn-over studies with administered isotopes, although it was difficult to work in the vicinity of a radium ward or other source of variable radiation.

In 1970 it was assumed that each Regional Isotope Centre at least, should have an apparatus of high sensitivity and several hospitals now have the expensive well-shielded counters. It seems, however, as though the use of such a facility has failed to hold the interest of clinicians except for a few specialised researchers.

An important development has been the use of *in vitro* techniques, not only to replace *in vivo* diagnostic tests but also to expand enormously the range and sensitivity of clinically important biochemical assay methods. In the early 1960s, for example, our own rather unsatisfactory attempts to use, first, red cells and then resin to estimate tri-iodothyronine as a thyroid diagnostic test were overtaken by the commercial introduction of very satisfactory kits. The normal uptake studies for thyroid function were thus no longer needed except where therapy procedures were to follow. The dramatic development at that time was that of saturation analysis or radioimmunoassay worked out independently in 1960 by Yalow and Berson in the States and Ekins in the Middlesex Hospital, the former for the measurement of plasma insulin and the latter for measurement of plasma thyroxine. These techniques have now reached a routine stage over a wide field of biochemical interest and they have opened up new avenues of diagnosis and research of clinical importance.

May I comment on the therapeutic applications of radioisotopes? The use of solid sources comes properly in a review of radiotherapy methods and I can leave this. Unsealed sources in therapy have played a relative minor role. The treatment of thyrotoxicosis and some forms of thyroid cancer are the outstanding successes—the early trials have led to established procedures which have, it seems, changed little over the past 20 years—perhaps still further long term study is needed to differentiate the value of the various therapy routines now used. The use of phosphorous-32 for treatment of blood diseases continues but also seems scarcely to have changed over the whole period of our review.

The search for a radioactive drug with highly selective localisation in neoplastic tissues has been a continuing one. Professor Mitchell has worked on this for many years and is still hopeful that a compound labelled with an alpha-emitter will provide an answer.

There are so many important things I have not mentioned in this brief review. I have not commented on the mathematical input, particularly in compartmental analysis, nor about the important work done in working out the problems of radiation dosimetry of administered isotopes. And I have not included any reference to the activities in the isotope work which were generated by the various codes of practice or legislation such as the Radioactive Substances Act of 1960, which had us all organising radioactive waste disposal and so on. Again most of these developments took place in the early half of the 40 years story.

I would like to end by commenting on the effect that the introduction of radioisotope techniques has had on the role of the professional physicist in the clinical field, especially perhaps in the UK. The specialist skills of the physicist, which first led to openings in radiotherapy, were also recognised as important when unsealed sources and the associated nucleonic equipment began to be used in hospitals. The new field of cooperation between physicist and clinician opened up for physicists opportunities to contribute in new ways to clinical science, biochemistry and physiology. This was an exciting time, comparable to the earlier period when radiation physicists made an impact on radiobology. At the present time other technological advances, notably in electronics and engineering, have more influence on the changing role of physics applied to medicine. The involvement of physicists in radioisotope procedures, however, remains of great importance and advantageous to medicine both in our own hospitals and in the third world. I mention this latter since from the early 60s a lot of useful work has been done by physicists in many parts of the developing world especially through the WHO and the IAEA. These organisations have also provided great stimulus in the isotope field by their symposia and specialist panels.

The development of nuclear medicine has not been without its problems. From 1960 onwards the entrance of the physicist into hospitals where there were no radiotherapy services was a step of which some medical consultants were a little suspicious, and the role of the physicist was not always clearly defined. The development of the organisational aspect of nuclear medicine took place in a variety of ways—with much discussed at both local and national level. Many reports appeared: Windeyer (1966), Zuckerman (1968), and later the Intercollegiate Standing Committee, the BIR, the Faculty (now Royal College) of Radiologists, and many local ones. These set out possible patterns of future organisation. In the end these could only be made to work well by good personal relationships, with arrangements to suit local circumstances. When physicists accepted their role as serving the patient via the clinical responsibility of the doctor interested in these new techniques, and the doctor accepted the physicist as responsible for a high quality of scientific and technical service, then a variety of arrangements developed under which unsealed radioactive sources proved of great value to clinical medicine. The growth of clinical isotope work continued, of course, but it had to take its place alongside other diagnostic techniques, and I guess that the balance has not yet been finally achieved.

Reprinted from *Physics in Medicine and Biology*, **29**, (2) pp. 163, by kind permission of The Institute of Physics (UK). ©1984

40 years of development in radioisotope imaging

N Veall DSc FInstP

Radioisotopes Division, Clinical Research Centre, Harrow, Middlesex

Neglecting the subject of autoradiography, which is important in its own right, the development of γ-ray imaging techniques can be traced from the pinhole camera with x-ray film recording of the image—a hopelessly insensitive technique. A major step towards improved sensitivity was to increase the absorption of incident γ-radiation in the detector by using a suitable crystal, such as calcium tungstate, soon replaced by thallium activated sodium iodide, which gave an output in the visible spectrum. Detection of the emitted light quanta improved from the photographic film by way of the image intensifier until high gain photomultiplier tubes became commercially available. This made possible the collimated single crystal probe detector still widely used today. During this period of development a good deal of useful clinical work, particularly thyroid studies, had been found possible using a collimated GM tube, which could be hand held, but this was completely superseded by the scintillation probe. It then became necessary to support the detector and carry out the scan mechanically. Thus, by about 1955 and for the next decade, the rectilinear scanner dominated the field of clinical γ-ray imaging; this was also the period when ^{131}I was by far the most widely used radionuclide in medicine. By 1958 Anger had already built the first gamma camera, probably five years before its time, closely followed by the positron camera.

The development of radioisotope imaging techniques has been closely linked with the development of radiopharmaceuticals. When radionuclides became generally available in the late 1940s the major clinical interest was in radioisotope therapy as a new cancer treatment. The 10:1 tumour/tissue uptake ratio was a major objective for this purpose and a good deal of metabolic data on radiolabelled compounds was obtained. Gradually this ratio became more important in connection with tumour localization by radioisotope imaging. Probably the most important advance in the 1960s was not the rapid technical developments in instrumentation but a modest paper presented by Harper in 1964 on the clinical use of ^{99}Tc^{m}. This made it possible to acquire the necessary bits of information to give useful images without a prohibitive radiation dose to the patient; equally important, the gamma camera and ^{99}Tc^{m} were obviously made for each other, and this made possible the explosive growth of 'nuclear medicine'.

The more recent technical developments in computer data reduction and processing are familiar to this audience. The dominant role of the gamma camera and 'region of interest' facilities has tended to encourage the superstition that the human body is a two-dimensional object. The increasing interest in tomographic imaging represents a return to reality, but the increased amount of information required means the use of even shorter lived radionuclides. Thus we have progressed to the stage where in some centres a cyclotron is regarded as a necessary attachment to an up-to-date imaging system, thus neatly reversing the situation which existed 40 years ago.

0031-9155/84/020163+01$2.25

320 PROCEEDINGS OF THE IEEE, VOL. 71, NO. 3, MARCH 1983

Single-Photon Emission Computed Tomography

GLENN F. KNOLL, SENIOR MEMBER, IEEE

Invited Paper

***Abstract*—The subject of single-photon emission computed tomography (SPECT) is generally reviewed. The basic interaction processes of gamma rays in matter are outlined, and the formation of conventional gamma-ray images is described. We then outline the extension of these concepts to the formation of three-dimensional or tomographic images. Of particular concern in emission tomography, the effects of gamma-ray attenuation and scattering are outlined. Several examples are given of practical SPECT systems, and representative results are given.**

INTRODUCTION

THE SUBJECT of computerized tomography in medicine, around which this special issue is centered, is conveniently divided into three major areas. The most widely known of these is transmission computed tomography in which an image is produced that displays the absorption coefficients for X-rays passing through the body from an external source. This technique, commonly known as CT scanning, is described in detail elsewhere in this issue.

The remaining two categories are positron-emission tomography (PET) and single-photon emission computed tomography (SPECT), sometimes grouped together under the common terminology of emission-computed tomography (ECT). In both these applications, the objective is quite different from that of CT scanning. The source of the electromagnetic radiation, rather than an external X-ray tube, is now a radioisotope that is distributed within the body. The objective is to provide a spatial map or image of this distribution, usually displayed as various two-dimensional sections or slices through the body. Therefore, while CT techniques generally display absorption properties that are related to anatomy, emission techniques reveal the distribution of radioactive tracers that can often indicate various aspects of physiological function. Tracers can also be used in CT in the form of radio-opaque contrast media, but the amounts that are required in order to create enhanced absorption are orders of magnitude greater than the amounts of radioactive material that can be detected in emission tomography. Therefore, interest in emission tomography is often centered about those applications in which the tiny amounts of tracer can follow functions that would be disturbed by the much larger amounts of contrast medium required for transmission studies. There has also been a long history of research in incorporating radioactive tracers with physiologically active agents. These radiopharmaceuticals can be used to single out certain functions or abnormalities with a high degree of specificity.

All three types of tomography require the recording of data along a large number of directional "rays." In most cases, data

Manuscript received June 14, 1982; revised December 8, 1982.

The author is with the Department of Nuclear Engineering, University of Michigan, Ann Arbor, MI 48109.

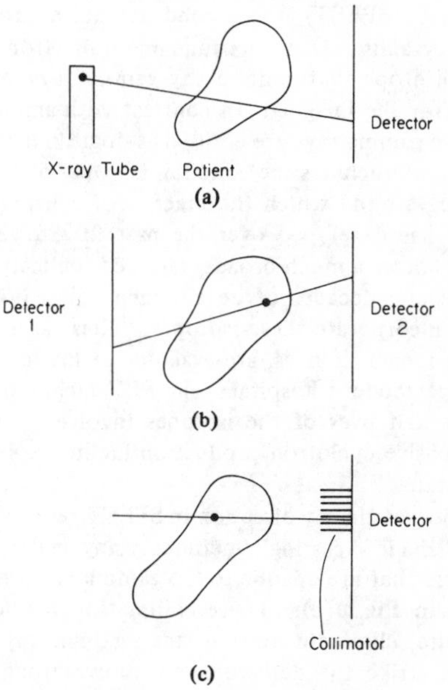

Fig. 1. Methods of defining the ray direction in three different types of tomography. (a) CT. (b) PET. (c) SPECT.

from many such rays lying within a single plane are collected, and the computational methods described elsewhere in this issue are applied to reconstruct the image in that plane or "slice." The three general methods of tomography outlined above differ greatly in the method used to define a ray.

In transmission tomography, a single ray is defined as shown in Fig. 1(a). The source of X-rays is a well-focussed spot on the anode of an X-ray tube, and this point, therefore, defines one end of the ray. On the opposite side of the patient, an X-ray detector is provided that can in some way sense the position of the detected X-ray photon. The source and detection points thus uniquely define the single ray of interest.

In emission tomography, the situation is made much more complicated by the fact that one does not *a priori* know the position of the source point. In fact, it is just this position that is the general objective of the measurement. While one can apply position-sensitive detectors similar to those used in transmission tomography, a second piece of information is required to define the *direction* of the ray. The two methods of emission tomography, PET and SPECT, differ in the methods used to define the ray direction.

The scheme used for PET is illustrated in Fig. 1(b). The technique is limited to those radioisotopes that decay by emitting a positron. This positron travels a short distance (a few millimeters at most) before stopping and annihilating with a normal

0018-9219/83/0300-0320$01.00 © 1983 IEEE

negative electron from the tissue of the patient. Annihilation radiation is formed at the site of this combination to carry away the lost mass energy. The unique property of this annihilation radiation is that it consists of two electromagnetic photons, each of energy 0.511 MeV, that are emitted in time coincidence and are oppositely directed along exactly the same line. Thus by placing two position-sensitive detectors on opposite sides of the patient, one can fully define the ray if the two detected events correspond to the photons from a single annihilation event.

The topic of this paper is single-photon emission computed tomography (SPECT), the second general approach to emission tomography. It differs fundamentally from PET in that any radioisotope that emits decay gamma rays may be used as the basis for the imaging. In contrast with annihilation radiation, these gamma rays are emitted as single individual photons. This category includes the common isotopes of Tc-99m, I-125, I-131, etc., around which the practice of conventional nuclear medicine has developed over the past 20–25 years. SPECT therefore allows a much broader range of applications that draw on the several decades of development of radiopharmaceuticals that incorporate these isotopes. These agents, with their relatively long half lives, are available at low cost in virtually every large modern hospital. The PET technique, because of the short half lives of the isotopes involved, requires that a very expensive cyclotron production facility be located within the hospital.

Definition of the ray direction in SPECT requires the application of "shadow-casting" methods using materials as spatial collimators that are opaque to the gamma rays. An example is sketched in Fig. 1(c). These collimation methods, by their very nature, eliminate most of the gamma rays that would otherwise strike the detector, and allow through only those that are incident in the prescribed direction. Because a typical collimator discards something like 999 out of every 1000 incident gamma-ray photons, this process is highly inefficient in its use of the information they carry. However, one must in general live with these inefficiencies in SPECT. It would take a detector that recorded both the position *and direction* of each incident gamma-ray photon to fully utilize all the information. Although some efforts have been made along these lines [1], there does not appear to be much prospect for the development of such detectors in the foreseeable future.

In summary, SPECT is characterized by the following:

1) A potentially wide utilization because of the availability and economy of the radioisotopes that can be employed.

2) A lower detection efficiency (by 1 or 2 orders of magnitude) when compared with PET methods, resulting in greater statistical noise in the derived images for equal radioisotope activities and imaging times.

3) As we will show in the following discussions, the instrumentation needed for SPECT is less complex than that for PET, and can often be applied to other nontomographic purposes in general nuclear medicine imaging.

Conventional Gamma-Ray Imaging

The remainder of this review will be limited to the instrumentation and techniques that have evolved for SPECT. As a background, it will be helpful first to review the simpler topic of conventional radioisotope imaging in nuclear medicine. Such methods lead to the recording of nontomographic or conventional images that are two-dimensional projections of three-dimensional intensity distributions. In nuclear medicine, the

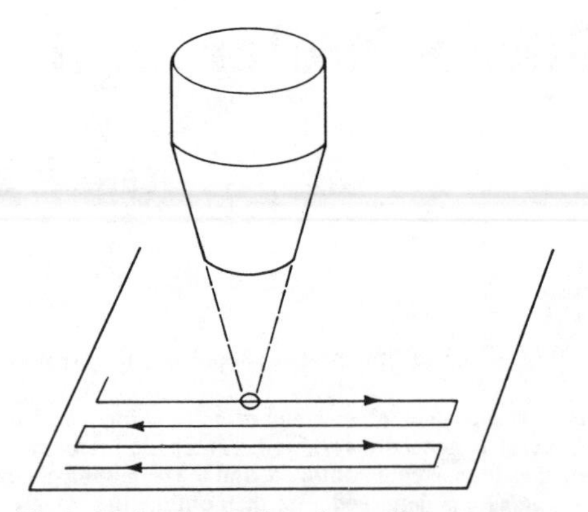

Fig. 2. Mode of operation of a rectilinear scanner.

three-dimensional distribution is that of the radioisotope within the patient. In some fraction of its decays, the radioisotope must emit a gamma-ray photon. If the energy of this photon is above about 100 keV, there is a reasonable probability of its escape from the patient without further interaction. The source is, therefore, analogous to a transparent self-luminous object, and the conventional two-dimensional image corresponds to a single view of this object from one specific perspective outside the patient. The recorded intensity at any point in this image is, therefore, the intensity of the source integrated along a line (or ray) that corresponds to a specific direction. As we will see later, the purpose of tomography is to derive an image that is free of these integration effects, and which represents the true intensity of the object at a specific point in space.

There are several general approaches to two-dimensional or conventional imaging. One is to choose a single detector, as in Fig. 2, and to fit that detector with a collimator made from lead or other dense material. The sensitivity of the detector is then limited to gamma rays emerging from a small local region of the source. If this detector is now moved or scanned in a systematic fashion across the area of interest, the counting rate variations from the detector will correspond to intensity variations in the source. The recorded image, therefore, consists of a two-dimensional map of the detector counting rate as a function of its position over the patient. Such "rectilinear scanners" were the first practical instruments for recording gamma-ray images, and are still in limited use today. The most commonly used gamma-ray detector is a crystal of sodium iodide operated as a scintillation counter. If the gamma-ray photon enters the crystal, it may interact in one of several ways to form a fast electron. As this electron loses its energy over a short track within the scintillation crystal, it leads to the emission of a flash of visible light. A fraction of this light is sensed by a photomultiplier tube optically coupled to one face of the crystal and is converted to an electrical pulse. The intensity of the image from a rectilinear scanner is proportional to the counting rate of these pulses at each position. When recorded in digital fashion, the image consists of an array of discrete picture elements or "pixels," and the content of each pixel is then the total number of pulses recorded from the detector during the time taken to scan across that specific pixel.

A scanner records information from different positions in the patient in a serial fashion, and information from all other areas is ignored while the scanner is located at a specific position. This inefficient use of information is a heavy penalty to pay because the quality of nearly all images in nuclear medicine

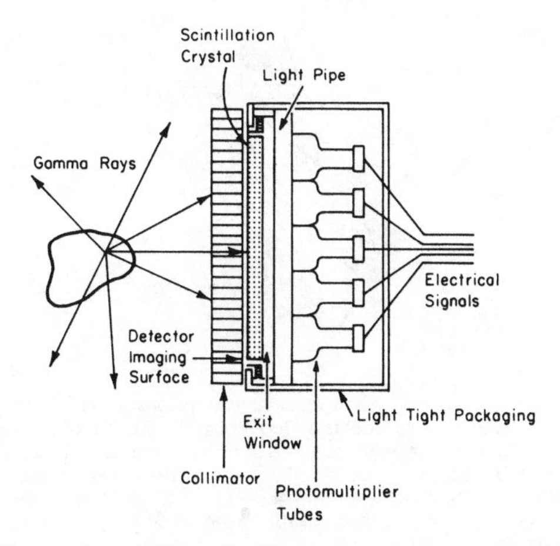

Fig. 3. Cross section of an Anger-type gamma-ray camera.

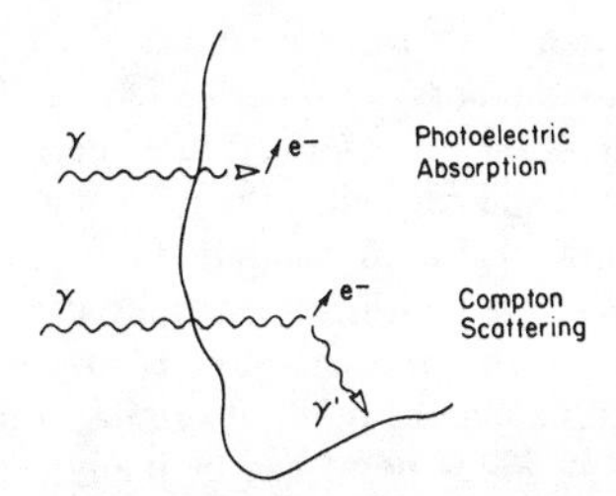

Fig. 4. The two important types of gamma-ray interactions.

is information limited. The number of photons available to form the image is never as large as one would like, and the statistical fluctuations in the relatively small number of recorded photons is a serious source of image degradation. Therefore, a device known as a gamma-ray camera that can record photons from all areas of the image *simultaneously* has far greater potential of retaining the maximum amount of available information. This advantage, coupled with the fact that the detector can now be stationary, has resulted in a general preference for cameras over scanners for modern conventional gamma-ray imaging.

A schematic of the most common type of gamma-ray camera is shown in Fig. 3. This device is often called an Anger camera, named after the inventor of the prototype [2], [3] first developed in the late 1950's. Gamma-ray cameras record a two-dimensional image by using a large scintillation crystal with diameter of 35 to 50 cm as the radiation detector. This stationary detector views the object through a collimator consisting of a large number of holes in a lead matrix. In the most common type of collimator, these holes are parallel and perpendicular to the face of the detector crystal. A parallel hole collimator defines rays for each element on the detector entrance surface that are close to the perpendicular ray at that point. In this way, the desired two-dimensional projection of the image distribution along each ray is registered as the pulse counting rate at the corresponding location within the crystal. It is, therefore, necessary to sense the position of each gamma-ray interaction. For this purpose, the opposite surface of the crystal is fitted with a large number (ranging from 19 to almost 100) of photomultiplier tubes. When a gamma-ray photon finds its way through a collimator channel and interacts in the crystal, the light that is generated is distributed among the nearest photomultiplier tubes. By sensing the relative output signal from all the tubes, the camera produces X and Y position signals for each event. In analog imaging systems, that event is recorded as a spot of light at a corresponding position on a sheet of film. In digital systems that are becoming more common, the X and Y signals are first digitized and then used to increment a corresponding address location in a two-dimensional image memory. In the latter case, the image that is accumulated is a digital image consisting of discrete counts in a two-dimensional array of pixels whose dimensions are determined by the "fineness" of the analog-to-digital conversion.

Several properties of gamma-ray detectors used in imaging are important both in conventional projection imaging and in SPECT. Because gamma-ray photons are always in short supply, the detection efficiency must be close to 100 percent to avoid further information loss. For low gamma-ray energies, that requirement corresponds to a minimum crystal thickness of about a centimeter. At higher energies of several hundred kiloelectronvolts, several centimeters of sodium iodide are required to provide a detection efficiency that approaches unity.

The energy of the incoming gamma ray is also of interest. Sodium iodide scintillators are capable of recording that energy because the light output per event is proportional to the energy deposited in the crystal. Therefore, the amplitude of the output pulse is an indicator of this deposited energy.

Interaction of Gamma Rays

Some basic properties of the behavior of gamma rays in matter are important when examining various approaches to medical tomography. In the energy range of interest for medical diagnosis, X- or gamma rays will interact in any matter through one of two principal interaction mechanisms: *photoelectric absorption* and *Compton scattering* (see Fig. 4). In photoelectric absorption, the incident gamma-ray photon interacts with an absorber atom and essentially disappears. Its energy is transferred to an electron in the absorbing material, and that electron travels only a few millimeters before being stopped near the place of its formation. In Compton scattering, the incident photon is only deflected in its interaction with an electron and transfers only a part of its original energy. The scattered photon continues on with reduced energy (or increased wavelength) and changed direction. The energy lost by the photon is transferred to the electron, which then deposits that energy locally near the scattering site.

In transmission tomography, the objective is to map the attenuation properties of a section of the body. In most applications, attenuation of the incident X-ray beam can take place through either photoelectric absorption or Compton scattering of the photon away from its original direction. Thus the local attenuation behavior reflects both these processes. The scattered X-rays, however, may still be recorded elsewhere in the detector at a more-or-less random location that is unrelated to their original direction. These scattered photons clearly are false information, and they simply decrease the contrast one would otherwise record in the absence of scattering.

In emission tomography, recall that we are interested in mapping source distributions rather than absorption properties. In fact, ideally there would be no interaction whatsoever of the emitted gamma rays before they emerge from the patient. However, such interactions inevitably occur even with the highest energy (or most penetrating) of gamma rays used in nuclear medicine. The attenuation of the gamma rays is characterized by an exponential drop off in their intensity with thickness of the overlying tissue. For the most commonly encountered isotope (Tc-99m at 140 keV), the original intensity is reduced to one-half its value in about 5 cm. Therefore, attenuation of the gamma rays cannot be ignored in most clinical situations. Correction for attenuation effects is an important aspect of SPECT, and we will return to this topic when discussing various practical implementations of the method.

Gamma rays scattered within the patient also pose something of a problem in emission tomography. If they emerge to strike the detector, their apparent point of origin is the scattering site within the patient. This point can be many centimeters away from the true position of the radioisotope source. If these scattered rays are recorded as part of the image, they again represent false information that only serves to add a diffuse background to the recorded image.

A simple method is widely applied both in SPECT and conventional projection imaging to eliminate much of this scattered gamma-ray background. Radioisotopes in common application emit only one or two prominent gamma-ray photons with discrete and well-defined energy. Because the gamma ray loses energy in the scattering process, the scattered photons can be distinguished from the unscattered or primary photons on the basis of their recorded energy. An "energy acceptance window" is set within the detector electronics to allow only those photons that deposit the primary energy to contribute to the recorded image. Scattered photons deposit less energy and are rejected through simple pulse amplitude selection.

This scatter rejection process depends on the assumption that the incident photons, whether they are primary or scattered, create a pulse from the detector with an amplitude that is proportional to the photon energy. Such will be the case if the gamma ray interacts within the detector by the photoelectric process, but not so if a Compton scattering interaction occurs instead. In the latter case, because all scattering angles are possible, a broad continuum of deposited energies are observed even for monoenergetic incident gamma rays. Therefore, detectors used for emission imaging are chosen from those materials in which photoelectric absorption dominates as much as possible. This is one of the criteria which has led to the widespread acceptance of sodium iodide as the material from which most scintillation crystals are fabricated. The high atomic number of the iodine constituent greatly enhances the photoelectric probability in sodium iodide compared with other materials of lower atomic number.

General Approaches to SPECT

The general philosophical approach to SPECT is similar to that for the other common methods of tomography described elsewhere in this issue. If one gathers sufficient information from a large number of recorded rays that span a wide enough range in direction and perspective, then it should be possible to reconstruct a full three-dimensional representation of the intensity of the object. In principle, one would like the set of collected data to include rays sampled at fine increments in the full 4π solid angle surrounding the object. It should be possible to then synthesize all these data simultaneously to derive the desired three-dimensional source distribution.

In practice, this synthesis is generally carried out in more limited steps. Nearly all the applications described in this review are based on reconstructing only one plane or slice through the object at a time. A full three-dimensional representation can then be built up by assembling a series of contiguous slices. This approach is dictated not only by practical limitations on the scope on the computations required, but also by restrictions on the possible orientation of detectors around the patient.

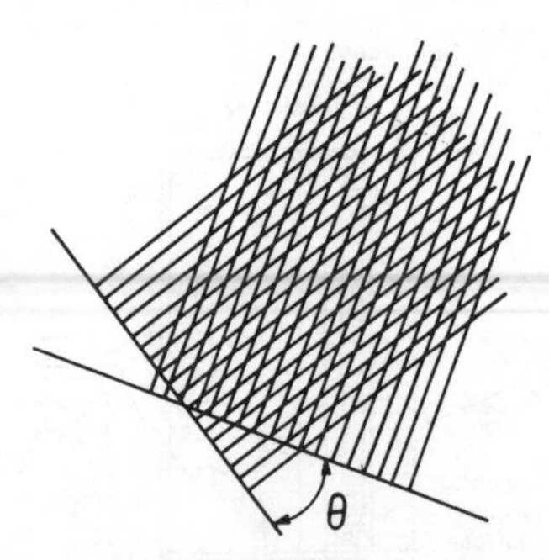

Fig. 5. Schematic of the method used by Kuhl and Edwards [4] to obtain projection data for the first clinical SPECT demonstration. A collimated scanner was moved to obtain one set of parallel ray projections in the plane of the slice. Then the detector orientation was changed by an angle θ, and a new set of data recorded. This process was repeated until a complete set of 360° perspective views was obtained.

There are two fundamental approaches to implementing SPECT. They are *transverse section imaging* in which the slice of interest is perpendicular to the long axis of the body, and *longitudinal section imaging* in which the image slice is parallel to the body axis. A more important distinction lies in the restrictions on data that can be obtained in the two cases. For transverse section imaging, data can generally be gathered in the plane of interest for a full 360° perspective around the patient. On the other hand, the realities of anatomy and patient shape make it virtually impossible to obtain data in the plane of a longitudinal section for more than a very limited range of angular samples. Transverse and longitudinal section imaging, therefore, differ more fundamentally in the density and type of data that can be gathered, and in the consequent mathematical reconstruction techniques. Because of their more widespread application, this review will concentrate primarily on transverse section tomographic methods. A brief mention of one longitudinal section technique will be given later for completeness.

Methods for Transverse Section SPECT

Historically, the first successful single-photon emission tomograph was developed by Kuhl and Edwards [4] in 1963. It anticipated many of the concepts that were necessary almost ten years later in the development of the immensely successful transmission tomography technique by Hounsfield *et al.* [5] and demonstrated an admirable degree of physical intuition.

This first emission system was based on the principle of a collimated scanning detector [6]. In a single scan across the patient, a set of parallel rays were recorded that defined the plane of interest. Next the scanner mechanism was indexed a few degrees so that the next scan produced a new set of parallel ray data that were oriented at a slight angle to the first, but in the same plane (see Fig. 5). By indexing the scanner through a full 360°, and recording consecutive scans after each small rotation, a full set of parallel ray projections were recorded for a single transverse section slice. In their early work, Kuhl and Edwards used simple backprojection techniques to reconstruct the tomographic image from these projections. Later they developed iterative algorithms that provide a more realistic reconstruction of the object [7].

Since that time, SPECT has rapidly evolved both in the instrumentation used to record the data and in the reconstruction methods. Tomographic reconstruction algorithms are reviewed elsewhere in this issue, and we will only mention several special problems that are unique to the single-photon emission tomography application. While there have been a large number of specialized instruments investigated for SPECT, modern approaches have tended to concentrate in two areas: those based on rotating Anger camera arrangements, and those that employ arrays of stationary detectors.

If a radioisotope camera fitted with a parallel hole collimator is rotated around the long axis of a patient, a series of images can be recorded at closely spaced angular intervals. In order to accommodate the large amount of computer processing that is necessary later, it should be understood that these images are inevitably recorded as digital images and stored in an appropriate computer system. As one example, data might be recorded for every 2° of rotation, corresponding to 180 images for a full 360° rotation. These data represent a wealth of information that can now be processed in a number of different ways. The most common approach is to recognize that a single plane in the patient perpendicular to the long axis of the body will be imaged as a specific line on each camera image. If we assume an orientation in which "up" corresponds to the side of the camera nearest the top of the patient, then each horizontal line in the camera image corresponds to a different transverse section through the patient. By selecting one of these lines, and collecting all the data from that line as the camera is rotated around the patient, a complete set of parallel ray projections are recorded for that slice. The beauty of the technique is that many such slices are recorded simultaneously corresponding to the set of such lines across the diameter of the camera.

There are several drawbacks to rotating camera systems. A typical camera head with its collimator and lead sheilding weighs several hundred pounds and is therefore a rather unwieldly object. In order to insure uniform rotation, rather elaborate mechanical drive systems are required. For greater detection efficiency, some systems incorporate two camera heads that are oppositely directed and provide some self-balancing of the rotating system. Nonetheless, the task of indexing the assembly through accurate angular increments is a formidable one. Furthermore, rotating the camera head can also create operational problems such as gain drift in the photomultiplier tubes. These effects arise because of the influence of stray magnetic fields on the electron trajectories within the photomultiplier tubes, and can be countered only through careful magnetic shielding of each of the photomultiplier tubes incorporated into the camera head.

Some recent attention has also focused on the effects of imperfections in the camera image on the reconstructed tomographic image [8]. Many such cameras display appreciable nonuniformity or nonlinearity in images of uniform or linear objects. Because these imperfections always occur in the same place in each image, their effects can be reinforced when many such images are synthesized into a single tomographic image. For example, if all camera images display a bright spot at some off-center position, then that bright spot will contribute to a bright ring in the reconstructed tomographic image. Because these ring artifacts are very noticeable visually, rather stringent demands are placed on the required degree of uniformity in the camera.

In principle, the difficulties mentioned above can be alleviated if the radiation detectors do not have to move. If a ring of individual gamma-ray detectors is provided, then it is possible to allow these detectors to remain stationary while recording the required set of ray data necessary for tomographic reconstructions. Furthermore, such detector rings have the potential of intercepting a large fraction of the gamma rays emitted from a particular slice, and can therefore be of attractive detection efficiency. We will describe one such ring system later in this review.

Spatial Resolution

An important property of any image is the degree to which fine detail or structure in the object can be reproduced. The most common specification of this spatial resolution is in terms of the *point-spread function*, or the image recorded for a single point source. The full width at half maximum of the recorded peak in the image is the quantity most often quoted as a measure of spatial resolution.

The resolution of a tomographic image is directly related to the resolution in each of the projection images. Because the reconstruction process is similar to a superposition of many of these projections, the tomographic image of a point source should have a width that is similar to the width of the image in each projection. This will then define the spatial resolution in the plane of the transverse tomographic slice. For typical recording systems, quoted figures range from 10 to 20 mm. These rather coarse resolution figures for emission systems should be contrasted with the very fine resolution (1-2 mm) obtainable with CT transmission devices.

The spatial resolution in the third dimension (perpendicular to the slice) is determined by the collimation properties in that dimension. In effect, it is a specification of the slice thickness that will be recorded from a source that is widely distributed along the axial dimension. This is somewhat of an arbitrary choice that may be varied to suit the application. Narrow collimation in the axial direction will result in thin slice definition, but only at the expense of sensitivity. Most present-day SPECT devices are designed for a slice thickness (or axial resolution) of 10-25 mm.

Device Sensitivity

One of the major considerations in SPECT is the efficiency with which emitted photons are detected and contribute to a reconstructed image. Because there is a need to minimize the amount of radiation dose to the patient, these photons are scarce. Even use of the most efficient systems result in imaging times of at least several minutes in order to produce images in which the statistical fluctuations are tolerable. In many potential applications, one would like this time to be as short as possible in order to follow the dynamics of time-dependent physiological processes. Hence there is a great interest in designing systems with high detection efficiency.

It is important to have a standard source for the intercomparison of the sensitivity of SPECT systems. One widely accepted standard consists of a 20-cm-diameter cylinder of radioactive solution whose axis is perpendicular to the transverse sections. Sensitivity is commonly quoted as the number of counts recorded per second per unit concentration (microcuries per cubic centimeter) of radioactivity in this solution. Because some systems image several different slices simultaneously while others record only single slices, it is also important to distinguish the total volumetric sensitivity from the single-slice sensitivity. The total volumetric sensitivity is based on summing the counts from all slices produced by the system, whereas the single-slice sensitivity is that for a single section only. In either case, these figures have significance only when other parameters such as the slice thickness and the spatial resolution of the reconstructed section image are also quoted. In most designs, sensitivity can be increased by relaxing resolution requirements, and *vice versa*. These efficiency and performance figures are also dependent on the energy of the gamma rays or the specific radioisotope used in the testing procedures.

Because emission tomographs designed specifically for single-slice sections often use some form of geometric focussing to expose a large area of detector to photons from that slice, their sensitivity per slice is usually greater than that obtained from multislice devices. However, the total volumetric sensitivity of a multislice system generally exceeds that of single-slice systems because the multiple slices are recorded simultaneously. The need for multiple slices will depend on the application. For those studies in which a large volume within the body must be resolved in three dimensions, then all the slices may be of interest and contribute to the total useful information. On the other hand, studies in which the process of interest is confined to a local region could, in principle, be investigated with only a single-slice section. As an intermediate case, only a few contiguous slices may be of interest. These could be obtained either as a portion of the data produced by a multislice system or by several sequential images from a single-slice machine with repositioning of the patient between images. It is clear that an intercomparison of sensitivity statistics must include an evaluation of the need for multiple slices in the application of interest.

Reconstruction Algorithms

The methods used for mathematical reconstruction of a tomographic image from its projections are given elsewhere in this issue. There are several factors in emission-computed tomography that are different from those in transmission tomography for which many of the algorithms have been developed. Nonetheless, the basic techniques are generally the same as those applied in transmission tomography, with differences accommodated through the application of corrections.

The algorithms themselves fall into general categories of either iterative techniques or filtered backprojection methods. Although successful examples of each of these approaches can be found in emission tomography, the filtered backprojection method has been most widely applied because of its rapid implementation on laboratory-scale computers.

One factor that is unique to emission tomography is the attenuation of the gamma rays. Without this attenuation, the line integrals become identical in form to those that apply in the transmission-tomography case and the transmission algorithms can be directly applied. Corrections for the attenuation can then be incorporated as described elsewhere in this review. The complications introduced by gamma-ray scattering will also be discussed later.

Another nonideality for the emission tomography case is the fact that any ray established by collimation will show some angular divergence. The backprojection algorithms assume that the width of each ray does not change as a function of distance from the detector. The divergent geometry in SPECT is at variance with this assumption. Nonetheless, this effect is usually ignored in applying the reconstruction algorithms. Things are not quite as bad as they might first appear, however, provided a full set of data recorded over a 360° orientation are available. Then the contribution of projections taken 180° apart are in some sense averaged, and the combined ray width is much more nearly constant.

Corrections for Gamma Attenuation

The basic reconstruction processes assume that the data taken along each ray represent the integrated intensity of gamma-ray sources along the same ray. One factor that causes a departure from this model is the inevitable attenuation of those gamma rays that have passed through some thickness of tissue before emerging from the surface. The fraction that escapes attenuation is roughly exponential with thickness. With common radioisotope gamma-ray energies, as many as half of the gamma rays emitted deep within the patient may be attenuated either by photoelectric absorption or by being scattered away from the detector before they emerge from the surface.

The effect of this attenuation is to cause an underestimation of the source intensity for deeply lying locations, unless some effort is made to correct for these losses. Thus all practical SPECT reconstruction algorithms include some analytic scheme to carry out approximate corrections for the effects of attenuation. In order to make the process tractable, all such correction schemes assume that the attenuation properties are uniform throughout the patient, and internal anatomical structures are generally ignored.

Some additional information beyond that ordinarily recorded in a SPECT study is required to carry out the attenuation correction. Data are required (or assumptions must be made) regarding the shape of the contour of the body within which the attenuation is taking place. One approach is to employ a supplementary measurement in which an external source of gamma rays is placed on the opposite side of the patient to the detector. The attenuation of these gamma rays is then recorded to yield data on the attenuation properties of the patient [9]. The position of the surface can then be deduced. For relatively simple geometries, such as a cross section through the head, direct physical measurements and modeling by simple shapes such as an ellipse can also be employed.

Another way to determine the body contour is to make measurements of the scattered radiation emerging from the patient. This scattered radiation is normally rejected by the energy acceptance criteria mentioned earlier since it is not a true indicator of source location. However, the scattering sites are broadly distributed throughout the patient volume, and the body contour can be deduced from the observed limits of scattering occurence [9]. Measurement of the scattered component requires that a second energy acceptance "window" be set electronically at an energy below that of the primary radiation, since scattering reduces the energy of the gamma-ray photon. In a number of commercial SPECT systems, such data are now recorded simultaneously with the primary data, and subsequent processing provides the contour information.

Once the surface profile has been established, it is possible to incorporate exponential attenuation into the analytic representation of the expected behavior of the tomographic projection data. The most common method uses an iterative procedure [10], [11]. A correction matrix is first defined for each point in the section plane that is related to the average attenuation calculated for that point. The first-order reconstructed image (without attenuation correction) is then multiplied by this correction matrix as a first estimate of the corrected image. Line integrals through this image are then computed and compared with the corresponding rays recorded in the original projection data. If the first-order correction were perfect, then these corrected line integrals should match the original data exactly. In general, the match will not be perfect, and the difference is derived between the original and derived data. This difference is then treated as an error projection and an error image is reconstructed using the same techniques applicable to general tomographic reconstruction. This error image (which may have both negative and positive elements) is then also corrected by the attenuation matrix and added to the first-

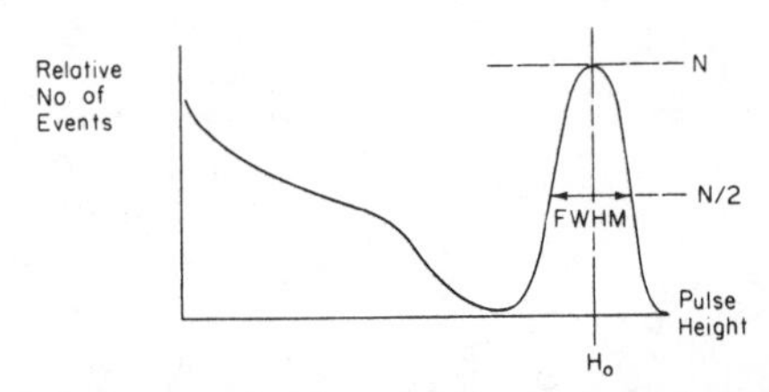

Fig. 6. A typical differential pulse-height spectrum for monoenergetic gamma rays. The energy resolution is defined as R = FWHM/H_0.

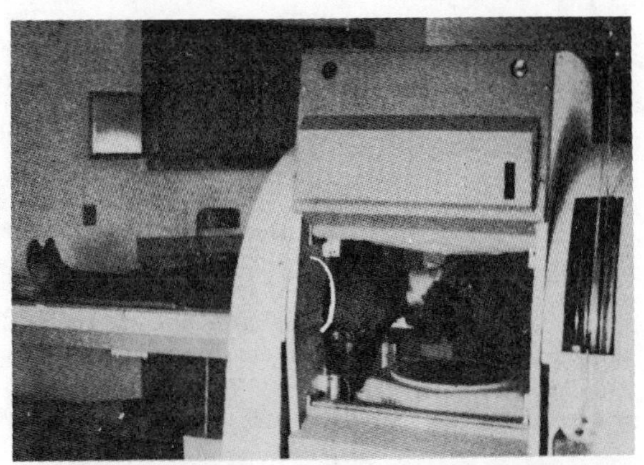

Fig. 7. A rotating gantry for carrying out SPECT with Anger cameras. Two cameras are mounted on opposite sides of the patient to increase detection efficiency (courtesy Dr. R. Jaszczak, Duke University).

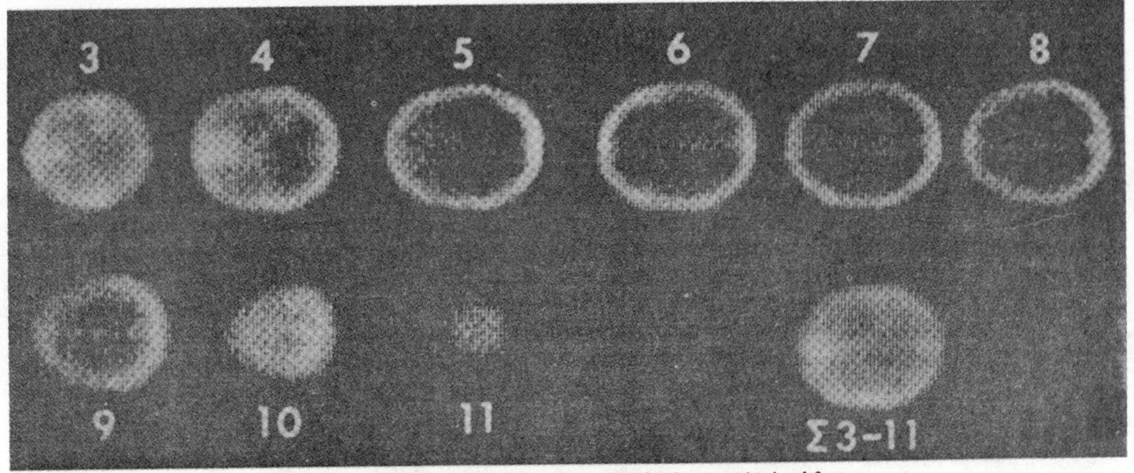

Fig. 8. Examples of reconstructed tomographic images obtained from a rotating Anger camera system. Images 3 through 11 represent slices through the patient's head from bottom to top. Also shown is the sum of the slice images, the approximate equivalent of a single projection image (courtesy Dr. R. Jaszczak, Duke University).

order corrected image. This step produces a second-order corrected image. The entire process could again be repeated, but it is found that usually one such iteration suffices. In practice, the process is sometimes truncated after the first-order correction. The reprojection step adds substantial complexity to the computation, and may also increase the noise in some reconstructed images.

Effects of Scattering in the Patient

The elimination of all scattered gamma rays would require that the detector be able to carry out infinitely fine energy discrimination. No real detector displays this perfect behavior, but rather will produce a spread in pulse amplitudes for a monoenergetic source. The energy resolution is defined as the full width at half maximum of the recorded pulse-height distribution for a monoenergetic source divided by the average pulse height (see Fig. 6). Expressed as a percentage, sodium iodide scintillators generally show an energy resolution of 10 percent or more for gamma-ray energies near 100 keV, decreasing to 5 percent at best for high-energy gamma rays in the vicinity of 1 MeV. Other types of detectors, such as germanium semiconductor diodes, are capable of an order of magnitude better energy resolution, but their high cost, requirement of cooling, and other difficulties have severely limited their application in SPECT.

Systems based on sodium iodide therefore inadvertently

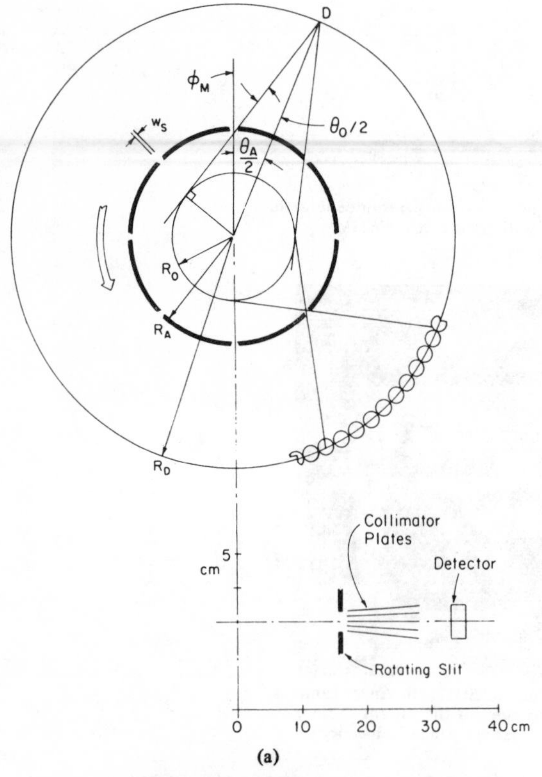

(a)

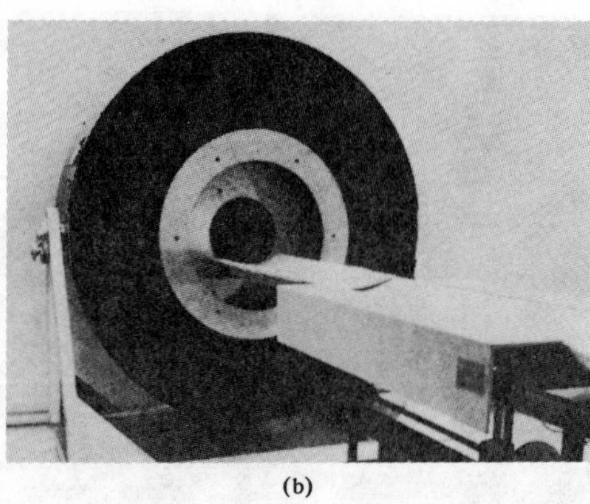

(b)

Fig. 9. The stationary ring device known as SPRINT developed at the University of Michigan. (a) Schematic diagram. The detectors are located at R_D, the rotating aperture at R_A, and the object to be imaged within R_0. A cross section through the plane of the slice is also shown. (b) A photograph of the completed device.

accept a significant fraction of the scattered gamma rays because the finite energy resolution does not allow their clean separation from the primary rays. The scattered gamma rays that are deflected the least will be the most troublesome since the energy lost in the scattering process increases with scattering angle.

The effects of scattering in conventional radioisotope imaging have been the study of several previous analytical and experimental studies [12], [13]. Because the degree of scatter acceptance will affect the accuracy of quantitative measurements, the influence of scattering has also been the subject of several recent investigations in SPECT applications [14], [15].

Sampling Reqirements

In any tomographic system, one must gather sufficient data so that the object is adequately sampled. At any given angular position or perspective, there is a requirement on the fineness with which the many rays making up the projection data must be sampled [16]. In many systems, there is a limit on the sampling spatial frequency that is set by the spatial resolution of the detector itself. There is little reason to take many samples over the smallest distance that can be resolved by the detection device. For typical systems, this translates into recording somewhere between 64 to 256 rays across a single projection.

There is also a requirement on angular sampling frequency. It can be shown that the number of angular views necessary for a full 360° perspective is approximately equal to the number of resolution elements across the image. In practice, rotating camera systems sample at intervals that are separated by not more than a few degrees. Therefore, 128 sets of projection data are typical for a full 360° rotation.

There have been several investigations of the results of undersampling [11], [17]. The most serious situations arise when a full 360° perspective cannot be recorded because of limitations of anatomy or the type of tomographic system in use. Methods have been developed to "fill in" the missing data under conditions of incomplete angular sampling. These problems are most severe in longitudinal section tomography where one is often restricted to angular perspective of less than 180°.

Example of a Rotating Camera System

In Fig. 7, an example is shown of the Searle Duke University SPECT system [9]. It is based on two large-field-of-view scintillation cameras mounted on a large gantry to permit their rotation around an axis coincident to the long axis of the body. The system is coupled to a large digital computer to allow implementation of the reconstructed algorithms. It also permits reorganization of the reconstructed data into sections that are different from the straightforward axial sections derived initially.

A number of commercial manufacturers are now marketing systems that are quite similar to this approach. Emission tomography based on rotating gamma-ray cameras has reached the point of extensive clinical evaluation, and there seems to be a high level of enthusiasm regarding the future potential of the technique. A representative tomographic image from actual clinical data is shown in Fig. 8.

The most direct approach to applying radioisotope cameras to SPECT is to retain the parallel hole collimator that is most often used with such cameras in conventional projection imaging. In that way, a given slice in the patient is seen only by a narrow band across the face of the camera. The sensitivity per slice is limited because the solid angle subtended is relatively small. However, slices at different axial positions through the patient are imaged simultaneously on corresponding parallel bands across the detector face. The multislice sensitivity is therefore quite high.

Other specialized configurations specifically designed for SPECT have also been investigated. Jaszczak *et al.* [18] have described a collimator with focussed channels fitted to a conventional gamma-ray camera which, when applied to SPECT, results in a reported increase in sensitivity while maintaining resolution in the reconstructed image. This collimator converges to a line of focus parallel to the axis of rotation, and when used with large-diameter gamma-ray cameras, more effectively uses the available detector for area imaging. It has also been proposed that a curved camera face be used, although the gains to be anticipated over the more conventional flat-face camera are not large.

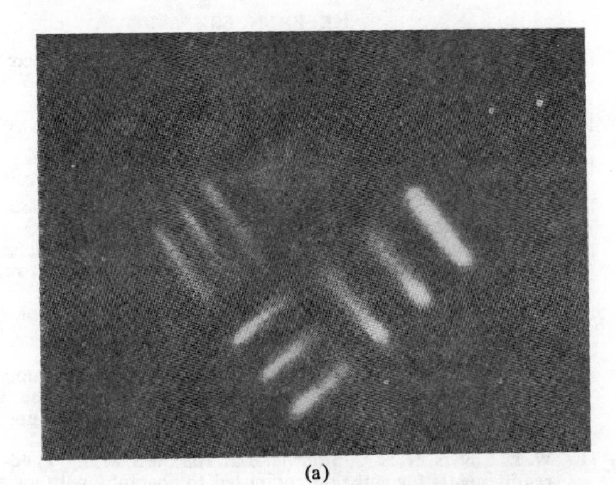

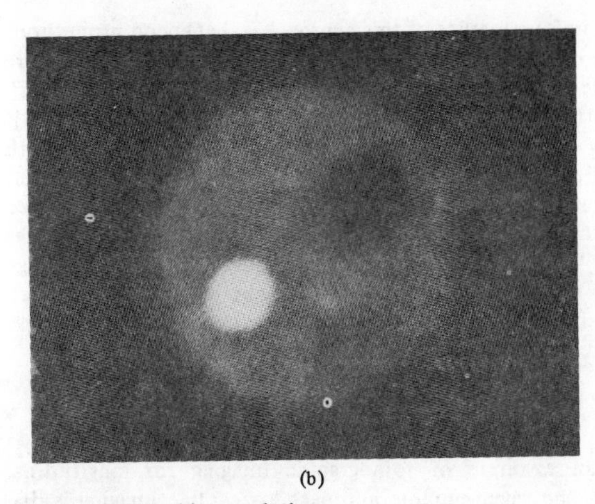

(a) (b)

Fig. 10. Tomographic images obtained with SPRINT. (a) A resolution phantom with the pattern shown of radioactive solution in the plane of the slice. The small bars are 6.3 mm wide and are separated by 6.3 mm. (b) A head phantom with 17.8-cm inside diameter. The hot and cold "lesions" are 3.8-cm diameter.

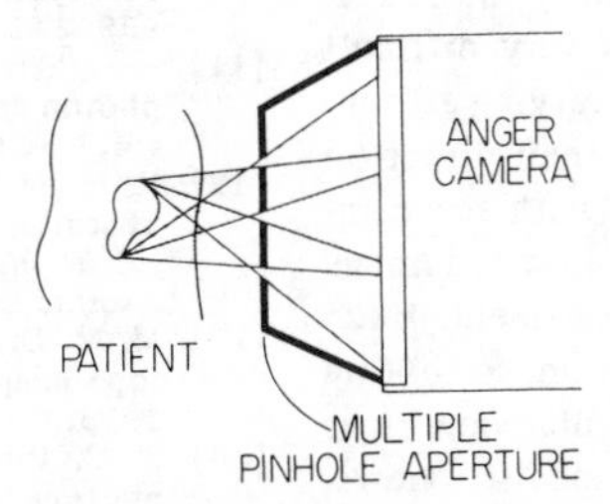

Fig. 11. Use of a multiple-pinhole aperture to simultaneously record several perspective views of the object. Each view is confined to a unique subsection of the total camera area.

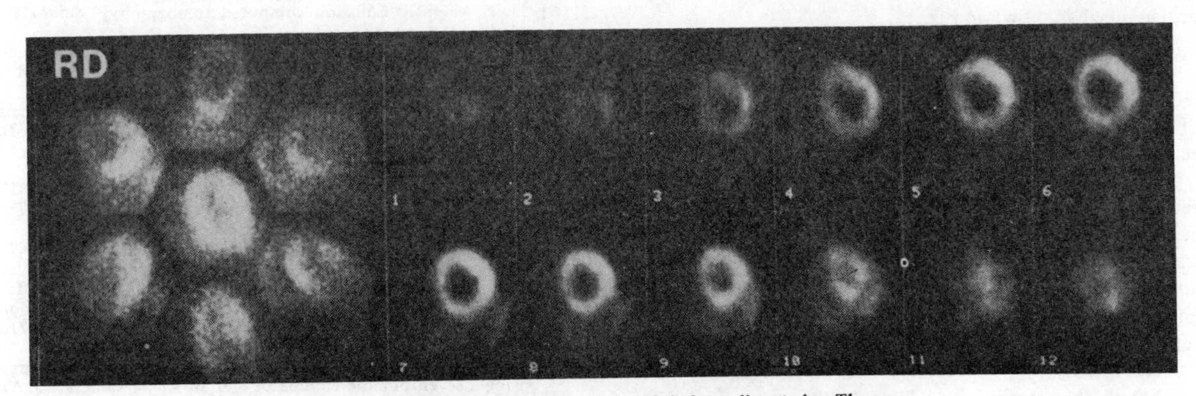

Fig. 12. Examples of data from a seven-pinhole cardiac study. The raw data are shown on the left, and reconstructed tomographic images at increasing depths are numbered 1 through 12 (courtesy of D. L. Kirch, University of Colorado–Veterans' Administration Hospital).

Example of a Stationary Detector Ring System for SPECT

An example of a dedicated device for transverse section tomography is SPRINT (Single Photon RINg Tomograph) developed at The University of Michigan [19], [20]. In this system, 78 individual sodium iodide detectors are arranged in a ring pattern that completely surrounds the patient and defines the plane of the tomographic slice (see Fig. 9). The detectors remain stationary during the data-gathering process, and each develops its individual perspective view of the object by recording a set of "fan-beam" rays emanating from the point of the detector. Each ray is defined by a slit in an inner cylinder that rotates to scan the entire field of view. One rotation of the cylinder allows each detector to gather a complete set of such fan-beam data. Conventional reconstruction methods are then applied to derive the tomographic image. An example is shown in Fig. 10.

One unique feature of this approach is that encoding methods can be used to define the direction of the ray. If many slits are provided in the rotating cylinder and arranged in an appropriate pattern, then the average counting rate of each detector is greatly increased over the situation that would prevail with one slit. The information must now be decoded, however. This can be carried out using correlation techniques provided the slit pattern is chosen to have certain statistical properties [21]. While coding has not demonstrated significant advantage in common applications of radioisotope imaging, it may well offer significant advantage in overcoming unmodulated background signals present in many applications.

Longitudinal Section Tomography

As an example of tomographic imaging for longitudinal planes, the seven-pinhole approach to cardiac imaging is diagramed in Fig. 11. This approach, pioneered by Vogel and Kirch [22], attempts to reconstruct cross-sectional images of the heart by placing a modified gamma-ray camera in an oblique orientation at the front of the patient's chest. It relies on limited angle sampling of those rays emerging only in the forward direction. Complete angular sampling is very difficult for cardiac applications because of the greatly varying attenuation of emitted gamma rays as a function of their direction of emission. The gamma-ray camera is fitted with seven individual pinhole apertures, each of which casts an inverted image of the object onto a corresponding section of the camera. Each of the pinholes is in a somewhat different position, so that its perspective of the object involves a different angular view than all the others. By synthesizing all these perspective views, reconstructions can be made that correspond to a number of assumed depths of the section. Each of these planes is parallel to the plane of the detector, and sections the heart at different depths relative to the chest wall. An example of a typical cardiac image is shown in Fig. 12.

Conclusion

There has been considerable recent progress in the development of tomographic methods for the imaging of conventional single-photon radioisotopes. These methods allow a high degree of specificity in localizing the position of radioisotopes in three dimensions, and clinical exploitation of this ability is already underway. Lesion detectability using SPECT has been shown [23] to be significantly improved compared with conventional projection imaging. Both moving and stationary detector systems have demonstrated the ability to form images of useful quality.

Although their sensitivity will probably never equal that of PET systems, the wider applicability of SPECT systems to conventional radiopharmaceuticals will be a strong positive influence in their practical utilization.

References

[1] D. Doria and M. Singh, "Comparison of reconstruction algorithms for an electronically collimated gamma camera," *IEEE Trans. Nucl. Sci.*, vol. NS-29, no. 1, pp. 447-450, 1982.
[2] H. O. Anger, "Scintillation camera," *Rev. Sci. Instrum.*, vol. 29, pp. 27-33, 1958.
[3] ——, "Scintillation camera with multichannel collimators," *J. Nucl. Med.*, vol. 5, pp. 515-531, 1964.
[4] D. E. Kuhl and R. Q. Edwards, "Image separation radioisotope scanning," *Radiology*, vol. 80, pp. 653-662, 1963.
[5] G. N. Hounsfield, "Computerized transverse axial scanning (tomography): Part 1, Description of system," *Brit. J. Radiol.*, vol. 46, pp. 1016-1022, 1973.
[6] D. E. Kuhl and R. Q. Edwards, "Cylindrical and section radioisotope scanning of the liver and brain," *Radiology*, vol. 83, pp. 926-935, 1964.
[7] D. E. Kuhl, "The current status of tomographic scanning," in *Fundamental Problems in Scanning*, A. Gottschalk and R. N. Beck, Eds. Springfield, IL: Charles C. Thomas, 1968, pp. 179-190.
[8] W. L. Rogers, N. H. Clinthorne, B. A. Harkness, *et al.*, "Field-flood requirements for emission computed tomography with an Anger camera," *J. Nucl. Med.*, vol. 23, pp. 162-168, 1982.
[9] R. J. Jaszczak, L. T. Chang, N. A. Stein, and F. E. Moore, "Whole-body single-photon emission computed tomography using dual, large-field-of-view scintillation cameras," *Phys. Med. Biol.*, vol. 24, pp. 1123-1143, 1979.
[10] L. T. Chang, "A method for attenuation correction in radionuclide computed tomography," *IEEE Trans. Nucl. Sci.*, vol. NS-25, pp. 638-642, 1978.
[11] ——, "Attenuation correction and incomplete projection in single photon emission computed tomography," *IEEE Trans. Nucl. Sci.*, vol. NS-26, pp. 2780-2789, 1978.
[12] R. N. Beck, M. W. Schuh, T. D. Cohen, and N. Lembares, "Effects of scattered radiation on scintillation detector response," in *Medical Radioisotope Scintigraphy*, vol. I. Vienna, Austria: Int. Atomic Energy Agency, 1969, pp. 595-616.
[13] M. M. Dresser and G. F. Knoll, "Results of scattering in radioisotope imaging," *IEEE Trans. Nucl. Sci.*, vol. NS-20, pp. 266-272, 1973.
[14] S. C. Pang and S. Genna, "The effect of Compton scattered photons on emission computerized transaxial tomography," *IEEE Trans. Nucl. Sci.*, vol. NS-26, pp. 2771-2774, 1979.
[15] R. J. Jaszczak, R. E. Coleman, and F. R. Whitehead, "Physical factors affecting quantitative measurements using camera-based single photon emission computed tomography (SPECT)," *IEEE Trans. Nucl. Sci.*, vol. NS-28, no. 1, pp. 69-80, 1981.
[16] M. E. Phelps, "Emission computed tomography," *Semin. Nucl. Med.*, vol. 7, pp. 337-365, 1977.
[17] R. A. Brooks, G. H. Weiss, and A. J. Talbert, "A new approach to interpolation in computed tomography," *J. Comput. Assist. Tomog.*, vol. 2, pp. 577-585, 1978.
[18] C. B. Lim, L. T. Chang, and R. J. Jaszczak, "Performance analysis of three camera configurations for single photon emission computed tomography," *IEEE Trans. Nucl. Sci.*, vol. NS-27, pp. 559-568, 1980.
[19] J. J. Williams, W. P. Snapp, and G. F. Knoll, "Introducing SPRINT: A single photon system for emission tomography," *IEEE Trans. Nucl. Sci.*, vol. NS-26, no. 1, pp. 628-633, 1979.
[20] W. L. Rogers, N. H. Clinthorne, J. Stamos, *et al.*, "SPRINT: A stationary detector single photon ring tomograph for brain imaging," *IEEE Trans. Med. Imag.*, vol. MI-1, no. 1, pp. 63-68, 1982.
[21] G. F. Knoll and J. J. Williams, "Application of a ring pseudorandom aperture for transverse section tomography," *IEEE Trans. Nucl. Sci.*, vol. NS-24, no. 1, pp. 581-586, 1977.
[22] R. A. Vogel, "A new method of multiplanar emission tomography using a seven pinhole collimator and an Anger camera," *J. Nucl. Med.*, vol. 19, pp. 648-654, 1978.
[23] R. J. Jaszczak, F. R. Whitehead, C. B. Lim, and R. E. Coleman, "Lesion detection with single-photon emission computed tomography (SPECT) compared with conventional imaging," *J. Nucl. Med.*, vol. 23, pp. 97-102, 1982.

Medical ultrasonics

©1984 IEEE. Reprinted, with permission, from *IEEE Spectrum*, **21**, (12), pp. 44–51, (December 1984).

As new techniques are perfected, ultrasound extends its usefulness from diagnosis to research, therapy, and even surgery

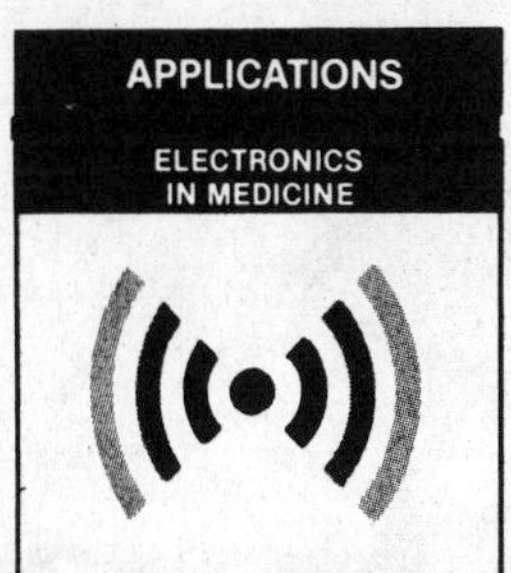

One of the most widely used electronic techniques in medicine today is ultrasound. Physicians using it can now measure the velocity of blood flow in high-risk pregnancies; examine the brains of newborn babies; take motion pictures of a patient's heart; and detect cancerous tumors and the plaques in blood vessels that cause arteriosclerosis. In obstetrics the use of ultrasound has become routine; it is believed that today as many as 40 percent of pregnant women in the United States have at least one such scan.

Although the primary use of ultrasound is in diagnosis, it is also being used increasingly in medical research, therapy, and even surgery. For example, ultrasound microscopes can offer a look at the structure of living cells unobtainable with light or electron microscopes; ultrasound-induced heating (hyperthermia) can be used to destroy some malignant tumors; and intense ultrasound shock waves have been applied to kidney stones and other calculuses to disintegrate them.

The basis for many of these new applications is the increasing technical perfection of a number of ultrasound techniques—in particular, Doppler ultrasound, which allows direct measurement of blood-flow vector velocities in patients. Ultrasound image resolution has also greatly improved. These improvements, in turn, have often been directly linked to the increased information-processing capabilities and lower costs of digital electronic circuits.

Technique based on sonar principle

Ultrasound scanning in medicine works on the same principle as sonar in undersea navigation. In medical applications, ultrasound signals, generally in the megahertz frequency range, are generated by a piezoelectric transducer, which converts RF electrical pulsed signals into mechanical vibration. These vibrations are coupled to the skin of the patient. When the ultrasound waves encounter any changes in characteristic impedance within the body, echoes are reflected back to the piezoelectric transducer, which acts as a detector. (The characteristic impedance of a material is equal to the product of its density and the speed at which ultrasound propagates within it.)

The time between the emission and detection of the echo gives, as in sonar, the range of the reflecting surface. When the returned signals are properly analyzed, they can be displayed as maps of the internal organs of the body.

In most applications of pulse-echo ultrasound, the emitted pulses are scanned across a plane and the echoes are displayed in two dimensions to produce a cross section of the patient. This is called the B-scan. Two other modes—the A-scan and M-mode—do not use beam scanning.

In the A-scan the echoes along a single line are displayed against time. This is used when accurate measurements of distance (as in measuring the axial lengths of components of the human eye) or amplitude of pulses are required. The M-mode also uses a fixed beam, but it records the echoes so as to outline the motion of intercepted structures away from or toward the source. When recorded on light-sensitive paper by a fiber-optic cathode-ray tube, such M-mode outputs are used in the study of heart dysfunctions.

In the B-scan, the ultrasound beam can be moved across the patient either by hand or by a mechanical scanning device or, in the case of a linear or phased array, electronically. In manual scanning the probe is mechanically constrained to movement in two dimensions. The probe is moved across the patient's skin. Coupling gel or oil is used to eliminate air. Echoes detected by the probe are arranged to modulate the brightness of the display.

Separate time-base generators for the horizontal and vertical deflection drives of the display are simultaneously triggered by the system clock. The starting points and amplitudes of the time-base signals are controlled by resolvers mounted on the scanning frame. These measure the position and angle of the probe, so the resulting time-base line on the display appears to be linked to the ultrasonic beam passing through the patient. Thus manual movement of the probe produces a two-dimensional image on the display.

Split-second scanning possible

Although with manual scanning an image is usually formed in about a second, much more rapid scanning is possible. The rate at which each line's echoes can be obtained is limited by the speed of sound in tissue. For example, for a typical penetration of 15 centimeters, it takes 200 microseconds for a pulse to penetrate and for the farthest echo to return. Thus the fastest possible scanning at a 15-cm penetration would be 5000 pulses, or lines, per

Defining terms

Aliasing—interference caused by beats or signals between the emitted and received signals of a Doppler ultrasound system.
Annular array—a group of transducers, antennas, or other detectors arranged in an annulus (ring).
Calculus—in medicine, an abnormal mass or deposit formed in the body, such as a kidney stone.
Endoscopic—refers to the examination of a body cavity by means of an instrument.
Histology—the study of the cellular structure of biological tissue; also, the structure of the tissues of an organism.
Piezoelectric transducer—a device for converting electrical energy into mechanical and vice versa.
Resolution cell—that region that is exposed to sound waves and from which echoes are received in an ultrasonic application.
Tomography—techniques, frequently aided by computer systems, of making cross-sectional images of a patient.

Peter N.T. Wells Bristol General Hospital

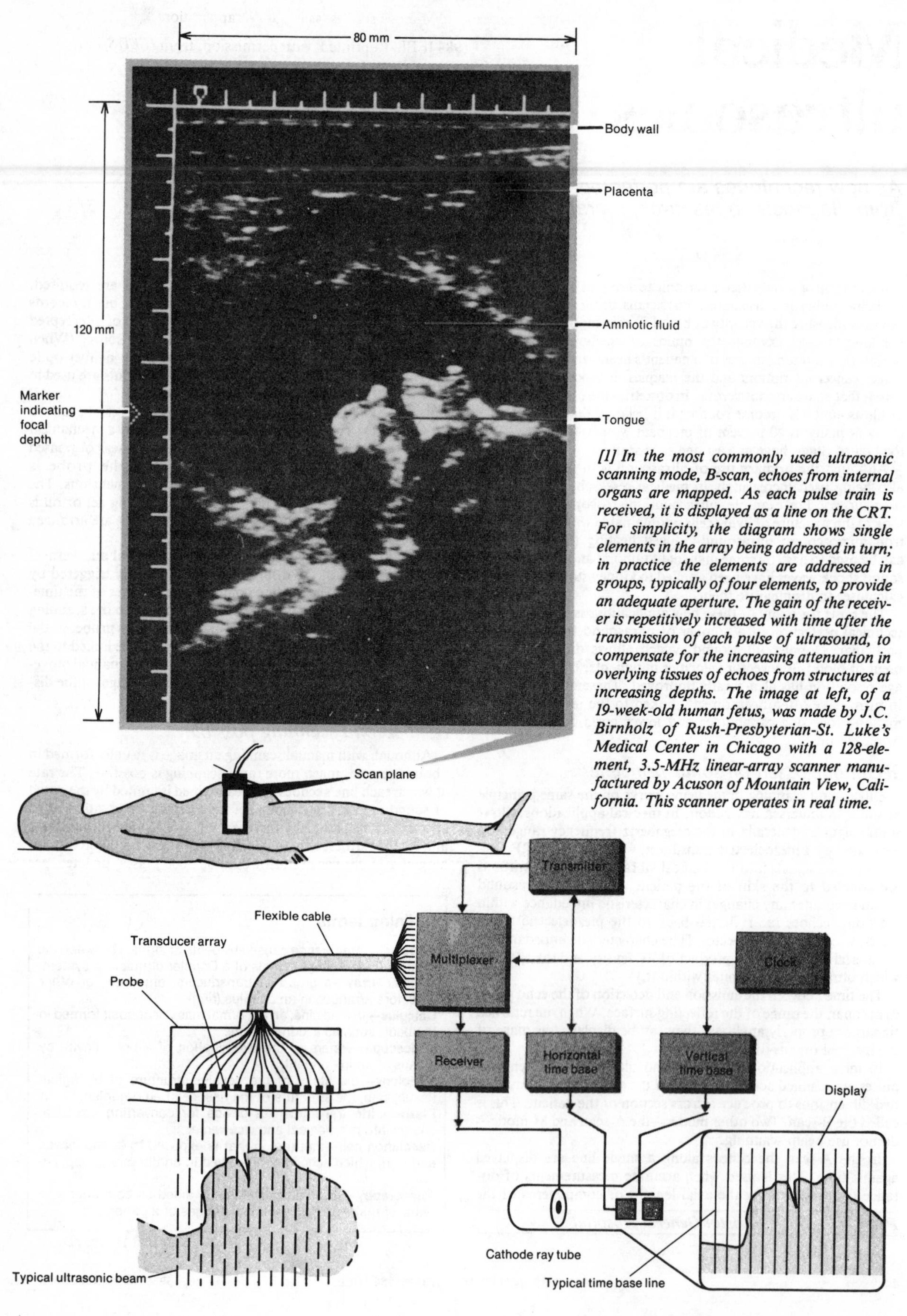

[1] In the most commonly used ultrasonic scanning mode, B-scan, echoes from internal organs are mapped. As each pulse train is received, it is displayed as a line on the CRT. For simplicity, the diagram shows single elements in the array being addressed in turn; in practice the elements are addressed in groups, typically of four elements, to provide an adequate aperture. The gain of the receiver is repetitively increased with time after the transmission of each pulse of ultrasound, to compensate for the increasing attenuation in overlying tissues of echoes from structures at increasing depths. The image at left, of a 19-week-old human fetus, was made by J.C. Birnholz of Rush-Presbyterian-St. Luke's Medical Center in Chicago with a 128-element, 3.5-MHz linear-array scanner manufactured by Acuson of Mountain View, California. This scanner operates in real time.

second. For a typical display of 100 lines, this is 50 frames per second. For such rapid scanning, mechanical or electrical devices are needed—a class of scanners called real-time.

Mechanical systems may control the beam direction of a transducer (or annular array) by oscillating the transducer, by rotating a wheel carrying one or more rim-mounted transducers with radially directed beams, or by oscillating a mirror reflecting the beam. In electronic systems a stationary array of small linear-transducer elements placed side by side is used to produce a scan with either a rectangular [Fig. 1] or a sector format. For sector scanning, the time grading of the excitation on transmission, as well as the delays on reception across the transducer elements in the aperture, are chosen to steer the beam through an angle in the plane of the scan. The approach uses the same principle as that of phased-array sonar or radar.

Real-time imaging has four main advantages over static scanning: rapid physiological movements can be studied; involuntary movements do not degrade the ultrasound image; the operator can quickly elucidate even complex anatomical relationships; and the two-dimensional image can be used to guide the instruments for either biopsy (the use of a needle to collect small samples of tissue for microscopic examination from selected sites) or M-mode studies (as in cardiac applications).

An image memory is needed to store the transient echo signals as the ultrasonic beam moves across the patient to form an image. Modern instruments use digital scan converters, which not only store the echo signals, but also convert the random ultrasonic-beam scanning format into the standard TV format of the display. A typical scan converter for this purpose has 512 by 512 picture elements, or pixels, each occupying 5 bits of random-access memory. The dynamic range of the 5-bit image is about 30 decibels.

The quality of the image—measured in terms of the ability of an observer to distinguish subtle differences in contrast and texture—depends on the gray-scale transfer characteristic. Many modern scanners give the operator choices of prestorage and poststorage gray-scale maps, but the scientific basis for optimal selection of these maps for particular clinical applications remains to be worked out.

B-scan use in obstetrics now routine

One of the earliest and still most common uses of ultrasound B-scans is in obstetrics. Here ultrasound images can guide clinicians in their use of a needle to draw samples of amniotic fluid.

Ultrasound in medicine: the engineering basics

The first step in the adaptation of ultrasound for medical purposes is the generation of the sound waves by a transducer. The most commonly used transducer material is lead zirconate titanate, a synthetic ceramic polarized during manufacturing to make it strongly piezoelectric. Lead zirconate titanates with wide bandwidths and high electromechanical coupling coefficients are most suitable for diagnostic applications, particularly those using short pulses. Their principal disadvantage is that they have a characteristic impedance about 20 times greater than that of biological soft tissue and water.

As in an electrical circuit, a large impedance mismatch causes energy to be reflected away from the interface rather than transmitted through it. Consequently impedance-matching layers are usually applied to the surface of the transducer in contact with the load, to improve the sensitivity. In its simplest form, a matching layer consists of a plate of material a quarter of a wavelength thick with characteristic impedance equal to the geometric means of the characteristic impedances of the transducer and the load. Because of the large mismatch that exists in practice and because matching is required over the frequency spectrum of the pulse—that is, over the corresponding range of wavelengths—multiple matching layers are generally used.

Once the signal is produced, it may be focused so the beam is as narrow as possible and the sources of the echoes are sharply defined. This can be achieved if the transducer is made concave or if the beam is reflected from concave mirrors by lenses or transducer arrays. The principle is similar to that used in phased-array radars.

Lenses are constructed from plastics and in the simplest form have to be concave, because the speed of ultrasound in such plastics is higher than that in biological soft tissues. It is sometimes desirable for the probe to have a flat or convex front surface so it makes good contact with the patient; such a surface can be provided if a strongly focusing concave transducer is combined with a defocusing convex plastic lens so that the resulting ultrasonic beam has the required degree of focus.

A limitation of fixed-focus devices is that only one range is in focus; the resolution of the system deteriorates at ranges other than the focused one. A phased-array focusing system can avoid this problem.

Focusing with an annular or linear array involves introducing electronic delays in the signal paths associated with each of the individual elements. This is to compensate for ultrasonic delays along the different ray paths between the transducer and the focus. Likewise the ultrasonic beam can be steered through any desired angle from the central axis of a linear array by appropriate timing of the electronic delays associated with the elements along the array.

On transmission, the position of the focus (and the direction of the beam) is fixed for each transmitted pulse. Usually a separate transmitter is associated with each transducer element, and the transmitters are triggered by voltage-controlled monostables, which are themselves triggered by a master clock. The time gradient across an array with a 20-millimeter aperture is about 10 microseconds when the beam is steered 40° off the central axis. On reception, the position of the focus can be swept along the beam axis if the individual delays associated with each element are continuously changed so the focus always coincides with the instantaneous positions of the echo-producing targets at steadily increasing beam depths. Moreover, simultaneous focusing and steering of the ultrasonic beam is possible with a linear array of transducers.

The outputs from all the delay circuits are summed to provide the signal input to the receiver. The delays can be generated by voltage-controlled analog circuits or by digital techniques. Apart from the need for accurate timing of the individual delays and the expense of providing the hardware—problems that are being solved by the application of VLSI circuit technology—the main difficulties are those of sideloops in the radiation pattern introduced when the beam is steered off axis and noise is introduced by the delay circuits.

The resolution of ultrasound images can in principle be improved if the wavelength of the emitted ultrasound is decreased. This is done by increasing the wavelength's frequency. However, since the attenuation of ultrasound by human tissue increases with increasing frequency, the frequency of ultrasound is limited by the depth to which the ultrasound must penetrate. Since the attenuation is generally around 1 decibel per centimeter per megahertz and since about 70 dB of attenuation in tissue can be tolerated, the maximum penetration is about 230 wavelengths.

For example, for abdominal and cardiac studies, penetrations of 10 to 20 cm are required, so frequencies of 2 to 3.5 MHz are used. Proportionately higher frequencies are used for smaller structures, and slightly higher frequencies may be used for structures with attenuation coefficients of less than 1 decibel per centimeter per megahertz or with high-gain instruments. Nowadays, for example, the best results in abdominal scanning are often obtained at 5 MHz.

—P.N.T.W.

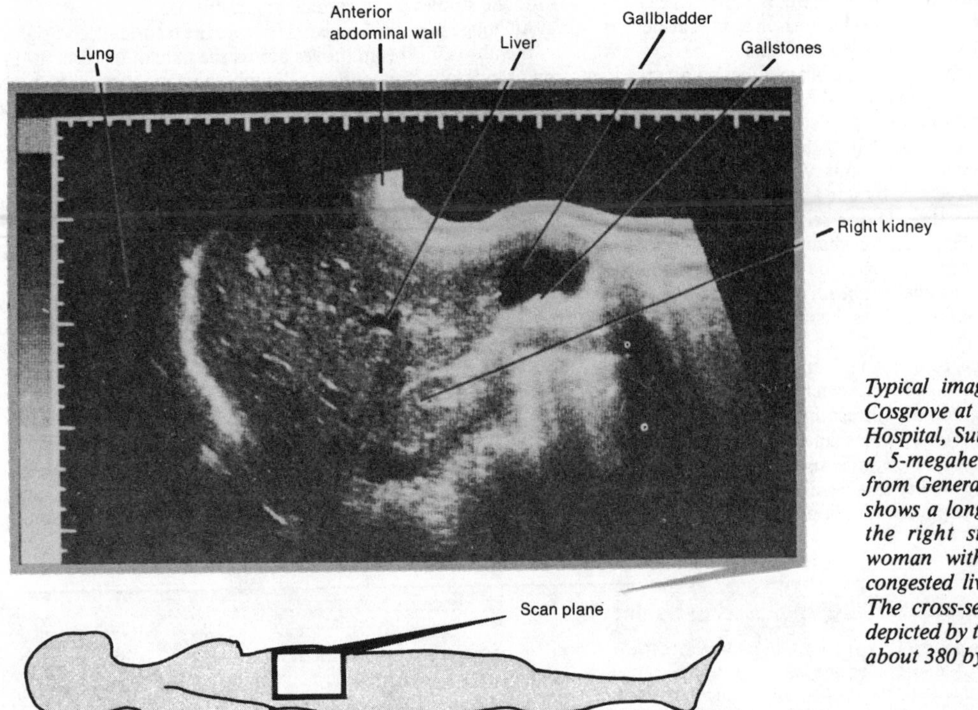

Typical image made by D.O. Cosgrove at the Royal Marsden Hospital, Sutton, England, with a 5-megahertz static scanner from General Electric. The scan shows a longitudinal section of the right side of an elderly woman with pleural effusion, congested liver, and gallstones. The cross-sectional scan plane depicted by this photo measured about 380 by 220 millimeters.

The fluid that surrounds the baby in the womb is then analyzed for chromosomal birth defects. Ultrasound guidance makes it possible to avoid the risk of the needle harming the placenta or fetus.

Ultrasound images of the fetus's head, body, and limbs give accurate assessments of the length of gestation and help to determine whether the fetus is suffering from abnormalities or retarded growth. Ultrasound can also show whether there is more than one fetus—a situation that previously could be determined with confidence only by X-rays, which are potentially harmful to both the expectant mother and the fetus.

A second major role of B-scan is in the detection and evaluation of cancer tumors. Frequently tumors growing in soft tissues are more obvious in ultrasound images than in X-ray pictures. The same is true for many other disease conditions of the body's organs. Since X-rays pass easily through most tissues, special contrast media, such as drinks containing barium for the X-raying of the gastrointestinal tract, must be used as aids in radiography. Ultrasound images, however, frequently show good contrast between various categories of soft tissue without artificial enhancement.

Ultrasound is also on the way to becoming a routine surgical aid, enabling the surgeon to see structures and organs that may be obscured from direct view. And recent technological advances in ultrasound are opening up still more applications. The most important of these advances include the perfection of Doppler techniques to study blood flow; the development of three-dimensional scanning technology; the use of ultrasound to characterize disease states in tissues through textures rather than image outlines; and the general improvement of ultrasound resolution and sensitivity.

The ins and outs of Doppler methods

Ultrasonic Doppler techniques depend on the change in frequency that occurs when a wave is reflected by a moving target. For example, the Doppler-shift frequency (the difference between the received and transmitted frequencies) is about 4 kilohertz when a 3-megahertz wave is backscattered by blood flowing in the direction of the ultrasonic beam at a velocity of 1 meter per second. If the directions of flow and of the ultrasonic beam do not coincide, the directions of the vectors determine the shift frequency; the instrument takes this into account by including the cosine of the angle between the two directions in the calculation of the Doppler-shift frequency.

It is significant that the Doppler-shift frequency usually lies in the audible range. The choice of optimal ultrasonic frequency depends on the properties of the target and the required penetration. For the study of blood flow, for example, a tradeoff between backscattered power and attenuation in the intervening tissue (both of which increase with increasing frequency) is necessary.

The simplest type of medical ultrasonic Doppler instrument employs continuous waves. The probe contains separate transducers for transmission and reception to minimize crosstalk. The frequency of the received signals equals that of the transmitted signals when the echoes are from stationary structures, but the frequency is shifted by the Doppler effect if the corresponding reflecting structures (or ensembles of scatterers, such as blood) have components of velocity that coincide with those of the ultrasonic beam.

Simple Doppler systems cannot distinguish between approaching and receding targets, and they cannot measure the depths of targets along the ultrasonic beam. Instead, they merely detect the frequencies and values of the amplitudes of the Doppler-shifted signals. The frequency of the received signal is greater than that of the transmitter if the target is approaching the ultrasonic probe, and vice versa. Therefore directional information can be obtained either by phase-quadrature detection (which determines whether the phase of the received signal leads or lags behind that of the transmitted reference signal) or by upper and lower sideband filtering.

Several applications involve continuous-wave Doppler ultrasound, especially in obstetrics. For example, with this technique, a fetal heartbeat can be monitored as early as the eleventh week of pregnancy, when the beat is too faint to be detected with a stethoscope.

For many applications, however, it is important to know the distance to the Doppler-producing blood flow—for example, in a vessel deep within a patient's body, relatively far from the measuring instrument. For this, pulsed-Doppler systems measure both the frequency of the returning echo and its timing. Pulsed-Doppler systems may have a single receiver gate or multiple gates fed to parallel processing channels, or the signals may be processed continuously by a swept or infinite gate.

The upper limit to the Doppler-shift frequency that can be detected without aliasing is equal to half the pulse-repetition rate. This places a further constraint on the maximum ultrasonic frequency that can be used in any given situation. Although in principle this limitation could be avoided by use of a random code on a continuously transmitted carrier, neither this nor related techniques have yet been applied clinically.

Audible signals often suffice

Because Doppler-shift signals usually fall within the audible frequency range, physicians can often obtain adequate clinical information simply by listening to the signals. This is all that is needed for example, to confirm that a fetal heart continues to beat or that blood continues to flow through an arterial graft. The acquisition of quantitative data, however, requires quantitative analysis of the Doppler signals.

Measurement of the zero-crossing (root-mean-square) frequency of the signals is adequate for some applications, but it can be misleading, particularly when forward and reverse blood flows exist simultaneously. Frequency-spectrum analysis of the directionally detected signals is preferable, and this can be achieved off line by use of a slow-speed, swept-frequency bandpass filter, or on line by fast-Fourier transform or other techniques. Inspection of the frequency spectrum display is not always convenient, however, and it is sometimes more advantageous to process the signals to extract either the maximum or the mean frequencies in the spectrum and to display these as time-varying waveforms.

The main use of pulsed-Doppler techniques at present is in diagnosing arterial diseases, such as arteriosclerosis. The aim is to determine how and to what extent the diseased artery may be affecting blood flow. For this purpose, the Doppler signals are studied to reveal the shape of the blood-flow pulse created each time the heart beats. By aiming ultrasound beams at the input and output of the suspect arterial segment, the physician can determine the arrival times of the blood-flow pulse and the variation with time in the velocity of the blood. If the artery is constricted, the pulse becomes damped and altered in shape. An index derived from the pulse shape, called the pulsatility index, can indicate the presence and degree of disease.

In the past few years, through more sophisticated computerized methods of analyzing the waveforms of blood-flow pulses observed with Doppler techniques, doctors have been able to detect arterial disease at earlier stages. One new mode of analysis involves comparing the waveform of the blood-velocity pulse with a data bank of stored waveforms, each known to be typical of a given arterial condition. By a mathematical technique known as principal-components analysis, each feature in the pulse waveform can be measured and the stored waveform most similar to it automatically selected. By use of Laplace transforms, the pulse waveforms can be used as a direct measure of how stiff the arteries have become and to what extent they impede the flow of the patient's blood.

Sometimes determining the range alone is not enough to locate accurately the diseased site in an affected artery, and then it is useful to have a two-dimensional map of the blood vessels. Such images can be obtained by a scanner in which the probe of a Doppler system is mounted on a two- or three-dimensional coordinate-measuring system. When the ultrasonic beam passes through moving blood, as the probe is manually scanned over the skin, the Doppler detector generates an output signal. This signal either switches on the writing beam of a storage cathode-ray-tube display (or its digital equivalent) or color-codes the display according to the Doppler-shift frequency (which is proportional to the blood-flow velocity). This produces a two-dimensional map of the blood-vessel lumen—the hollow portion through which blood flows. When obtained with a continuous-wave Doppler system, such an image is essentially a plan view; sectional images can be obtained using range-gated pulsed-Doppler systems.

Duplex scanning: a sophisticated approach

The most sophisticated contemporary form of Doppler investigation involves combining pulse-echo B-scan imaging with pulsed-Doppler techniques [Fig. 2]. With this combination, called duplex scanning, a physician can locate a vessel with the imager and then probe within it or along it to determine blood-flow characteristics.

In principle, the simplest type of duplex scanner has separate transducers for imaging and for Doppler studies. A typical arrangement consists of an electronically switched linear-array, real-time scanner with a Doppler transducer mounted at one end, so the Doppler beam intersects the real-time scan plane at either a fixed or an adjustable angle. A common arrangement in use at the present time consists of a mechanical real-time sector scanner with an ultrasonic beam that can be held stationary in a desired position for Doppler studies; real-time imaging and Doppler measurement cannot be accomplished simultaneously with this method.

Simultaneous imaging and Doppler measurements are possible, however, with the "agile" ultrasonic beam of a steered-array scanner, since the position of the beam can be changed suddenly without having to overcome mechanical inertia. (The same approach is used for simultaneous pulse-echo imaging and M-mode studies.) Although the Doppler beam can be electronically steered off the central axis, noise in the delay lines may limit the performance of the system, so some of these instruments only operate with the Doppler beam aimed straight ahead from the array.

Duplex methods, which have come into general use only in the 1980s, have already proved valuable in a number of applications. For example, it is possible to locate and assess shunts communicating between the left and right sides of the heart. There is a quantitative relationship between the blood-flow velocities through the cardiac valves and the pressure drops across them; duplex scanning may greatly reduce the need for invasive cardiac catheterization. Even cardiac numbers can be studied by observing blood-flow turbulence.

Measuring blood-flow rate

Perhaps the most advanced application of Doppler-imaging ultrasound is in the measurement of blood-flow rate, an important parameter in many medical situations. There are three approaches to the problem.

The first is based on the multigated Doppler imaging system, arranged to determine the orientation of the blood vessel (to allow correction for the velocity vector) and the blood-flow velocity profile, which is assumed to have circular symmetry. The physician calculates volume flow by integrating the contribution of all the annuli forming the velocity profile.

In the second method, the diameter of the blood vessel is measured by pulse-echo ultrasound, and the average velocity is estimated from Doppler signals collected at known angles. It is assumed that the blood vessel is uniformly exposed to sound waves and that consequently the power density of the frequency spectrum is proportional to the blood-flow volume.

In the last method, the total backscattered power and spectral distribution of the Doppler signals from all the blood moving in the vessel are measured (again assuming uniform sound exposure). The instrument computes the volume flow by correcting for the attenuation in the intervening tissue. This attenuation is estimated from a measurement of the backscattered power from

[2] An ultrasonic duplex scanner combines real-time two-dimensional imaging with a pulsed-Doppler system for blood-flow studies. In this instrument (A), manufactured by Johnson & Johnson Ultrasound of Englewood, Colo., the direction of the ultrasonic beam is controlled by hand. Diagram (B) shows the general principles of the system. The region of interest identified by the operator on the real-time display is linked to the scanner motor servo control by means of the video overlay electronics, and Doppler information is then obtained by "freezing" the real-time image.

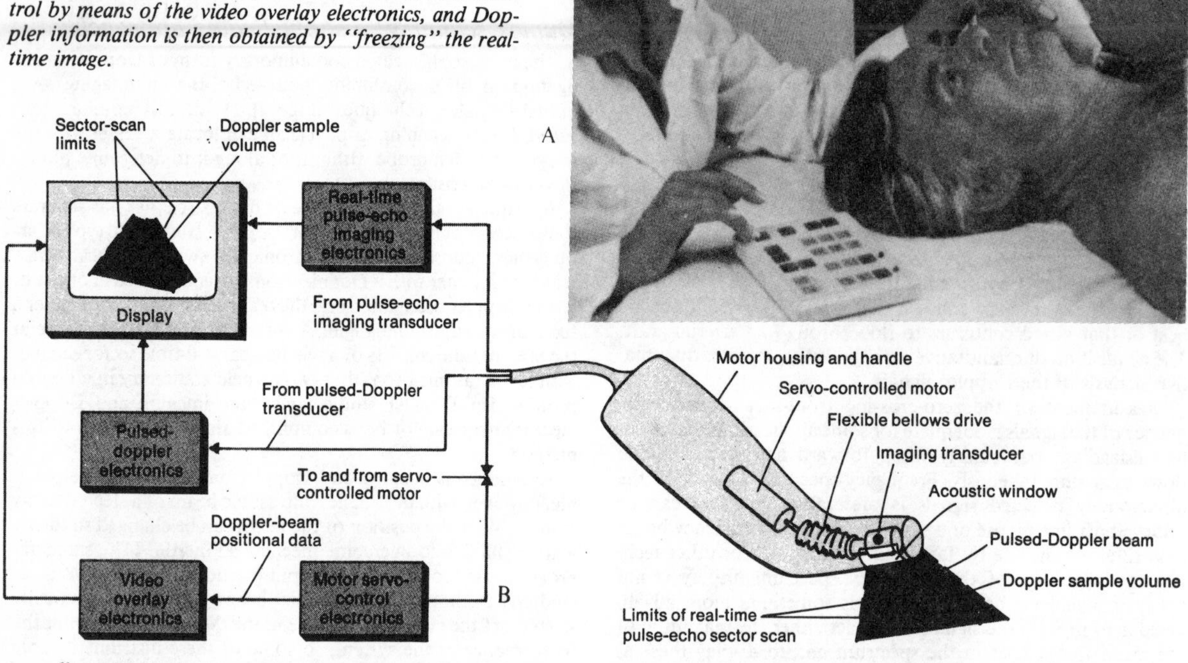

a small sample volume positioned entirely within the blood vessel.

The first two methods have been developed for clinical application. The errors associated with the measurements, however, are rather large. An error of 10 percent in the diameter measurement leads to an error of more than 20 percent in the volume estimate, and a similar error arises if the angle is 10° in error when it is thought to be 45°. Lack of uniform flow velocity and of uniform sound exposure is a further source of error with the first two methods. The third method has not yet been fully implemented even in the laboratory, but it should give good results with intermediate-size blood vessels.

Blood-flow-volume measurements may turn out to be useful in monitoring high-risk pregnancies. By measuring the amount of blood flowing to the fetus in the umbilical vein, an obstetrician can get advanced warning of abnormal blood supplies that could, if prolonged, lead to low birth weights in infants. Vascular surgeons can also benefit from blood-flow-volume measurements. Research is beginning into the measurement and characterization of blood flow in many other vessels. For example, flow patterns in the abdominal vessels promise to give insight into stomach, liver, and kidney functions.

3-D techniques being explored

While Doppler techniques are now being used clinically, another advance—three-dimensional imaging—is still in the research stage. There are two approaches to 3-D imaging, and there are difficulties with both. One simply generalizes the two-dimensional pulse-echo scanning technique, using either a series of 2-D scans next to each other or a two-dimensional array of transducers. Since each scan is made up of a series of range-gated echoes, the volume of data to be processed is substantial, as are the difficulties in displaying it.

For 3-D imaging from 2-D scans, the simplest approach is to scan in parallel planes separated by small increments. Anatomical access is sometimes limited, however, by overlying structures, such as the lung, which covers much of the heart. In such cases, the 2-D scan planes cannot all be parallel, but 3-D images—at least of structure outlines—can be displayed isometrically by computer processing of 2-D scan planes. The planes are randomly oriented and measured by sensors, such as those employing spark generators and orthogonal linear microphones.

The other approach is to use ultrasound not for reflection echoing, but for transmission imaging, much as X-rays are used. X-ray computer axial tomography (CAT) has already led to the production of 3-D images and, at least theoretically, ultrasound transmission measurements can be used in the same way. In transmission imaging, the amount of ultrasound transmitted through the body or the time it takes to penetrate the body are measured, not the time of echo return.

The results of early experiments on transmission imaging of soft tissues were disappointing. Apparently this was because coherent ultrasound was used with phase-sensitive detectors; the images were greatly degraded by interference caused by small differences in propagation speed along different ray paths in the beam. (Partly for the same reason, acoustical holography has failed in medical imaging.) This problem is significantly reduced by using spatially and temporally incoherent ultrasound produced by a random array of transducers excited by uncorrelated transmitters. It is clear that the method now has potential clinical value, particularly in the examination of the skeleton and joints of the newborn child and in the diagnosis of mammary disease.

Ultrasonic computer axial tomography involves acquiring a complete set of transmission profiles—time-of-flight or attenuation—through the plane of interest and subsequently reconstructing the two-dimensional cross-sectional image representing propagation speed or attenuation for these profiles. The reconstruction process is related to the filtered back-projection technique developed for X-ray CAT scanning. Although only a few structures, such as the female breast, are accessible to this approach—penetration in every direction is ideally required—encouraging results are being obtained.

Determining histology

Another area of active research is in the use of ultrasound to characterize tissues and determine parameters other than those obtained by imaging. Ultrasonic pulse-echo images are qualitatively characterized by a granular pattern known as a "speckle" that varies from one location to another and fluctuates in time as a result of motion. This speckle is derived from coherently forming the echo from many small scatterers within the resolution cell. Consequently the texture of the image of a given tissue structure depends on several properties of the imaging system, such as the frequency, the scanning pattern, and the geometry of the resolution cell.

The texture does not bear a direct relationship to the properties of the tissue, although different tissues generally give rise to different image textures. With experience, users become skilled in interpreting the textures produced by scanners in different clinical conditions. This is a qualitative form of what is known as tissue characterization, the process of identifying histology from ultrasonic information.

Ultrasound imaging techniques

Technology	First research	First clinical use	Clinical applications	Comments
Transmission imaging	1947	—	—	Under development
A-scan	1950	1956	Distance measurements	Adjunct to other techniques
Manual B-scan	1952	1963	Internal medicine, obstetrics, urology	Declining in use
M-mode	1954	1961	Cardiology	Used with real-time imaging
CW Doppler	1957	1964	Blood-flow studies	Simple system; high velocities of blood flow
Real-time scanning	1967	1974	Cardiology, internal medicine, obstetrics, ophthalmology, urology	Mechanical and electronic systems
Directional Doppler	1967	1967	Blood-flow studies	Mainly in vascular laboratories
Transducer arrays	1968	1972	Cardiology, internal medicine, obstetrics, urology	Real-time imaging
Pulsed Doppler	1969	1973	Blood-flow studies	Provides distance measurements
Acoustic microscopy	1936	1974	Histopathology	Under development
Doppler imaging	1971	1971	Blood vessels, cardiology	Under development
Computer axial tomography	1974	—	Breast	Under development
Duplex scanning	1977	1978	Blood vessels, cardiology	Next generation under development

Quantitative approaches to texture identification are just now being tried. So far the method is only applicable to relatively large areas of more or less homogeneous tissues. Such regions of interest can be identified by the clinician, and a computer can be used to determine statistical descriptors of the textures. A learning program is used to assemble a data base of the textures corresponding to normal tissues and to tissues with known pathologies. Comparison of suspect textures with the data base then allows identification of the corresponding histology within defined limits of probability.

Three properties of tissue are candidates for quantitative tissue characterization. These are the propagation speed (which tends to increase with increasing protein content and with decreasing water content), the attenuation (which increases with increasing propagation speed), and the scattering characteristics (which reflect the inhomogeneity and structure of the tissue). Speed can be measured either by transmission techniques, such as ultrasonic CAT, or by spatial cross-correlation of sets of pulse-echo images showing identifiable features scanned from different directions. Attenuation can also be measured by ultrasonic CAT. But when through-transmission cannot be achieved, measuring attenuation by reflection methods is the only practicable approach.

There are essentially two techniques for doing this. In the first, the decrease of echo amplitude with increasing range is measured, and the calculation assumes that the scatterers are uniformly distributed along the ultrasonic path. The second involves measuring the downward shift of the echo-pulse-frequency spectrum. This downward shift takes place because the attenuation of the pulse is proportional to the frequency; the ultrasonic pulse is assumed to have a Gaussian frequency spectrum. These methods are just beginning to emerge from the laboratory and to enter clinical evaluation.

Better resolution, sensitivity spur applications

Ultrasound applications have recently been growing simply because of the incremental improvements in ultrasound resolution and sensitivity. For example, ultrasound images of fetuses now allow obstetricians to make detailed examinations of the face and to detect defects, such as cleft lip, as early as the twenty-eighth week. Advanced knowledge of some defects can be of value in planning termination of pregnancy or delivery and early postnatal corrective treatment.

Although the main use of ultrasound is in diagnosis of abnormalities of structures and organs in the living body, it is also of use in research through ultrasound microscopy. One advantage of ultrasound microscopes is that they can visualize living cells. The staining necessary for optical microscopy and the metal coating needed for electron microscopy both kill the cells being studied. Ultrasound microscopes exploit the differences in attenuation of various parts of the cell, which, at high frequency, depend on differences in viscosity. No contrast media are necessary and thus living cells can be studied.

Ultrasonic microscopes typically operate at frequencies in the 10-to-2000-MHz range. The corresponding wavelengths in water and soft tissues are 150 to 0.75 micrometers. Thus the resolution of a 2-GHz ultrasonic microscope using a wavelength of 0.75 micrometer is comparable to that of a light microscope (the eye is most sensitive to light with a wavelength of about 0.55 micrometer).

Two kinds of ultrasonic microscopes have been developed. The scanning acoustic microscope (SAM) is the simpler. The specimen is mounted on a thin membrane fixed to a support, which is moved in a raster so that every element of the specimen passes in turn through the focal region of an ultrasonic beam. The beam is focused without aberration by a sapphire lens, and a liquid coupling medium is used. The microscope can operate in the reflection mode, with the same transducer acting as both transmitter and receiver, or in the transmission mode, with separate transmitting and receiving transducers on each side of the specimen and with coincident foci.

The second instrument is the scanning laser acoustic microscope (SLAM). This approach uniformly arranges the transmitted ultrasound waves so that they fall on one surface of a transparent plastic plate. The specimen can be anything from a thin tissue section to a small organ or even a small animal. The other side of the plate is scanned in a raster by a fine beam of laser light, and the ultrasonic disturbance of the surface of the plate is detected by optical interferometry to produce an electrical video signal that drives a TV display. An optical micrograph can be displayed simultaneously through a CCTV system by means of transmission of light through the specimen.

In principle, the image frame rate of the SLAM is faster than that of the SAM but the system cannot operate at frequencies higher than about 100 MHz because of the attenuation in the relatively long ultrasonic path.

Applying ultrasound to treat ailments

One of the most widespread applications of ultrasound in medicine is in physical therapy. Ultrasonic waves are thought to be helpful in relieving the symptoms of many unpleasant ailments, such as joint contracture, acute back and shoulder pains, and scar-tissue aches. Physical therapists decide on the best treatment for any particular clinical condition by applying rules that have been built up over a long time. In general, the advantage of ultrasound is that it can produce heat distributions not easily achievable by radio frequency or microwave radiation, the main alternative methods of deep heating. A typical course of ultrasonic physical therapy consists of a week of daily treatments, each lasting about 15 minutes, with frequencies between 1 and 3 MHz and intensities between 0.25 and 3 watts per square centimeter.

Recently interest in the treatment of cancer with hyperthermia has revived. Applying heat, either alone or simultaneously or in sequence with ionizing radiation or cytotoxic drugs, often enhances the damage to the tumor. There is only a narrow range of temperature, about 42° to 44°C, over which this is effective; normal body temperature is 37°C. The challenge is to design an ultrasonic irradiation scheme that heats the tumors to more or less uniform temperature but does not dangerously raise the temperature of neighboring normal tissues. Ultrasonic beams of megahertz frequency, with flat profiles, scanned focused beams, and multiple entry ports, may be used. The main problem, as with other methods of producing localized hyperthermia, is that of tumor thermometry for feedback control of the heating conditions.

'Trackless' surgery possible

The applications of ultrasound to surgery can broadly be considered in two classes, according to the ultrasonic frequency employed. First, ultrasound of megahertz frequency can be used to deposit energy in localized sites in much the same way as in hyperthermia treatment. The ultrasound may be focused, and the region of damage, or lesion, produced may be beneath intervening tissue that remains undamaged, because it is exposed to intensities lower than those at the focus. Thus it is possible to produce "trackless" lesions—lesions with no associated track caused by the passage of an instrument through surrounding tissue.

When arranged to produce lesions in the brain, the frequency is usually in the range of 1 to 3 MHz and the exposure conditions at the focus are chosen at between 20 kilowatts per square centimeter for 300 microseconds to 200 watts per square centimeter for 10 seconds. At the higher intensities, the process of lesion production includes a contribution from direct mechanical action, such as cavitation, in addition to heat.

Although focused ultrasonic surgery does have unique advantages, it is complicated, inconvenient, and expensive. The original incentive for the development of the method was for the treatment of certain behavioral disorders and for the destruction of deep-seated structures in the brain, such as the pituitary gland, which sometimes controls the development of secondary cancer tumors. Nowadays, however, neurosurgeons prefer to use more conventional techniques to treat their patients, and drugs have been developed that can control the symptoms of some previously intractable diseases.

Unfocused 3-to-10-MHz ultrasound has some specialized surgical applications. Direct application to the round window via the middle ear, or to a semicircular canal exposed by surgery, can control some of the symptoms of Meniere's disease, which is characterized by deafness and vertigo, possibly because of excessive pressure in the semicircular canals of the internal ear. Ultrasound can eradicate residual papilloma in the larynx. Other possibilities under investigation include the treatment of glaucoma and myopia, temporary contraceptive sterilization of the male, and termination of pregnancy.

Endoscopic ultrasonic disintegration of kidney, ureteric, and bladder stones is emerging as an attractive alternative to open surgery. A rigid metal rod of about 3 millimeters in diameter is used to apply ultrasound in the frequency range of 22 to 30 kHz directly to the calculus by means of endoscopic access—if necessary through a small incision in the abdominal wall. The excursion amplitude of the tip of the vibrating rod is too small to damage soft tissue with which it may come into contact. The stone is methodically broken up into fragments small enough to be removed by aspiration through a tube mounted beside the endoscope or to be passed naturally.

As an alternative to endoscopic ultrasonic disintegration, a machine called an ultrasonic lithotripter has been developed that can disintegrate kidney stones by focused shock waves applied through the skin under X-ray imaging control. The shock waves are produced by an underwater spark generator mounted at the near focus of an elipsoid; the machine is positioned so that the stone is at the far focus. A course of treatment may last 20 to 30 minutes, during which time 500 to 1000 shocks are applied to break the stone into fragments small enough to be passed without acute pain.

To probe further

A book that comprehensively describes the development of all aspects of ultrasound through the end of 1975 is *Biomedical Ultrasonics*, by P.N.T. Wells, published by Academic Press in 1977. For descriptions of progress that has been made since then, see the proceedings of the last World Congress of Ultrasound in Medicine and Biology, *Ultrasound '82*, edited by R.A. Lerski and Patricia Morley, published by Pergamon Press in 1983. *Ultrasonics in Clinical Diagnosis*, ed. by B.B. Goldberg and P.N.T. Wells, published by Churchill Livingstone in 1983, is an up-to-date overview of the most important medical applications.

The basic principles behind ultrasound are also covered in *Diagnostic Ultrasonics: Principles and Use of Instruments* by W.N. McDicken (published by Crosby Lockwood Staples of London in 1976) and in *Ultrasonic Imaging and Holography: Medical, Sonar, and Optical Applications*, edited by G.W. Stroke, W.E. Kock, Y. Kikuchi, and J. Tsujiuchi and published by Plenum Publishing Corp. of New York City in 1974. Plenum also publishes the series *Acoustical Imaging*, of which the latest book, Vol. 13, was published last August (Kaveh, Muller, and Greenleaf, eds.).

The IEEE journals *Biomedical Engineering* and *Sonics and Ultrasonics* often carry articles on medical ultrasound. Specialized journals include *Ultrasound in Medicine and Biology*, *Ultrasonic Imaging*, and Churchill Livingstone's serial publication, *Clinics in Diagnostic Ultrasound*, which comes out regularly with definitive reviews.

About the author

Peter N.T. Wells is a Fellow of the Institute of Physics and of the Institution of Electrical Engineers. He was elected a Fellow of the Fellowship of Engineering in 1983 and is an honorary Fellow of the American Institute of Ultrasound in Medicine. He was professor of medical physics in the Welsh National School of Medicine from 1972 to 1974 and is at present chief physicist to the Bristol and Weston Health Authority in England. He is the author or editor of nine books and of more than 140 articles, mainly on the medical applications of ultrasonics. He holds Bachelor of Science, Master of Science, and Ph.D. degrees in, respectively, electrical engineering, physics, and zoology. ◆

Reprinted with permission from *Medical Physics*, **11**, (1) pp. 1–14, K.F. King and P.R. Moran, "A unified description of NMR imaging, data-collection strategies, and reconstruction."

REVIEW ARTICLE

A unified description of NMR imaging, data-collection strategies, and reconstruction

Kevin F. King[a)]

Departments of Physics, Radiology and Medical Physics, University of Wisconsin, Madison, Wisconsin 53706

Paul R. Moran

Department of Radiology, Bowman Gray School of Medicine, Winston–Salem, North Carolina 27103

(Received 10 December 1982; accepted for publication 24 May 1983)

In nuclear magnetic resonance (NMR) imaging by the zeugmatographic methods, there is a common and unified theoretical description. All forms of two-dimensional and three-dimensional imaging involve NMR data which trace various geometric representations, in reciprocal transform space, of the subject's spatially blurred "effective" spin density. The effective density is proportional to the physical density modulated spatially by the several factors of receiver coil (B/I) ratio, rf pulse excitation terms, T_1-relaxation terms, and T_2-relaxation terms. These factors depend upon the rf pulse sequence and field-gradient modulation sequence, and they may be calculated according to some model or directly measured. From this viewpoint, all different imaging modes appear as variations in data-collection and image-reconstruction strategies. The results are used here to describe slice-oriented polar and Cartesian strategies, three-dimensional Cartesian and two forms of spherical strategies, and multiecho strategies of the "planar-echo" type.

I. INTRODUCTION

In nuclear magnetic resonance (NMR) experiments, it had long been known that temporal magnetic field modulations[1] were useful in phase encoding the detected signals. It also was shown that phase modulation of exciting rf fields was identical with field modulations, not only classically, but even for rigorous quantum descriptions.[2]

The concept of establishing a *spatially* varying field was proposed by Lauterbur[3] to link temporal frequencies to spatial encoding and thereby to enable the formation of NMR images of a subject in a high-resolution, direct, and relatively rapid NMR scan. For this linking, or "yoke," between temporal code and spatial information, the term "zeugmatography" arose[4] to denote this form of NMR imaging.[5] A comprehensive bibliography of references, reviews, and descriptive articles has been given by Thomas.[6]

In the current applications, one finds many variations and modulations of Lauterbur's original concept. An excellent description of basics is given by Mansfield and Morris.[7] It would seem, at first, from the current literature that various techniques of NMR imaging are quite truly physically and mathematically different and that each must be understood in its own specific context. The purpose of this paper is to present a development to make explicit a single unified description of all NMR imaging, from which the different methods all may be understood in a simple common viewpoint. According to this unified description, the various NMR-imaging "methods" are shown to be different data-gathering strategies and minor algorithmic variations of a single, general, transform inversion of reciprocal-space data to reconstruct a spatial image.

In the following sections, we review some basic NMR phenomena and develop notation, then describe the general nature of NMR-imaging signal detection. Next, we develop the fundamental NMR-imaging equations as a general reciprocal-transform expression. Finally, we apply this viewpoint to describe two-dimensional (2-D) polar, 2-D Cartesian, three-dimensional (3-D) spherical (single and double pass), 3-D Cartesian, and 2-D and 3-D multiecho data-collection and image-reconstruction strategies.

II. THE IMAGING EQUATIONS

A. Review of basic relations

In the NMR-imaging system, the subject is placed in a relatively strong external magnetic field $\mathbf{H}$,

$$\mathbf{H} = H_0\hat{z} + \mathbf{G}\cdot\mathbf{r}\hat{z}. \tag{1}$$

In Eq. (1), H_0 represents a dominant static homogeneous field, and three orthogonal coil sets produce small field gradients represented by the $(\mathbf{G}\cdot\mathbf{r})$ term. These fields are directed along the same axis in space (which we denote the z direction) as H_0, but their change-of-field sense (the gradient directions) and relative strengths are controllable

$$\mathbf{G} = G_x(t)\hat{x} + G_y(t)\hat{y} + G_z(t)\hat{z}, \tag{2}$$

so that $\mathbf{G}$ can be modulated temporally both in magnitude and direction.

The nuclear spin system of the subject evolves toward thermodynamic equilibrium after application of H_0. A particular spin isotope gains net magnetization along the field direction with the relaxation rate $W_1 = (1/T_1)$,

$$m_z(t) = m_{\text{init}} \exp(-W_1 t) + m_{\text{EQ}}[1 - \exp(-W_1 t)], \quad (3)$$

where m_z is the magnetization (spin magnetic moment per unit volume), m_{init} is any initial component existing at the instant defining $t = 0$, and m_{EQ} is the thermodynamic equilibrium value,

$$m_{\text{EQ}} = 2\pi\gamma\hbar[I(I+1)/3][\hbar\omega_L/kT]\rho. \quad (4)$$

In Eq. (4), ρ is the physical number density of spins, and ω_L is the precessional angular Larmor frequency, related to the spin isotope's gyromagnetic ratio γ (Hz/T) and the field strength H by

$$\omega_L = 2\pi\gamma H. \quad (5)$$

The gradient field values (G·r) are small compared with H_0 (typically 10^{-3} across the subject) so that

$$[\hbar\omega_L/kT] \simeq [\hbar\omega_0/kT] = [2\pi\gamma\hbar H_0/kT]. \quad (6)$$

The rms spin-level count $[I(I+1)]^{1/2}$ in Eq. (1) involves the isotope's spin number I; the entropy of alignment is the Boltzman factor $[\hbar\omega_0/kT]$ [typically 10^{-5}–10^{-6} for frequencies and temperatures (T) of interest in medical imaging].

Carefully timed pulses of rf magnetic field from an exciting transmitter can be applied to the system. The Larmor sense rotating component,

$$H_1[\cos(\omega_{\text{rf}} t)\hat{y} - \sin(\omega_{\text{rf}} t)\hat{x}],$$

resonantly influences the spin system when ω_{rf} is close to ω_L and generates, from m_z, a magnetization component, $m_\perp$ directed transverse to H_0, i.e., in the (x,y) plane, where it precesses subsequently,

$$m_x(t) = m_\perp \cos(\omega_L t + \phi) \quad (7)$$

and

$$m_y(t) = m_\perp \sin(\omega_L t + \phi). \quad (8)$$

The initial phase ϕ is defined by the specifics of the excitation pulse. Since $m_\perp$ is a nonequilibrium component, internal processes will act to cause its "dephasing" decay at some rate $W_2 = (1/T_2)$,

$$m_\perp(t) = m_\perp(0)\exp(-W_2 t). \quad (9)$$

Any internal process which restores m_z at rate W_1 also contributes to the decay of m_z. All W_1 processes require energy transfer from the environment in quanta of $\hbar\omega_L$, but other processes can also contribute to W_2 so that

$$W_2 = (1/T_2) = W_1 + \text{"other"}. \quad (10)$$

In general, (i.e., in solids) these "other" processes do not give the simple exponential "free-induction-decay" (FID) shape expressed in Eq. (9). However, in the "motionally narrowed case" $\exp(-W_2 t)$ is a very good approximation for homogeneous materials. In the liquid-state systems typifying subjects of medical interest, motional-narrowing limits are reached and Eq. (9) applies to any homogeneous region.

It is important that both W_1 and W_2 are exquisitely sensitive to both the mean-squared strengths and the "correlation-time" dynamics of environmental interactions. Different tissues in a biological subject, and even different microscopic locales (e.g., intracellular or extracellular environments) in a given uniform tissue, will exhibit different T_1's and T_2's. Thus, macroscopically observed relaxation and decay will be the composite, e.g., the net magnetic *moment* M_z from a macroscopic volume relaxes via Eq. (3) (with $M_{\text{init}} = 0$) according to

$$M_z(t) = \Sigma_v\{m_{\text{EQ}}(i)[1 - e^{-W_{1i}t}]\Delta V_i\}, \quad (11)$$

where ΔV_i is a volume element of uniform $m_{\text{EQ}}(i)$ and characterized by a particular T_1 rate W_{1i}.

B. Signal detection

The precessing magnetizations of Eqs. (8) and (9) generate Larmor-frequency dipolar field variations which link the turns of a receiving coil. This coil is part of a hi-Q-tuned receiving circuit, and the voltages V induced across the circuit are related to the coil's passive field-to-current ratio (B/I),[8] the circuit Q, the mean Larmor frequency, and the *total* magnetic moment in the coil volume by

$$V(t) = Q\omega_0(B/I)M_x(t), \quad (12)$$

where we have taken the x axis to be the receiving coil axis. From Eqs. (8) and (9), we can write the circuit voltage as

$$V(t) = Q\omega_0(B/I)\int m_\perp(\mathbf{r},t)\cos[\omega_L t + \phi]\,d^3r. \quad (13)$$

The circuit signal is input to an rf receiver amplifier of low-noise figure, and is amplified and passed to phase-coherent demodulators. The reference input to the demodulator is taken from the stable oscillator circuit which generates ω_{rf}. Two such demodulators (also called "product" detectors, single-sideband detectors, or heterodyne detectors) have their reference phase adjusted at 90° to each other; the two[9] form a quadrature detection pair phase adjusted so that

$$S_1(t) = AQ\omega_0(B/I)\int m_\perp \cos[(\omega_L - \omega_{\text{rf}})t]\,d^3r, \quad (14)$$

and

$$S_2(t) = AQ\omega_0(B/I)\int m_\perp \sin[(\omega_L - \omega_{\text{rf}})t]\,d^3r, \quad (15)$$

where A is the net active electronics amplification factor. It is convenient to combine S_1 and S_2 in complex number notation

$$S(t) \equiv S_1 + iS_2$$
$$= AQ\omega_0(B/I)\int m_\perp(\mathbf{r})e^{i(\omega_L - \omega_{\text{rf}})t}\,d^3r. \quad (16)$$

C. Image-reconstruction relation

If we now return to Eqs. (1) and (5), so as to express ω_L in terms of the "static" fields applied, we obtain from Eq. (16),

$$S(t) = [AQ\omega_0(B/I)]\int m_\perp(\mathbf{r},t)e^{i(\omega_0 - \omega_{\text{rf}})t}\,e^{2\pi i\gamma\mathbf{G}\cdot\mathbf{r}t}\,d^3r. \quad (17)$$

In Eq. (17), $m_\perp(\mathbf{r},t)$ is written explicitly time dependent to denote intrinsic T_2 decays or other dephasing factors due to motion of spins in the applied gradient fields G·r. It is possible to adjust and maintain H_0 so that the off-resonance phase term $(\omega_0 - \omega_{\text{rf}})t$ is kept negligibly small over the time duration of the measurement. Any residual time dependence arising from $(\omega_0 - \omega_{rf})$ detuning, we shall simply absorb implicitly into the $m_\perp(\mathbf{r},t)$ behavior. Then we have the detected signal $S(t)$ expressed by

$$S(t) = [AQ\omega_0(B/I)] \int m_\perp(\mathbf{r},t) e^{-2\pi i \mathbf{q}\cdot\mathbf{r}} d^3r, \quad (18)$$

where

$$\mathbf{q} \equiv -\gamma \mathbf{G} t. \quad (19)$$

We observe the fundamental relation of NMR imaging in Eq. (18); the phase-coherently detected signal [the net free-induction-decay $S(t)$ which may be directly observed on an oscilloscope trace following rf excitation] is proportional to the *spatial* Fourier transform $\tilde{m}(\mathbf{q})$ of the magnetization,

$$\tilde{m}(\mathbf{q}) \equiv \int m(\mathbf{r}) e^{-2\pi i \mathbf{q}\cdot\mathbf{r}} d^3r. \quad (20)$$

That is, if we define $\mathbf{q} \equiv -\gamma \mathbf{G} t$ and use the notation

$$\tilde{S}(\mathbf{q}) \equiv S(t) \quad \text{with } t = (q/\gamma G), \quad (21)$$

then Eq. (18) gives

$$\tilde{S}(\mathbf{q}) = \text{detected signal pair} = [AQ\omega_0(B/I)]\tilde{m}_\perp(\mathbf{q}). \quad (22)$$

By obtaining a sufficiently dense set of all $\mathbf{q}$ values for $\tilde{S}(\mathbf{q})$, the Fourier transform can be inverted numerically to produce an image of $m_\perp(\mathbf{r})$. One accomplishes this by repeating many cycles of $S(t)$ measurement, with $m_\perp$ excited identically in each and $[AQ\omega_0(B/I)] = constant$, but with $\mathbf{G}$ changed from cycle to cycle in effective direction.

In practice, $S(t)$ passes to an "integrate-sample-and-digitize" analog-to-digital converter (ADC) stage. $\tilde{S}(\mathbf{q})$ is sampled on a discrete mesh of $\mathbf{q}$ such as that the increment, Δq obeys the sampling theorem by being less than the inverse dimension D^{-1} of the subject to be imaged. That is, along any imaged dimension,

$$|\Delta q| < D^{-1}. \quad (23)$$

In practice also $\mathbf{q} = \gamma \mathbf{G} t$ is neither measured nor used in calculation to infinitely long times (infinitely large q). In hardware and software, the data set is apodized by some bandlimiting function $\tilde{\mathscr{H}}_A(q)$ to give an image-data set $\tilde{\mathscr{I}}(\mathbf{q})$,

$$\tilde{\mathscr{I}}(\mathbf{q}) = \tilde{\mathscr{H}}_A(q)\tilde{S}(\mathbf{q}). \quad (24)$$

The general principle then is that numerical Fourier inversion of the data set reconstructs the image,

$$\mathscr{I}(\mathbf{r}) = \int \tilde{\mathscr{I}}(\mathbf{q}) e^{2\pi i \mathbf{q}\cdot\mathbf{r}} d^3q = \int \tilde{\mathscr{H}}_A(q)\tilde{S}(\mathbf{q}) e^{2\pi i \mathbf{q}\cdot\mathbf{r}} d^3q. \quad (25a)$$

Thus $\mathscr{H}_A(r)$, the transform of $\tilde{\mathscr{H}}_A(q)$ is a spatial convolving or system point spread function, and Eqs. (22) and (25a) imply that

$$\mathscr{I}(\mathbf{r}) = \int \mathscr{H}_A(\mathbf{r} - \mathbf{r}') m_\perp(\mathbf{r}') d^3r' = \mathscr{H}_A \otimes m_\perp(\mathbf{r}). \quad (25b)$$

Consequently, the only differences between the various zeugmatographic imaging techniques are data collection strategies, whether $+q$ and $-q$ sets are obtained by appeals to the mathematical Hermitian properties of $\tilde{S}(\mathbf{q})$, by gradient reversal, or by spin-echo techniques, how the excitation and data-collection sequence sensitizes $m_\perp$ to dynamic quantities (such as relaxation rates), and finally whether the data set resulting is appropriate to spherical coordinate or Cartesian coordinate inversion or is truncated for two-dimensional planar reconstruction by the 2-D polar or 2-D Cartesian Fourier inversions. The terminology "back-projection technique" is used for polar transform numerical inversion, "(direct) fast Fourier transform (FFT)" is used for Cartesian inversion, and planar echo used in a technique which collapses all 2-D values for $\mathbf{q}$ onto a single temporal line for inversion; all methods actually are Fourier transform algorithms and can be intercompared and understood easily and powerfully from a unified Fourier transform viewpoint.

III. MAGNETIC MOMENT DYNAMICS

A review of nuclear spin dynamics in the presence of applied static and rf magnetic fields is helpful. The nuclear magnetic moment $\boldsymbol{\mu}$ of a nucleus is a dipole moment, and is directly related in direction and magnitude with the intrinsic angular momentum $\mathbf{L}$ of that isotope. An excellent description of the basic physics of NMR was given recently by Dixon and Ekstrand,[15] which is recommended for detail beyond the following abbreviated discussion. Any component of L takes on discrete values, e.g.,

$$L_z = I\hbar, (I-1)\hbar, \cdots, -I\hbar, \quad (26)$$

where I is the spin quantum number of the nucleus in question. The mean-squared value $\langle \mathbf{L}\cdot\mathbf{L}\rangle$ is always

$$L^2 = \langle \mathbf{L}\cdot\mathbf{L}\rangle = 3\langle L_z^2\rangle = 3\langle L_y^2\rangle = 3\langle L_x^2\rangle = I(I+1). \quad (27)$$

The moment $\boldsymbol{\mu}$ relates to $\mathbf{L}$ and the gyromagnetic ratio γ according to

$$\boldsymbol{\mu} = 2\pi\gamma \mathbf{L}. \quad (28)$$

The "forces" on $\boldsymbol{\mu}$ in magnetic fields $\mathbf{H}$ are torques and since $\boldsymbol{\mu}$ and angular momentum are linked, the resulting dynamics of $\boldsymbol{\mu}$ are gyroscopic precessions. When $\mathbf{H}$ is a static field, then the motion of $\boldsymbol{\mu}$ is simple. The component of $\boldsymbol{\mu}$ along $\mathbf{H}$ remains constant, and the component of $\boldsymbol{\mu}$ perpendicular to $\mathbf{H}$, (i.e., $\mu_\perp$), rotates around H at the Larmor frequency, $\omega_L = 2\pi\gamma H$. Consequently, when we view $\boldsymbol{\mu}$ in a frame of reference so that $\mathbf{H}$ in that frame has no explicit time dependence, then the spin dynamics are very easily described.

A. Rotating frame

We are concerned with large static fields, e.g., $H_0\hat{z}$, and a smaller (but resonant) rf magnetic field $H_1(t)$. The rf field which influences $\boldsymbol{\mu}$ is the Larmor sense rotating component of H_1 lying in the (x,y) plane. If we view the spin system in a frame of reference rotating about $H_0\hat{z}$ and phase locked to the rf frequency ω_{rf}, then neither H_0 nor H_1 have explicit time dependence. The effective $\hat{z}$ field h_0 (see Fig. 1) transformed to the rotating frame is

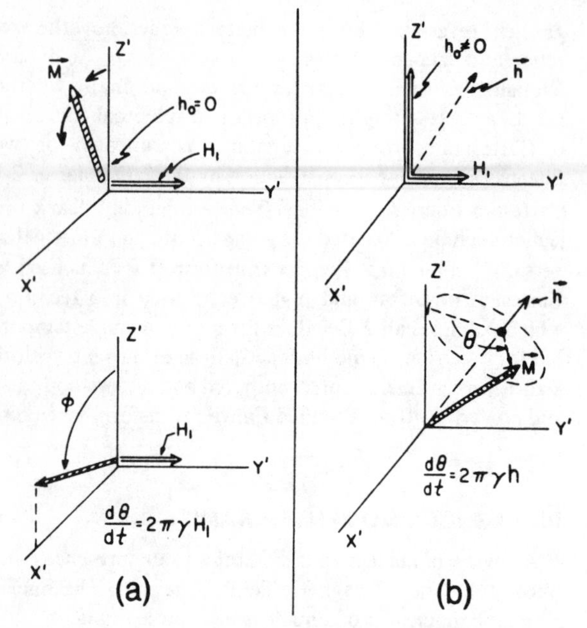

FIG. 1. (a) Motion of magnetization vector in rotating-frame-of-reference when $\omega_{rf} = \omega_L$ and (b) when $\omega_{rf} < \omega_L$.

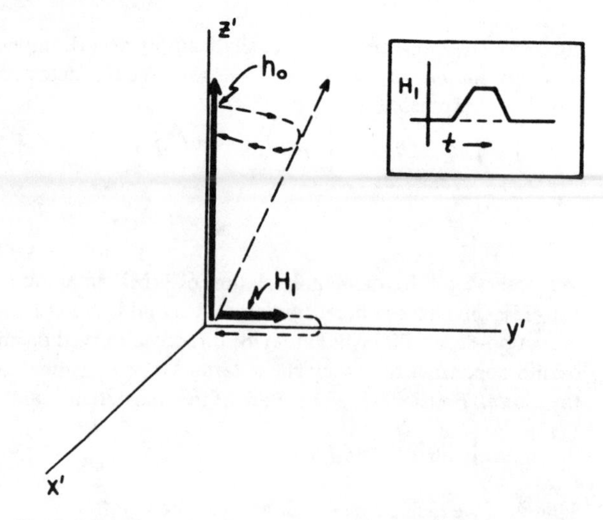

FIG. 2. Adiabatic motion of magnetization following **h** in the rotating frame.

$$h_0 = H_0 - (\omega_{rf}/2\pi\gamma), \tag{29}$$

and $\mathbf{H}_1$ is

$$\mathbf{H}_1 = H_1 \hat{y}', \tag{30}$$

where we simply assign the rotating coordinate axis $\hat{y}'$ to be defined by the direction of $\mathbf{H}_1$.

For example, suppose the magnetization **m** is originally in equilibrium along H_0, and H_1 is switched on at $t = 0$. In the rotating frame, if exact tuning gives $\omega_{rf} - \omega_L = 0$, then and the situation $h_0 = 0$ of Fig. 1(a) obtains. The magnetization vector precesses in the (z', x') plane about H_1 at a frequency ω_1,

$$\omega_1 = 2\pi\gamma H_1; \tag{31}$$

when $\omega_1 t = (\pi/2)$, **m** lies along x', and this condition is called a 90° pulse of rf. A pulse length for which $\omega_1 t = \pi$ is a 180° pulse, whereby **m** is inverted by rotating from $+\hat{z}$ to $-\hat{z}$.

Consider the nonresonant case where $\omega_L \neq \omega_{rf}$. Then the field **h** in the rotating frame is

$$\mathbf{h} = [(\omega_L - \omega_{rf})/2\pi\gamma]\hat{z}' + H_1 \hat{y}', \tag{32}$$

as shown in Fig. 1(b). The magnetization precesses in a cone about **h**; the dynamics in the laboratory frame of reference have also a rotational motion of frequency ω_{rf} about the z axis, i.e., the vector **h** defining the center of the cone precesses about the $\hat{z}$ axis at frequency ω_{rf} in the lab system.

A case of special interest in NMR imaging is where the excitation is off resonance, $\omega_{rf} \neq 2\pi\gamma H_0$, as in Fig. 2, but H_1 is "adiabatically" turned on and then off, as illustrated in the inset graph. In this case, the effective field **h** starts at $h_0 \hat{z}$ and slowly moves out to $h_0 \hat{z} + H_1 \hat{y}$, and then back. If this occurs on time scales which are long compared with $[2\pi\gamma h]^{-1}$, then **m** simply follows **h** in a very small precessing cone. When H_1 is maximum **m** lies along $h_0 \hat{z} + H_1 \hat{y}$, and when H_1 returns to zero, **m** is completely restored to its initial $\hat{z}$-direction location.

B. Spin-echo patterns

We now consider "spin-echo" phenomena. If different spins, at different spatial locations, experience different fields (e.g., due to gradient field application or intrinsic inhomogeneity of H_0), then low-field spins precess in the rotating frame accumulating negative relative phase, and high-field spins accumulate positive phase. This is illustrated in Fig. 3(a). The resulting NMR signal shows a free-induction-decay pattern (whose spectrum is the field-variation distribution) as sketched in Fig. 3(b). If the individual spins do not change location, then this field-inhomogeneity (T_2^*-decay) dephasing is reversible. If, for example, the rf transmitter is phase shifted by 90°, then H_1 will appear along the $\hat{x}'$ axis (instead of $\hat{y}'$), and a 180° pulse [which we will call a (180°)* pulse] of H_1 can be generated. Each dephased spin precesses about H_1 in a cone so that, if its accumulated phase was "$+\Theta$" before the pulse, then its phase immediately after is "$-\Theta$." Since the same field inhomogeneities exist as before, the system rephases to give a symmetric echo pattern reproducing the original FID. Such spin-echo sequences can be repeated indefinitely until the "true-T_2" irreversible phase decays obliterate $m_\perp$.

When the field inhomogeneity is purposefully introduced, e.g., by gradient field coils, then another spin-echo procedure is possible simply by switching the gradient-coil current to opposite gradient polarity. The phase accumulation rate everywhere reverses in sense and the spin system "refocuses" to give an echo pattern. The two procedures are almost equivalent; the echo from the (180°)* rf pulse reproduces the FID phase intact, whereas the odd-numbered echoes in gradient reversal give the "complex conjugate" of FID (recalling that the in-phase and quadrature-phase NMR signals are treated as a complex number pair). Field gradient reversals cannot recover an echo pattern from any static intrinsic field inhomogeneities, whereas the (180°)* rf pulse echo recovers all time-independent inhomogeneity phase scrambling. We note, here, that various combinations

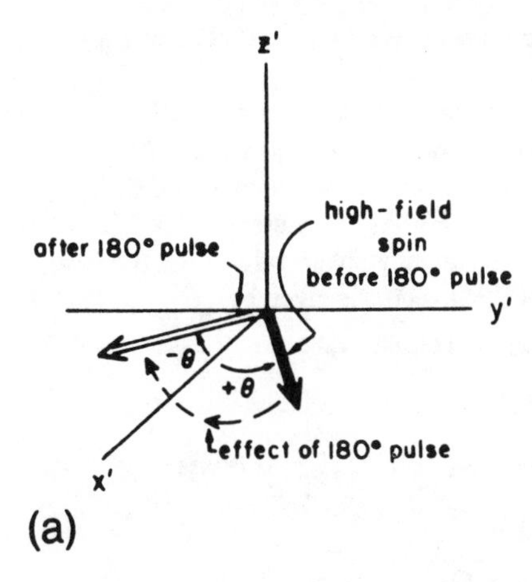

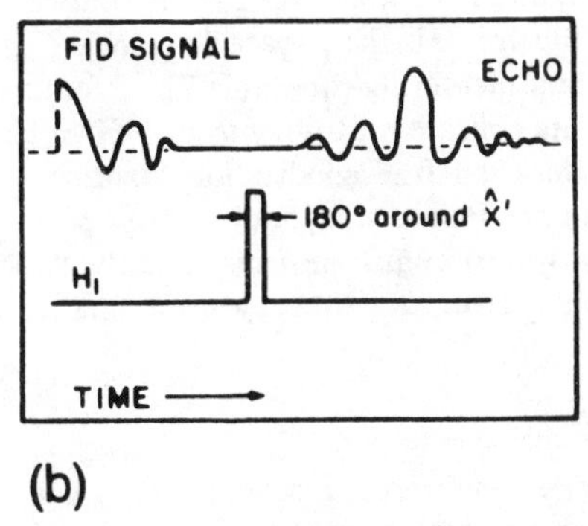

FIG. 3. (a) Motion of magnetization vectors off resonance and experiencing a 180° spin-echo rf pulse and (b) corresponding direct FID followed by echo FID.

of temporal sequence of gradient fields, field reversals, 180° rf pulses, etc., are possible. The number of ways the resulting NMR free-induction-decay pattern can thereby be encoded to extract physical and spatial information is limited apparently only by the imagination, experience, and patience of the system's operator. With the background of the preceding sections, one can categorize the families of NMR-imaging data collection and reconstruction strategies, and one can intercompare them for their relative strengths and weaknesses.

IV. TWO-DIMENSIONAL (PLANAR) STRATEGIES

Two-dimensional NMR-imaging (or planar data) strategies image a single slice by acquiring a two-dimensional data collection set. This gives a slice image similar in nature to that of x-ray computed tomography (CT) methods. Many slices can be acquired sequentially to produce stacked images for reformatted imaging, in principle, in any plane. In all slice-oriented NMR imaging, an initial slice selection sequence is applied such that the transverse magnetization $m_\perp$ is excited (and subsequently detected) only in a single tomographic plane. In discussions here, we denote the axis perpendicular to the image slice as the z axis, and the image plane to be the (x,y) plane. These arbitrary coordinates need not, of course, correspond to those used previously to discuss magnetic field directions.

Figure 4 illustrates one example of a workable slice selection sequence. First, ω_{rf} is set, relative to H_0, to determine a particular on-resonance plane. Next, a field gradient G_z is applied and an adiabatically shaped H_1 pulse generated. From the discussion of the previous section, we note that spins far from the on-resonance plane sense large values of effective **h** in the rotating frame. They will be tipped out and then restored to original m_z conditions at the H_1 termination. There will be a plane (adjustable by tuning ω_{rf} for fixed H_0) which has $\omega_{rf} = \omega_0$; for these spins $h_0 = 0$, only H_1 appears, and they are rotated into the transverse precessing state, $m_\perp = m_{z(\text{initial})}$.

By adjusting the G_z applied during this "adiabatic" shaped H_1 pulse, by controlling the maximum H_1, and the rf pulse-shape characteristics, a particular slice width, whose breadth we call W, and the distribution of $m_\perp$ within it can empirically be selected. Since this selection process occurs in a field gradient (i.e., G_z) inhomogeneity, not only will $m_\perp$ show a distribution in magnitude across W, but also some dephasing distribution. This dephasing can be recovered to some degree, producing a maximum in-phase distribution, by a "field-gradient-reversal-spin-echo." This is the purpose of the opposite-sense field gradient added to the trailing edge of G_z shown in Fig. 4. Usually, it is adjusted experimentally to recover the largest possible amplitude of the "in-phase" FID signal. In practice, the slice width can be adjusted from tenths of millimeters for W up to many centimeters. At this point, $m_\perp$ is processing at ω_0 and excited only in a single $\hat{z}$-axis slice of width W; all spins outside the slice have been returned adiabatically to their original m_z condition. We may also note here a modification; the shaped rf pulse may

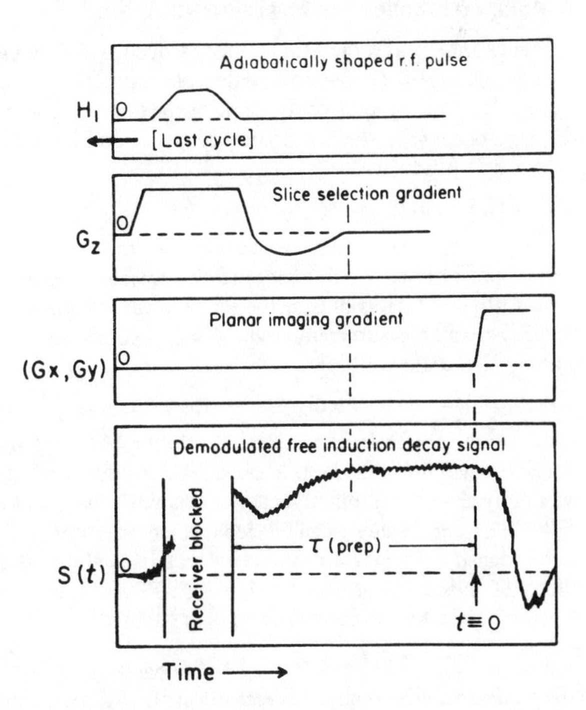

FIG. 4. A typical "slice-selection" sequence for slice oriented, or planar 2-D imaging.

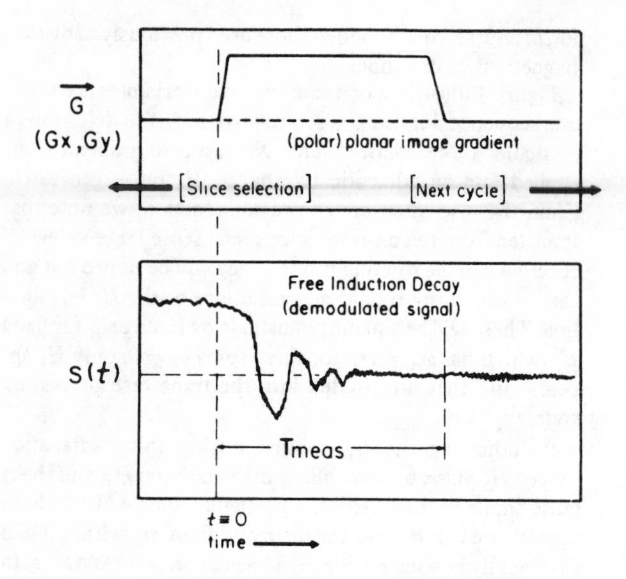

FIG. 5. One data cycle for 2-D imaging with "polar" data collection.

be modified to give, say, a 45° turn; then $m_\perp = \sqrt{1/2}\, m_{z(\text{initial})}$, and there remains $\sqrt{1/2}\, m_{z(\text{initial})}$ aligned *along* H_0, even *within W*.

The system now is prepared for spatial encoding [by (G_x, G_y) gradients] of the $m_\perp$ signal arising from a single slice. Differences in data collection strategies for the different slice-oriented NMR-imaging modes all occur subsequent to a common slice-selection procedure such as the example described above.

A. Polar data collection and reconstruction

Figure 5 illustrates the field-gradient sequence (following slice selection) for simple polar data collection. Note that some delay time, τ(prep), controllable by the system's operator, occurs between the preparation sequence and the turn on of a field gradient **G**.

$$\mathbf{G} = G_x\hat{x} + G_y\hat{y}. \tag{33}$$

During τ(prep), the excited $m_\perp$ decays via "true" T_2 processes (and residual H_0 inhomogeneities, which we ignore henceforth as artifacts outside the scope of this treatment); the FID-decay measurement origin, $t = 0$, occurs at the beginning of **G** and so

$$m_\perp(t \equiv 0) = [m_\perp(\text{excited})]\exp[-\tau(\text{prep})/T_2]. \tag{34}$$

In biologically interesting samples, T_2 is in the 10 to 100 ms range. We will not consider this effect explicitly any further; we simply absorb it implicitly into the meaning of $m_\perp(t = 0)$ and will express it later as an imaged (protocol-sensitive) effective density of spins ρ_{EFF}. Similarly, we simply absorb the effects of chosen spin-rotation pulses (45°, 90°, 30° preparations, etc.) into the effective density, ρ_{EFF}; that is

$$m_\perp(t = 0) = 2\pi\gamma\hbar[I(I+1)/3](\hbar\omega_0/kT)\rho_{\text{EFF}}. \tag{35}$$

We return to a more complete description of ρ_{EFF} later. It is sufficient here to observe that varying τ(prep) in a series of complete-data-collection runs is one way to make the resulting images sensitive to spatial W_2 (or T_2) variations in the subject.

The application of **G**, depicted in Fig. 5, marks $t = 0$ for data collection. In typical practice, G is chosen so that the field deviation across the diameter of the subject, is on the order of $10^{-3}H_0$. As "t" increases, the FID signal traces out $\tilde{S}(\mathbf{q})$, the frequency-space values, for the imaged slice. The angle of the **q** vector is given by

$$\tan(\Theta_q) = [G_x/G_y], \tag{36a}$$

and

$$\tilde{S}(q,\Theta) = [AQ\omega_0(B/I)] \int\int Wm_\perp(x,y)e^{2\pi i(G_x x + G_y y)t}\,dx\,dy;$$

with

$$q = -\gamma(G_x^2 + G_y^2)^{1/2}t. \tag{36b}$$

The magnitude of **q** evolves radially outward from the origin, as illustrated in the **q**-space diagram (e.g., q_1) of Fig. 6. There is some measuring-time limit, T_{meas}, for this particular cycle of data collection. It is controlled either by the algorithmic truncation [the apodization function, $\mathscr{H}_A(q)$, described in a previous section] or by intrinsic effects such as T_2 decay. This maximum measuring time (typically ~50 ms intrinsically) gives a frequency-space spatial bandlimit, *F*:max,

$$F\text{:max} \equiv \gamma|G|T_{\text{meas}}. \tag{37}$$

More precise definition of *F*:max and T_{meas} can be derived, but that is beyond the current scope.

Recalling that the image $\mathscr{I}(\mathbf{r})$ is reconstructed from the Fourier transform $\tilde{\mathscr{I}}(q) = \tilde{\mathscr{H}}_A(q)\tilde{S}(q)$, we can express the general inversion in polar coordinates:

$$\mathscr{I}(\mathbf{r}) = \int_0^{2\pi} d\Theta \int_0^\infty \mathscr{H}_A(q)\tilde{S}(q,\Theta)e^{2\pi i\mathbf{q}\cdot\mathbf{r}}\,|q|\,dq, \tag{38}$$

If $\hat{u}_\theta$ represents a unit vector directed along the particular **q** line, at angle θ, we can write Eq. (38) as

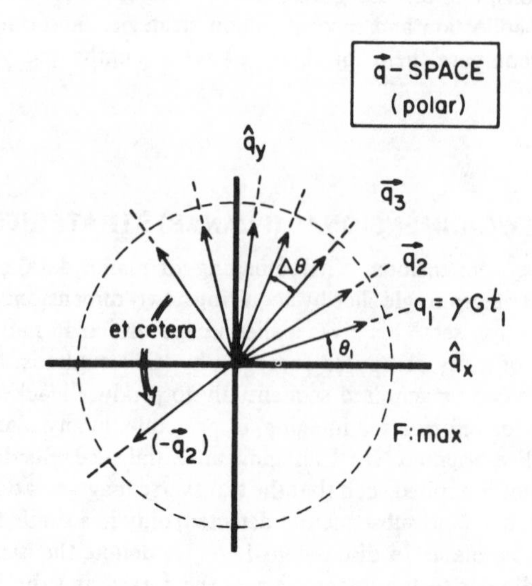

FIG. 6. Sampling of **q** space in 2-D imaging with polar data strategy.

$$\mathscr{I}(r) = \int_0^{2\pi} d\theta \int_0^{\infty} \mathscr{H}_A(q)\widetilde{S}(q,\hat{u}_\theta)e^{2\pi i q \hat{u}_\theta \cdot r}\,|q|dq$$
$$= \int_0^{\pi} d\theta \int_{-\infty}^{\infty} |q|\mathscr{H}_A(q)\widetilde{S}(q,\hat{u}_\theta)e^{2\pi i q \hat{u}_\theta \cdot r}\,dq. \quad (39)$$

This is the "back-projection" expression familiar in x-ray CT algorithms. It is simpler to work with when Eq. (39) is written in the form

$$\mathscr{I}(\mathbf{r}) = \int_0^{\pi} d\theta \left\{ \int_{-\infty}^{\infty} \left[\int_{-\infty}^{\infty} |q|\mathscr{H}_A(q)\widetilde{S}(q,\hat{u}_\theta)e^{2\pi i q x}\,dq \right] \delta(x - \hat{u}_\theta \cdot \mathbf{r})dx \right\}. \quad (40)$$

Starting within the "[ ]" bracket integrand we see the one-dimensional transform of a single line of apodized $\widetilde{S}(q)$ data, "filtered" by $|q|\mathscr{H}_A(q)$, into a spatially filtered projection along a line "x" parallel to $\hat{u}_\theta$. These values are "back projected" by the integral within "{ }" brackets onto all image coordinates $\mathbf{r}$ which project onto the particular value $\hat{u}_\theta \cdot \mathbf{r} = x$. The outer integral represents a summation of this back-projection process over all angular views. Thus a complete data collection involves many individual cycles; in each subsequent cycle the angle Θ is varied so as finally to acquire a complete set of data covering positive q values with Θ spanning 2π radians, or covering $+q$ and $-q$ values with Θ spanning π radians.

The direct method simply uses two cycles with $\mathbf{G}$ reversed to give the angles Θ and $\Theta + \pi$. An algorithmic alternative employs the Hermitian property of $\widetilde{S}(\mathbf{q})$; since the subject distribution is a "real" function then its transform $\widetilde{S}(\mathbf{q}) = S(t = [q/\gamma G])$, must obey

$$\widetilde{S}(q) = S_1(q) + iS_2(q) \quad (41a)$$

with

$$\widetilde{S}(-q) = \widetilde{S}^*(q) = S_1(q) - iS_2(q). \quad (41b)$$

A third way to obtain $\pm q$ values is by a spin-echo measurement; a spin-echo FID generated by gradient reversal is sketched in Fig. 7(a) (a 180° pulse spin-echo method is possible also). An illustration of the time evolution of $\widetilde{S}(q)$ measurement for the echo technique is given in Fig. 7(b). Two aspects of the echo data strategy should be noted; first, a greater interval τ elapses between preparation and the defined origin $t \equiv 0$ at the echo center, than occurs for direct [Fig. 6(a)] FID measurement. Hence, explicit T_2 decay can more strongly influence ρ_{EFF} for the echo case than for the direct method. Second, for a fixed intrinsic value of T_{meas} the spin-echo requirement for (γG) must be increased by a factor of two in order to acquire the same limits on q, i.e., $-(F\text{:max}) \lessdot q \lessdot +(F\text{:max})$, whence

$$\gamma G T_{\text{meas}} = 2(F\text{:max}). \quad (42)$$

Twice as many q values are acquired in a single echo measurement than in a single direct measurement, but only half as much time[11] is available to sample and digitize each value.

Another, subtler, effect also can play a role. In the expression of Eq. (18) relating the FID signal $\widetilde{S}(q)$ to the transverse magnetization $m_\perp$, we note that $m_\perp(\mathbf{r},t)$ is time dependent due to T_2 processes occurring *during* the T_{meas} period. That is

$$m_\perp(\mathbf{r},t) = m_\perp(\mathbf{r},0)\exp[-W_2 t], \quad (43a)$$

from which we observe that the T_2 decay *during* measurement gives an intrinsic apodization,

$$\mathscr{H}_2(q) \equiv \exp[-W_2(\mathbf{r})(q/\gamma G)], \quad (43b)$$

which enters multiplicatively with the algorithmic-limiting term, $\mathscr{H}_A(q)$, described earlier. If $W_2(\mathbf{r})$ is rapidly varying in space compared with $\mathscr{H}_A(\mathbf{r})$, then the result is not describable only as a blurring point spread of $m_\perp(\mathbf{r})$, and spatial T_2 distortions occur. However, when $W_2(\mathbf{r})$ has small spatial variation over a few resolvable volumes, then $\mathscr{H}_2(q,r)$ gives only a spatially slowly varying blurring function in the image, one which is locally T_2 dependent. Returning to the

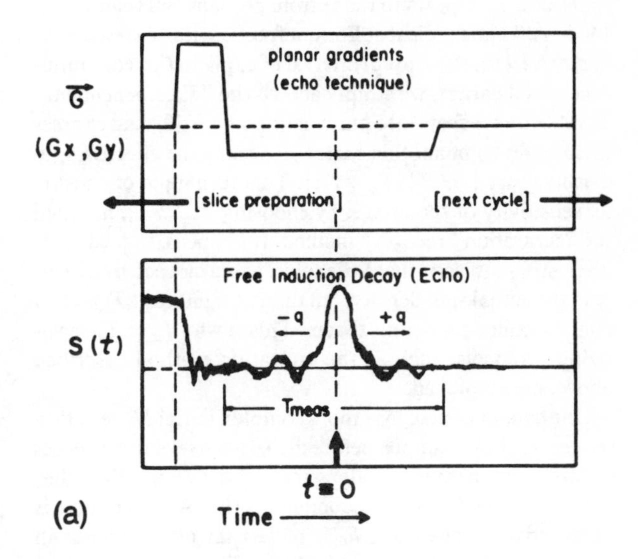

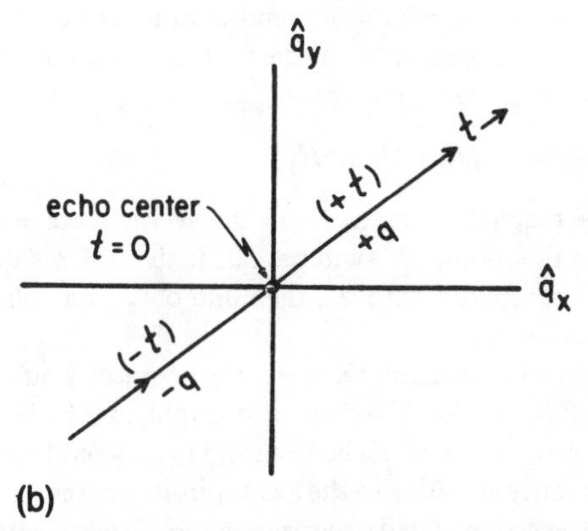

FIG. 7. (a) Spin-echo data sequence for gradient-reversal echo in polar strategy and (b) evolution of q-space data for echo FID.

spin-echo pattern of Fig. 7, in comparison with the direct FID of Fig. 6, one can see that this T_2 blurring tends to cancel (to linear terms) across the $\pm q$ values reconstructed into for the echo pattern, but remains in full force for the direct (or the algorithmic) methods of obtaining $\pm q$ values.

Another potential advantage of spin-echo measurement, especially in systems where the field gradient control circuits are not well stablized, is that the crucial $\tilde{S}(q)$ values at $q = 0$ (i.e., $t = 0$) occur outside the range of transients in the leading edges[12] of the G switching employed (as well as allowing $t = 0$ to be more removed from residual receiver recovery effects induced by the large H_1 pulse). Other modifications, e.g., multiple spin echoes within a longer intrinsic T_{meas} interval, obviously also are possible.

After each measurement cycle, the G-direction controls modify to give a new "Θ", a delay period ensues, and a new cycle finally begins with a repeat of the *slice-selection* sequence. We denote the total intercycle time between successive rf pulses, as T_{rep}. During this time, m_z relaxes toward m_{EQ} with a local T_1 characteristic. Since T_{rep} is not $\gg T_1$, m_z does not reach its full equilibrium value between 90° pulse and

$$m_{z(\text{initial})} = m_{\text{EQ}}\{1 - \exp[-W_1(n)T_{\text{rep}}]\}. \qquad (44)$$

In tissues, T_1 appears to range from perhaps less than 100 to 1400 ms. This is typical at Larmor frequencies of a few megahertz. As with the τ(prep) interval of explicit T_2-decay influence noted earlier, we simply absorb the "T_{rep}- dependent" T_1 relaxation effects into our definition of $m_\perp(0)$, and express them in terms of another factor governing the effective spin density, $\rho_{\text{EFF}}[(\tau/T_2);(T_{\text{rep}}/T_1)]$. This technique of causing T_1 sensitivity of the image, by choosing $T_{\text{rep}} \lesssim T_1$, is called the "saturation" recovery method. It happens, in medically interesting subjects, that little subject contrast occurs directly in the actual spin density, but there is significant T_1 and T_2 contrast among different tissues. This is why T_2 or T_1-sensitizing protocols, such as the saturation method described above, are employed.

One can, of course, use more complex initial H_1 selection sequences. For example, before the W-slice selection process occurs, one can invert m_z along the $-\hat{z}$ direction. We earlier noted that a 180° rf pulse accomplishes this. An alternative is "adiabatic-fast-passage"; h_0 is pulsed far off resonance in one sense, and H_1 is switched on. Then h_0 is returned through resonance, $\gamma H_0 = \omega_{\text{rf}}$, and continued far off resonance in the opposite sense. If the variation is adiabatically slow, then spin precession about

$$\mathbf{h} = [H_0 - (\omega_{\text{rf}}/2\pi\gamma)]\hat{z} + H_1\hat{y}$$

causes the magnetization to follow the vector sense of $\mathbf{h}$ and invert. At this point H_1 switches off; if the process occurs "rapidly" compared with T_1, then one obtains a complete 180° inversion of m_z.

In the case of inversion recovery, the operator would then set a relaxation period T (relax), before applying the W-slice-selection process for $m_\perp$. The resulting ρ_{EFF} would thus become explicitly sensitive to the local spin-inversion recovery [again governed by $T_1(\mathbf{r})$] occurring in the T (relax) interval. In the inversion-recovery sequence, when each tissue is assumed to be characterized by a unique T_1, then the expression analogous to that in Eq. (44) is

$$m_{z(\text{initial})} = m_{\text{EQ}}\{1 - 2\exp[-T(\text{relax})W_1] + \exp[-T_{\text{rep}}W_1]\}. \qquad (45)$$

The inversion-recovery method giving a T_1-sensitive image also is called the "180-τ-90" method. An important consideration, yet remaining, is to determine how many different angular views, or discrete values, must be accumulated in successive cycles in order to adequately reconstruct $\mathscr{I}(\mathbf{r})$.

We can calculate the minimum increment $\Delta\Theta$ in angular step necessary to avoid "angular-aliasing" artifacts. At the edge of the zone, where $q = F$:max, one must satisfy the Fourier sampling requirement by obtaining q-value increments perpendicular to $\mathbf{q}$ with a maximum permissible increment $\Delta q_\perp < D^{-1}$ as shown in Fig. 8, where D is the diameter of the subject. Consequently,

$$(F\text{:max})\Delta\Theta = \Delta q_\perp \lesssim D^{-1}, \qquad (46)$$

or

$$\Delta\Theta \lesssim [D(F\text{:max})]^{-1}; \qquad (47)$$

this means that the number of cycles necessary to subtend 180° in Fourier space (or in image space) is

$$J_\pi = (\pi/\Delta\Theta) \gtrsim \pi D(F\text{:max}). \qquad (48)$$

If one performs a direct simple measurement of $+q$ and $-q$ by gradient reversal, then the total number of cycles spanning 2π rad is

$$J(\text{direct}) = 2J_\pi \geqslant 2\pi D(F\text{:max}). \qquad (49)$$

If one uses the Hermitian property algorithmically setting $\tilde{S}(-q) = \tilde{S}^*(+q)$, or uses the spin-echo measurement to obtain $\pm q$ data from a single cycle, then

$$J(\text{echo or Hermitian}) = J_\pi \geqslant \pi D(F\text{:max}). \qquad (50)$$

The digitized sampling along the t axis, that is, along the radial direction of a given q, must be a minimum increment of

$$\Delta q_t \lesssim D^{-1}; \gamma G\Delta t \lesssim D^{-1}. \qquad (51)$$

Consequently, from $(-F\text{:max}) \lesssim q \leqslant (+F\text{:max})$, we must obtain "$d$" samples,

$$d \gtrsim 2(F\text{:max})D; \qquad (52)$$

this strategy requires J_π lines to cover q space, so the grand total sampling number is

$$\text{data-samples-polar} \gtrsim 2\pi[D(F\text{:max})].^2 \qquad (53)$$

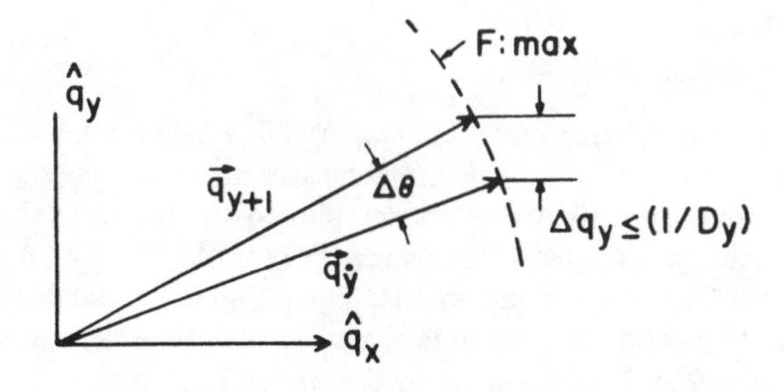

FIG. 8. Polar coordinate diagram for determining maximum nonaliasing angular increment in polar-data-strategies.

The Fourier sampling theorem shows that the distance p between independently reconstructable points in the image is given by

$$(F\text{:max}) = (1/2p). \tag{54}$$

We may use this relation to re-express the number of polar-strategy samples of Eq. (53) as

$$\text{data-samples-polar} = 2[(\pi/4)(D/p)^2]. \tag{55}$$

Since (D/p) is the number of reconstructable points along a diameter, $(\pi/4)(D/p)^2$ counts the number of "independent" pixels in the circular field of view. We note that the number of polar strategy data samples in Eq. (55) must be a factor of two larger than the number of reconstructable points. In polar-sampling strategies, the polar data *must* always be redundant by a factor of two, because of moments constraints on the projections of the two-dimensional subject. Of all polar data collected in the fashion described, only "half" are independent values; the other half are necessary to satisfy the moments rotation constraints. It is, nevertheless, necessary to measure all the polar-data values and to accept the redundancy. The benefit of this twofold redundancy is elsewhere[13] shown to be *not* increased resolvable points, but a twofold decrease in image noise variance.

B. Planar Cartesian data

In this data collection strategy, one employs independent temporally separated applications of G_y and G_x.[10] A schematic representation is shown in Fig. 9(a); the G_y pulse induces a phase accumulation, so that at $t = 0$, defined by G_x application, we have

$$S(t=0) = \int\int Wm_\perp(x,y)e^{2\pi i\gamma G_y y\tau_y}\,dx\,dy. \tag{56}$$

Subsequently, with G_x applied, $S(t)$ evolves according to

$$S(t) = \int\int [Wm_\perp(x,y)e^{2\pi i\gamma G_y y\tau_y}]e^{2\pi i\gamma G_x xt}\,dx\,dy, \tag{57}$$

which we can re-express by defining

$$q_y \equiv -\gamma G_y\tau_y, \tag{58}$$

and

$$q_x \equiv -\gamma G_x t, \tag{59}$$

in the form

$$\tilde{S}(q_x,q_y) = \int\int Wm_\perp(x,y)e^{-2\pi i(q_x x + q_y y)}\,dx\,dy. \tag{60}$$

As shown in Fig. 9(b), the measurement evolution of q_x is continuous as t increases, while q_y is defined discretely for each new cycle. Usually, G_y is varied in integral steps positively and negatively, so that

$$q_y = n(\Delta q_y);\Delta q_y \lesssim D^{-1}, n = \text{integer}. \tag{61}$$

After the q_x values are sampled and digitized, and q_y is cycled to cover $(-F\text{:max})$ to $(+F\text{:max})$, as shown in Fig. 9(b), one obtains a Cartesian data set of discrete (q_x,q_y) values. The image is obtained from a Cartesian inversion,

$$\mathscr{I}(x,y) = \int\int \mathscr{H}_A(q_x,q_y)\tilde{S}(q_x,q_y)e^{2\pi i(q_x x + q_y y)}\,dq_x\,dq_y; \tag{62}$$

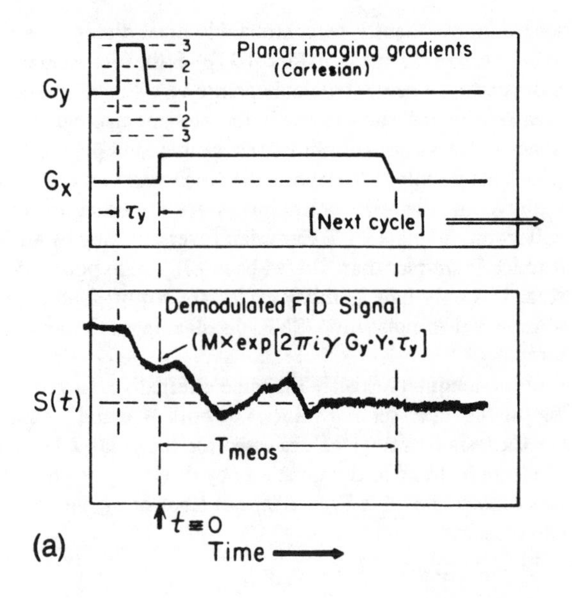

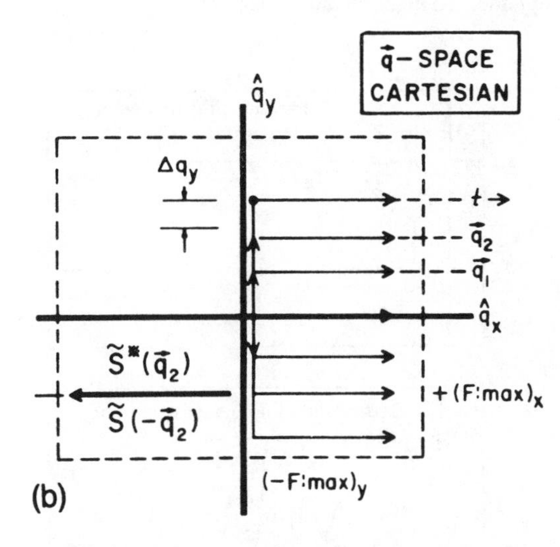

FIG. 9. (a) Separate G_x and G_y gradient applications for 2-D Cartesian-data-strategy and (b) evolution of **q**-space data.

usually this is implemented in software by a double pass, or 2-D, fast-Fourier-transform algorithm. We note, in particular, that the relation in Eq. (62) corresponds exactly with that in Eq. (38); there is no difference in essential method and the terms "back-projection technique" and "Fourier technique" imply only different algorithms for doing the Fourier-inversion calculation.

As with polar data, we can count data samples. Both q_x and q_y range over a span of $2(F\text{:max})$ in increments as large as D^{-1}; the number of cycles covering π radians about the x axis is

$$J_\pi(\text{Cartesian}) = 2(F\text{:max})D. \tag{63}$$

Along a temporal line (i.e., q_x axis) there are also $2(F\text{:max})D$ samples required. The sample total is

$$\text{data-samples-Cartesian} = 4[(F\text{:max})D]^2 = (D/p)^2, \tag{64}$$

where independently reconstructable pixel dimensions are as before, $p = (2F\text{:max})^{-1}$. Since $(D/p)^2$ is the total number of independently reconstructable points in a subject plane of area D^2, Eq. (64) shows there is not the same twofold redundancy in Cartesian sampling as encountered in polar strategies. That is, only half as many FID-measurement cycles are required in Cartesian collection as are necessary in polar collection. While a FFT Cartesian inversion may be algorithmically simpler than filtered-back-projection polar inversion, and only half as many cycles are required, there are some signal-to-noise ratio (SNR) disadvantages to Cartesian sampling.[13]

One encounters exactly the same alternatives in generating the full "2π" set of q values, i.e., both $+\mathbf{q}$ and $-\mathbf{q}$, for the Cartesian strategy as discussed for the polar 2-D case. They are handled in the same way by direct G_x reversal, by spin echo as shown in Figs. 10(a) and 10(b), or by the Hermitian property

$$\widetilde{S}(-q_x, -q_y) = \widetilde{S}^*(+q_x, q_y),$$

implemented algebraically in software.

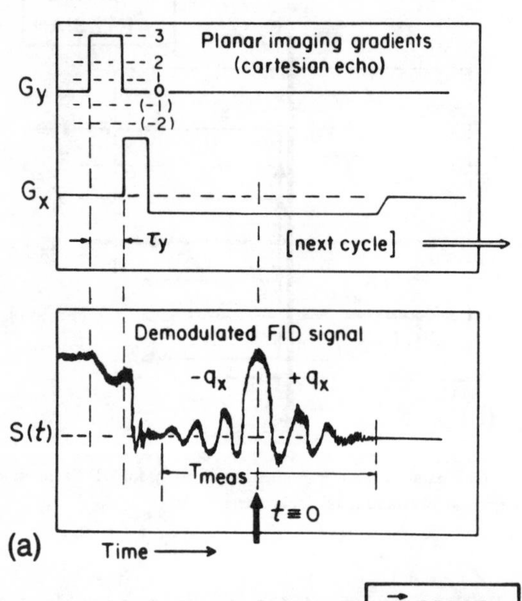

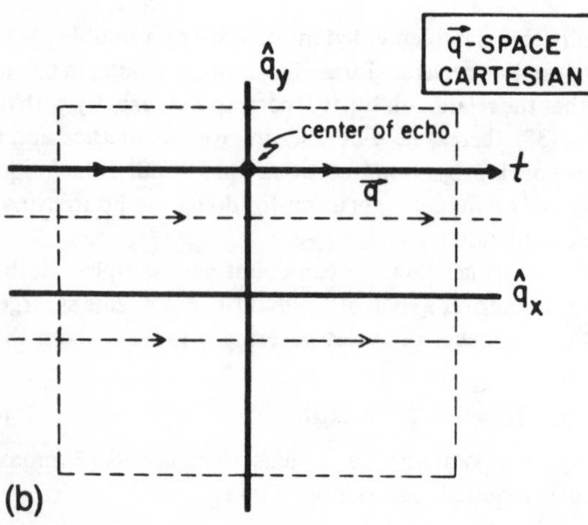

FIG. 10. (a) Gradient sequence for Cartesian strategy echo data and (b) evolution of q-space data from echo FID.

V. THREE-DIMENSIONAL IMAGING

The field-gradient sequences for 3-D NMR imaging occur after an initial excitation sequence. In three-dimensional techniques, $m_\perp$ is induced over the whole active volume of the subject (rather than in a single slice), and a $\mathbf{G}$ vector in three dimensions varies to give spatial volume coding. As in the planar imaging modes described earlier, the same kinds of T_1-factor and T_2-factor decays and relaxations occur and, by similar variation of $\tau(\text{prep})$, T_{rep}, T (relax) intervals, images combining various combinations of true density ρ, T_2 sensitivity, and T_1 sensitivity are produced and describable in terms of the effective density ρ_{EFF}.

Similarly, along the temporal axis of measurement, one has a certain q evolution $q = -\gamma G t$; one determines $\pm q$ values with the same alternatives as for planar techniques.

A. Cartesian 3-D

Cartesian 3-D strategies follow by simple extension of the planar Cartesian method, as illustrated in Figs. 11(a) and 11(b). Temporally separated $(G_z\tau_z)$, $(G_y\tau_y)$, and finally $G_x t$ sequences are implemented in a single cycle. The path of the measured $\widetilde{S}(\mathbf{q})$ in reciprocal space is shown in Fig. 11(b). Using the same notation as before,

$$q_z = -\gamma G_z \tau_z, \tag{65}$$

$$q_y = -\gamma G_y \tau_y, \tag{66}$$

and

$$q_x = -\gamma G_x t, \tag{67}$$

we can reconstruct $\mathcal{I}(\mathbf{r})$ from

$$\mathcal{I}(\mathbf{r}) = \int\int\int \mathcal{H}_A(q_x, q_y, q_z)\widetilde{S}(q_x, q_y, q_z) \times e^{2\pi i(q_x x + q_y y + q_z z)}\, dq_x\, dq_y\, dq_z, \tag{68}$$

by a three-pass FFT in Cartesian coordinates. In Eq. (68), $\widetilde{S}(q_x, q_y, q_z)$ is given by

$$\widetilde{S}(q_x, q_y, q_z) = [AQ\omega_0(B/I)] \int\int\int m_\perp(x,y,z) \times [\exp(-|q_x|/\gamma G_x T_2)] e^{2\pi i \mathbf{q}\cdot\mathbf{r}}\, dx\, dy\, dz, \tag{69}$$

and the T_2 decay during the encoding and measurement processes provides, as before, an added image blurring.

Again we can determine the minimum number of measurement cycles:

$$J_y \gtrsim 2(F\text{:max})D_y \tag{70}$$

and

$$J_z \gtrsim 2(F\text{:max-}z)D_z. \tag{71}$$

In Eq. (71), we explicitly note that we need not have the same z-axis resolution (e.g., patient axial dimension) as we use for transverse dimension, which also, of course, may be chosen with different resolving limits for (F:max). From Eqs. (70) and (71), the number of cycles is

$$J_\pi(\text{3-}D) = 4(F\text{:max})(F\text{:max-}z)D_y D_z = (D_y D_z / p_y p_z), \tag{72}$$

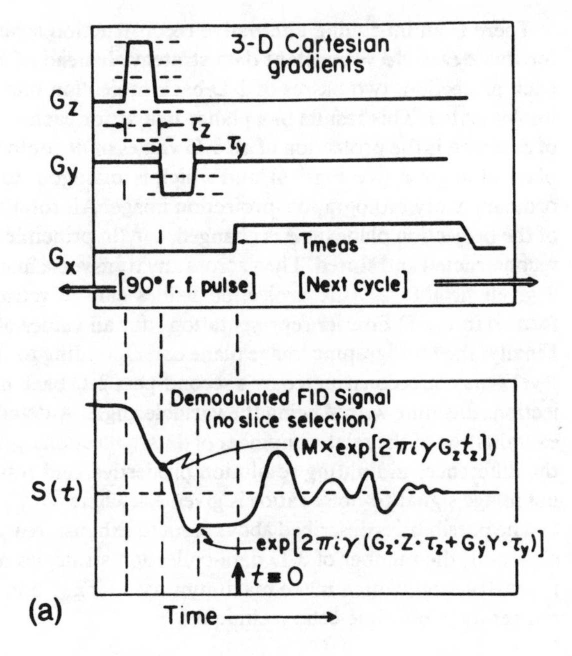

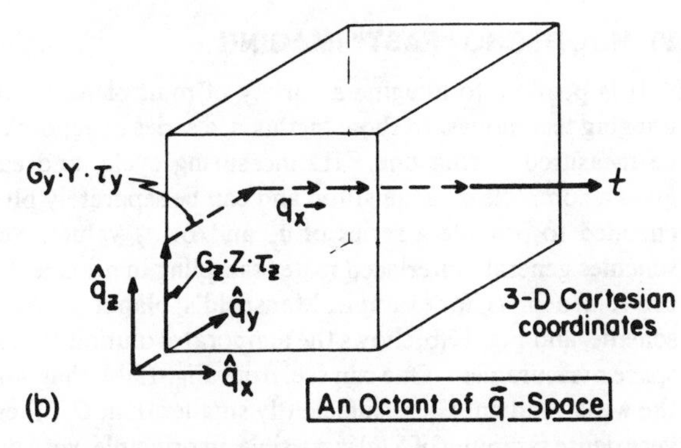

FIG. 11. (a) Gradient sequence for Cartesian 3-D data-collection strategy and (b) the q-space data trace.

and twice that number if $+q$ and $-q$ are to be directly measured without an echo technique or algorithmic Hermitian processes. With a total number of (D_x/p_x) samplings on the temporal line,

$$[-(F{:}\max - x)] < q_x < [(+F{:}\max - x)],$$

the total data count is

$$\text{data-samples-Cartesian-3D} = [D_x D_y D_z / p_x p_y p_z]; \quad (73)$$

which also, as anticipated, is the total count of independently reconstructable pixels in a rectangular cylindrical subject of sides D_x, D_y, and D_z.

B. 3-D spherical strategy

The data strategies here involve simultaneous G_x, G_y, and G_z application to give a resultant radial G vector. During the intracycle delay, the gradient components are readjusted so that a complete series of cycles completely samples both in [see Figs. 12(a) and 12(b)] azimuthal increments Θ and polar increments ϕ. If $+q$, along the temporal axis, is determined directly by reversed gradient cycles then the (Θ,ϕ) combinations subtend 4π steradians of solid angle. If spin-echo or Hermiticity methods are used, then the (Θ,ϕ) cycles need subtend only 2π steradians. The Fourier inversion expression in spherical coordinates is

$$\mathscr{I}(\mathbf{r}) = \int_0^{\pi} d\theta \int_0^{2\pi} \sin(\theta)d\phi \int_0^{\infty} \tilde{\mathscr{H}}_A(q) \times \tilde{S}(q,\hat{u}_{\Theta\phi})e^{2\pi i q \hat{u}_{\Theta\phi}\cdot\mathbf{r}}\,|q^2|dq; \quad (74)$$

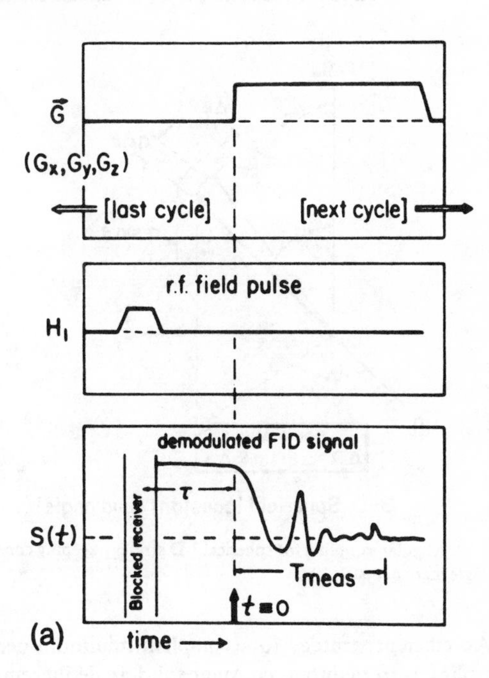

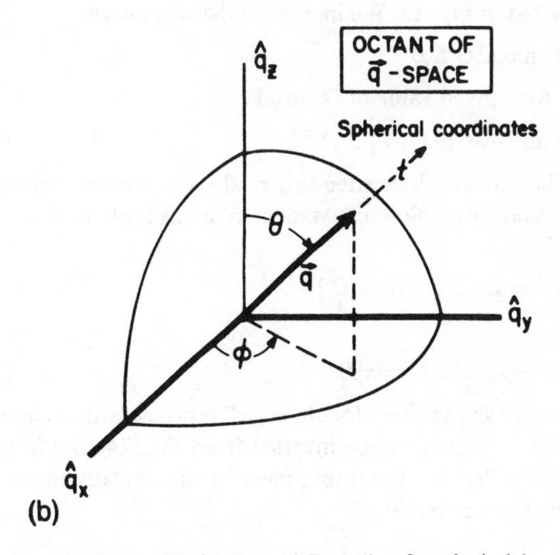

FIG. 12. (a) Excitation rf pulse and 3-D gradient for spherical data strategy and (b) data evolution in q space.

we note this is a spherical back-projection expression. The data along the line $\mathbf{q} = -\gamma\mathbf{G}t$ are filtered by $[\tilde{\mathscr{H}}_A(q)\cdot q^2]$ and transformed to a spatial axis. All points $\mathbf{r}$ in the plane normal to $\mathbf{q}$ in the 3-D volume, which project onto the spatial variable given by $\hat{u}_{\Theta\phi}\cdot\mathbf{r}$ receive the tranform's value. This is then summed over all polar angles, with the solid angle step $\sin(\Theta)d\phi$ (from the spherical coordinates Jacobian), and resummed over all azimuthal views Θ.

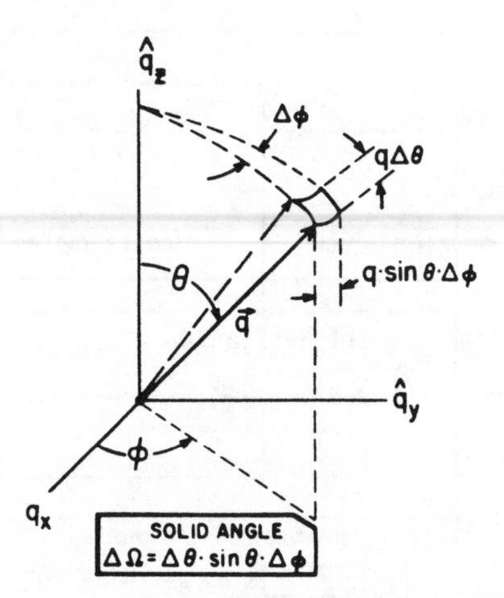

FIG. 13. Angular sampling for spherical 3-D strategy keeping constant solid-angle increments.

An efficient strategy to accomplish minimum necessary sampling is to maintain constant solid-angle increments as depicted in Fig. 13. We increment θ always with

$$(F{:}\max)\Delta\Theta \lesssim D^{-1}, \tag{75}$$

and for a given value of Θ we take

$$(F{:}\max)\sin\Theta\,[\Delta\phi\,] \lesssim D^{-1}. \tag{76}$$

There is an alternative spherical data strategy possible. This scheme employs constant increments for both Θ and ϕ, e.g.,

$$\Delta\Theta = \Delta\phi \lesssim [(F{:}\max)D\,]^{-1}, \tag{77}$$

and

$$J_\Theta = J_\phi = \pi(F{:}\max)D. \tag{78}$$

Figure 14 shows this "double-polar" representation schematically. The data can be inverted from Eq. (74); but in this case, the "sin θ" weighting must be numerically inserted, since $\Delta\phi$ = constant.

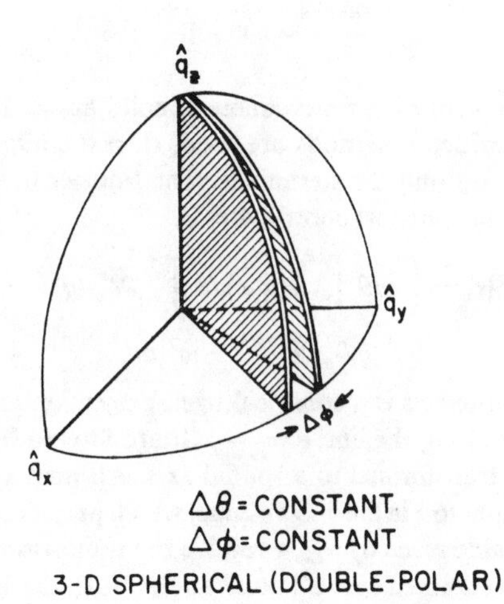

FIG. 14. Angular sampling for spherical 3-D strategy of "double polar" type showing two "projection" data planes.

There is an intriguing alternative reconstruction scheme for this $\Delta\Theta = \Delta\phi$ = constant data strategy. Instead of 3-D back projection, two passes of 2-D back projection may be implemented. This results in a planar image, for each value of ϕ, which is the protection of all 3-D values of $m_\perp$ onto the plane at angle ϕ (see Fig. 14), and which is analogous to an ordinary x-ray radiographic projection image. All rotations of the projection plane, as ϕ is changed, can (in principle) be reconstructed and stored. Then across any transverse line, at a given height "z," the projection values can be retransformed into 1-D Fourier representations, for all values of ϕ. Finally, the tomographic image plane corresponding to that level z may be reconstructed by a second-pass 2-D back projection, this time with ϕ being the variable angle. A detailed examination of the relative number of data-collection cycles, the differences in limiting resolution properties, and resulting image signal-to-noise ratios is given elsewhere.[8]

The possibilities described above seem to exhaust, reasonably well, the number of 3-D data-collection strategies and reconstruction, which make maximum use of T_{meas} and do not resort to multiple-echo techniques.

VI. MULTIECHO "FAST" IMAGING

It is possible to imagine a variety of multiple-echo, fast imaging techniques. In these methods, a series of echoes can be measured during one FID measuring cycle, and each gives a complete q_x acquisition and can be separately phase encoded to provide a series of q_y and/or q_z values. Such schemes generate interlaced raster sampling in q space. Figure 15(a) depicts, for example, Mansfield's[7] planar-echo data scheme, and Fig. 15(b) shows the temporal evolution of the q-space measurement. One can see, from Fig. 15(b), that when the weak gradient G_y, is sufficiently smaller than G_x, then a very dense sampling of $\tilde{S}(q)$ is possible, in principle, with only a single FID cycle. As "t" increases, of course, the T_2 decay will increasingly attenuate the data for increasing q_y values, but this is an acceptable intrinsic apodization. Values of $\tilde{S}(q_x,q_y)$, for q_y having negative values, are represented by the dotted lines in Fig. 15(b); they may algorithmically be generated by using the Hermitian-property requirement,

$$\tilde{S}(-q_x,-q_y) = \tilde{S}^*(+q_x,q_y).$$

A subtlety exists here in the practical Fourier inversion of such data; note that even-numbered echo patterns correspond to $(+G_x)$ FID signals, while odd-numbered echo patterns correspond to $(-G_x)$ FID signals. This opposite sense of the q_x evolution in alternating echoes is easily accounted for in the inversion mathematics, provided one is aware of its existence.

Moreover, Mansfield and colleagues[7] pointed out an interesting special case, which allows relatively simple reconstruction mathematics. When G_y is chosen so small that several G_x echoes occur within a range $\gamma G_y t \simeq D_y^{-1}$, then a helpful redundancy of q_x data results. In this case, one need only do a single one-dimensional Fourier temporal inversion of the entire data string; the result gives a periodically spaced set of image value "lines" which are interpretable as viewing the subject along discretely sampled lines of $\{x_i$

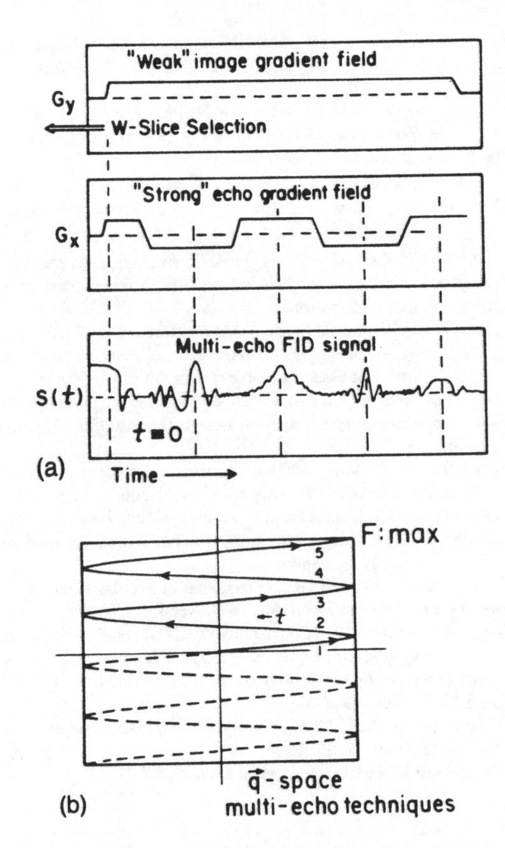

FIG. 15. (a) A multiecho data strategy for 2-D "planar-echo" example and (b) raster sampling of q-space data by multiecho methods.

= (integer)$\cdot\Delta x$}, but at a slight angle so that the y variable, $y = (x - x_i)$, traces out the "y" variation along that line. While this certainly is an ingeneous Fourier-inversion reconstruction scheme, it is not yet clear whether it actually offers any advantage, in image quality or computational simplicity, over an explicit two-dimensional inversion.[14]

In principle, this multiecho data collection scheme can be extended to 3-D imaging by adding a series of G_z pulses. After sampling one q_z-plane of $\tilde{S}(q_x, q_y, q_z = n\Delta q_z)$, as described above (but without slice selection of a particular W thickness), then reversal of G_y and an interspersed G_z pulse will increment the data collection to

$$\tilde{S}[q_x, q_y, q_z = (n+1)\Delta q_z].$$

In practice, the difficulty encountered is in generating and controlling sufficiently large G_x gradients (and G_z pulses for the 3-D extension) to reliably sample sufficient data within a T_{meas} period limited by the subject's shortest T_2's.

The limitations of these multiecho, fast imaging schemes are not difficult to imagine. They require the acceptance of low resolution and high noise; this is exactly analogous with comparing fluoroscopy imaging (with its high-noise, low-resolving-power characteristics) to the very low noise and high resolving power of an ordinary screen–film radiograph. It is, of course, the real-time "fluoro" nature realizable with multiecho NMR images (Mansfield and Ordidge have demonstrated TV-frame rates for planar echo) that constitutes their value.

Moreover, it should be clear from the preceding discussion, that a hybrid combination of multiple-cycle and multiple-echo techniques is possible. Thus, one may achieve any intermediate trade off of imaging time versus noise and resolution which one desires in NMR imaging.

VII. SUMMARY

In summary, for all NMR imaging modes, three orthogonal field gradient modulations provide spatial-position encoding of the NMR free-induction-decay signals. Different modes of NMR imaging correspond to different strategies of data-collection dependency on the specific temporal sequence of gradient field activation. When one gradient field occurs *during* rf pulse excitation, then the subject volume is truncated to a selective excited slice, and subsequent cycles employ the remaining gradient-direction modulations for 2-D slice-oriented imaging. One may then choose simultaneous 2-D gradient activations to collect Fourier-space data sets in polar-coordinate form, or one may separately pulse the planar gradients to collect data in Cartesian coordinate geometry. When no slice selection occurs in the rf excitation phase, then 3-D imaging occurs; all three gradient directions are activated during the data collection part of the cycle. In 3-D, one may choose spherical-coordinate geometry, "two-pass" double-polar strategies, Cartesian coordinate strategies or, in principle, a combination. There are no physically significant differences in the subsequent image reconstruction from these various Fourier-space data sets; in all cases, the image results from a Fourier inversion calculation. Differences in nomenclature, e.g., "zeugmatography back-projection" or "Fourier imaging," refer primarily to the names of algorithms written to accomplish the Fourier inversion appropriate to the geometry of the data strategy employed. In all cases, also, the quantities measured are the spatial Fourier-transform components of precessing magnetization and the quantities imaged by reconstruction is the spatial distribution of precessing magnetization during the data window used. The precessing magnetization is proportional to the physical density of precessing spins, but also may be modulated locally by rf excitation factors, T_1-relaxation factors, T_2-decay factors and, in principle, other physical effects such as chemical-shift spectral dispersion, spin-motion effects, and so on. These sensitizations may be generated in a variety of ways by controlling intercycle repetition times, intracycle delay times, rf pulse sequences, etc. Thus, nomenclature such as "inversion-recovery-imaging" or "spin-echo imaging" does not reflect any physical difference in what is imaged as it always is the distribution of precessing magnetization in the subject. These names imply that a specific intracycle method generates the appropriate sensitization and thereby modulates the physical spin density in producing that distribution of precessing magnetization.

In practice, the limit on time interval of a data-collection window is more often determined by base-magnet inhomogeneity than by intrinsic T_2 factors. To achieve equal limiting resolutions, higher *absolute* inhomogeneity requires higher field-gradient levels, and shorter data-window times; to some extent, one may overcome disadvantages of worse homogeneity, in principle, by using rf pulse spin-echo techniques and acquiring multiecho data. Other forms of multiecho data may be used in producing fast "fluoro-mode" NMR scans.

ACKNOWLEDGMENT

The authors acknowledge Fellowship support of K.F.K. on NIH Training Grant No. NIH-NCI-P3014520.

[a)] Present address: General Electric Medical Systems, Milwaukee, Wisconsin 53201.

[1]A. Abragam, *Principles of Nuclear Magnetism* (Oxford, London, 1961).
[2]P. R. Moran, J. Phys. Chem. Solids **30**, 297 (1968).
[3]P. C. Lauterbur, Nature (London) **242**, 190 (1973).
[4]P. C. Lauterbur, Pure Appl. Chem. **40**, 149 (1975).
[5]We confine discussions here to the modern pulsed-gradient imaging techniques. Lauterbur's first experiments used narrow-band cw NMR in a dc gradient. Some early systems employed rapid ac gradients to define "sensitive lines" or "sensitive points" which were rastered spatially to acquire image data. These methods were useful in early research prototypes as they required much less ambitious data rates and less sophisticated digital electronics. For practical medical imaging, however, they suffer extreme disadvantages.
[6]S. R. Thomas, Refresher Course in NMR, AAPM-1982 New Orleans; available from compiler at Department of Radiology, University of Cincinnati, Cincinnati, Ohio.
[7]P. Mansfield and P. G. Morris, "NMR Imaging in Biomedicine," in *Advances in Magnetic Resonance, Supplement 2*, edited by John Waugh (Academic, New York, 1982).
[8]K. F. King, "Signal-to-Noise Ratios in NMR Imaging," Ph.D. thesis (University of Wisconsin, 1982) (University Micro-films; Ann Arbor, Michigan); also available.
[9]An alternative is to run ω_{rf} at an off-resonance frequency, e.g., $\omega_{rf} < 2\pi\gamma(H_0 + \mathbf{G}\cdot\mathbf{r})$. Then the output of a single electronic demodulator multiplexes at a rate $(2\pi\gamma H_0 - \omega_{rf})$ between "S_1-cosine" and "S_2-sine" phase signals. This method suffers some SNR disadvantages and, because of initial phase uncertainties, often allows one to reconstruct only the modulus of the spin distribution.
[10]A. Kumar, D. Welti, and R. Ernst, J. Magn. Reson. **18**, 69 (1975).
[11]In practice, however, effects other than intrinsic may dominate actual data-gathering time. For example, in the case of limiting by the inhomogeneity of the base field, which also is "echoed" in a 180°-rf-pulse spin–echo, *each* side of the echo pattern may be as long as that limiting a direct-FID measurement.
[12]In practice, the ideal "non-echo" method (Fig. 5) is little used because of the rise-time limitations of the imaging-gradient pulses. Most scanners, consequently, actually always employ some spin–echo [Fig. 7(a)] method to circumvent this difficulty, even though the name used for such modes might imply a nonecho sequence.
[13]A full description of the need for data redundancy in polar angle collection was given by one of the authors (KK) in a University of Wisconsin Medical Physics report, WMP-130, "Constraints on CT Data Due to Moments Transformations of Bounded Objects" (1981). The question of SNR improvement generated by the polar-strategy data redundancy is treated in Dr.King's Ph.D. thesis (Ref. 8).
[14]I. R. Young and M. Burl "Nuclear Magnetic Resonance Systems," U. S. Patent 4,355,282, Oct. 19, 1982.
[15]R. L. Dixon and K. E. Ekstrand, Med. Phys. **9**, 807 (1983).

IEEE TRANSACTIONS ON BIOMEDICAL ENGINEERING, VOL. BME-26, NO. 9, SEPTEMBER 1979 497

Magnetocardiography: An Overview

DAVID B. GESELOWITZ, FELLOW, IEEE

Abstract–Since 1963 when recording of the human magnetocardiogram (MCG) was first reported, the number of clinical studies has been limited. High-quality tracings can now be easily obtained in special shielded chambers, but problems remain if records are to be made in hospitals without such chambers. Measurement techniques, theory of the MCG, and model studies are discussed. Configuration of the MCG waveform is much the same as that of the electrocardiogram (ECG). Measurements of dc currents of injury in dogs by means of the MCG have been reported, and are potentially of great clinical interest. Effects of magnetic susceptibility changes in the torso associated with blood movement may contribute to variations in the external field. The questions of what new diagnostic information is available in the MCG and what lead systems are most appropriate remain to be answered.

Introduction

Electromotive forces in the heart muscle have their genesis in electrical excitation and the subsequent recovery of cardiac cells. Since the surrounding tissues are conductive, currents spread through the body and give rise to potential differences of the order of millivolts on the body surface. Records of these potentials are called electrocardiograms (ECG's) and provide the clinician with important information concerning the heart.

In 1820, Oersted demonstrated the existence of a magnetic field surrounding a wire carrying a current and Ampere subsequently developed quantitative laws relating magnetic fields to electric currents. Currents existing in the body because of the activity of nerve cells or muscle cells will therefore create magnetic fields outside the body. These magnetic fields are extremely small, but modern electronic technology enables us to detect and record them.

It is now more than 15 years since Baule and McFee reported the inscription of a human magnetocardiogram (MCG) [1]-[3]. They measured the extracorporeal magnetic field of the intact heart using two coils in opposition. One coil was placed near the body, the other one was placed a small distance away (see Fig. 1). Since the ambient magnetic field does not vary greatly over this distance, while the magnetic field of the heart decreases rapidly with increasing distance from the heart, the difference in the voltages induced in the two coils is proportional to the rate of change of the cardiac magnetic field at the proximal coil. MCG's showed a peak amplitude of about 5×10^{-7} G. Thus, the magnetic field of the heart recorded near the surface of the body has an intensity only one-millionth that of the earth's magnetic field.

Manuscript received March 1, 1979; revised May 15, 1979. This work was supported in part by the National Science Foundation under Grant GK 36608, and the National Heart Lung and Blood Institute under Grant HL 21283.

The author is with the Bioengineering Program, The Pennsylvania State University, University Park, PA 16802.

The work of Baule and McFee was later confirmed by Safonov *et al.* [4], who used a similar technique except that the detector and the subject were placed in a shielded chamber to attempt to reduce the ambient magnetic fields. About the same time Cohen [5] used a single magnetometer and a heavily shielded chamber to record the MCG, and subsequently reported the distribution of the magnetic field vector around the torso of normal subjects during *QRS* [6].

In the intervening years from these early investigations to the present time, the major experimental contributions have come from the laboratory of Cohen at the National Magnet Laboratory in Cambridge, MA, although within the past several years at least four other laboratories in various parts of the world have become interested in clinical MCG. Concurrently, the original theoretical contributions of Baule and McFee have been supplemented by analyses provided by Geselowitz [7], Plonsey [8], and Grynszpan and Geselowitz [9].

In spite of the period of more than a decade since the original publication by Baule and McFee, the MCG literature is relatively small. The number of clinical studies is much more limited. It is probably accurate to state that, as of this writing, specific situations where the MCG will provide new or improved diagnostic information are yet to be firmly demonstrated.

Instrumentation and Measurements

Basically, there are two techniques which have been used for detecting the MCG. One technique involves the use of a coil consisting of many turns of wire. A voltage e will be induced in the coil whenever it is placed in a time varying magnetic field H. Specifically,

$$e = \mu N A \cos\theta \frac{dH}{dt} \tag{1}$$

where N is the number of turns, μ is the permeability of the material inside the coil, A is the area of the coil, and θ is the angle between the magnetic vector and the axis of the coil. Therefore, the voltage induced in the coil will be proportional to the component of the magnetic field along the axis of the coil.

It should be noted that the magnetic field intensity H is a vector and that its complete description at each point in space requires registration of its three orthogonal components, e.g., H_x, H_y, and H_z. Simultaneous detection of the three components would require three coils arranged with their

0018-9294/79/0900-0497$00.75 © 1979 IEEE

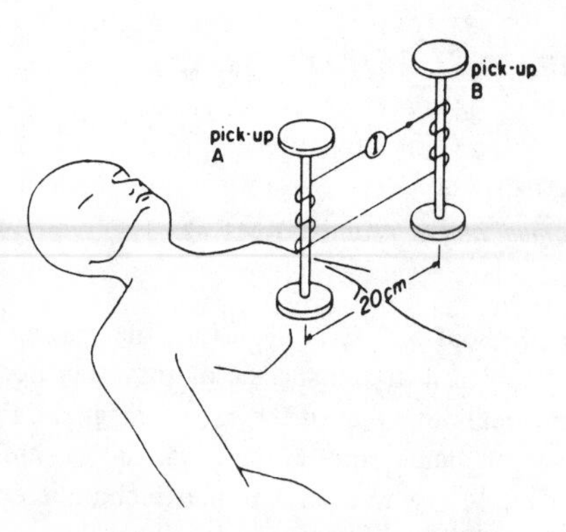

Fig. 1. Two coil arrangement for detecting the MCG. The cardiac field at coil B is very small, while the extraneous field is ideally the same at the two coils which are connected in opposition. (Reprinted from Baule and McFee [3] with permission of the *American Heart Journal.*)

axes orthogonal to each other in space. Alternatively, three orthogonal components could be obtained sequentially with a single coil by rotating its axis to be parallel in turn to the x, y, and z axes. With the sequential technique, the individual components are obtained for different heartbeats. The fact that the magnetic field is a vector rather than a scalar, as is electric potential, poses an added difficulty for the recording and display of MCG's. This point will be discussed later in this paper.

Since H is extremely small, N and μ must be very large in order to obtain reasonable values of induced voltage. Baule and McFee used coils with 2 000 000 turns. A high permeability was achieved by wrapping the coil around a ferrite core.

The second technique uses the SQUID magnetometer which operates on the basis of the Josephson effect in superconductors [10], [11]. The Josephson effect is observed at liquid helium temperatures. Therefore, this device requires a Dewar and associated cryogenic technology. Once again the measurement of the vector field requires either a three coil arrangement in the Dewar or the sequential rotation of the detector.

SQUID magnetometers have the property of responding to dc magnetic fields. This fact is of considerable interest in cardiology and will also be discussed later in this paper.

With modern electronic technology, amplification of very small signals, such as those associated with biomagnetic fields, poses no insurmountable problem. The significant factor is not the amplitude of the signal, but rather the noise level. In order to be usable, the signal as finally displayed must have a large signal-to-noise ratio.

Noise includes extraneous or extracardiac magnetic fields existing in the vicinity of the torso together with electric fluctuations inherent in the detector and amplifiers. SQUID magnetometers have noise levels which are well below MCG signal levels, and hence, intrinsic instrumentation and magnetometer noise is not a problem. The major problem is ambient noise fields associated with urban sources, especially electric currents. Cohen estimates that urban background noise is typically 10^{-4}-10^{-3} G.

Ambient noise can be reduced either by shielding or by using two magnetometers in opposition. Consider the two coil arrangement. Let the fields at the two coils be H' and H'', and let $H' = H'_h + H'_n$ and $H'' = H''_h + H''_n$ where the subscript h refers to the heart signal and the subscript n refers to the ambient noise field. The signal-to-noise ratio in the difference is then $(H'_h - H''_h)/(H'_n - H''_n)$. If the heart field is say 5×10^{-7}, and a 50:1 signal-to-noise ratio is required, then $H'_n - H''_n$ must be of the order of 10^{-8}. Hence, in a moderately noisy environment $(H'_n - H''_n)/H'_n = 10^{-5}$, which is extremely difficult to achieve, especially if the separation of the coils is large in order to make H''_h negligible. If the coils are brought closer together, H''_h is no longer negligible, and the arrangement approaches that of a gradiometer which responds to the spatial derivative of the magnetic field. Baule and McFee pointed out that a greater cancellation of the extraneous field introduced by a remote source could be achieved with a four coil arrangement. Use of such a second order gradiometer is discussed by Saarinen *et al.* [12]. It appears that it will be extremely difficult to achieve adequate signal-to-noise ratio in the hospital environment without shielding, at least without signal averaging, which is another technique for reducing the effect of noise [13].

Cohen *et al.* have constructed a five-layer shielded room which provides very effective shielding [11] (see Fig. 2). MCG's published by Cohen *et al.* are remarkably free of noise, including muscle tremor artifact (see Fig. 3). The shielded room, however, is quite costly, although less expensive modest shielded closets might be adequate in a hospital setting. Therefore, there are still several problems to be solved before MCG becomes a generally available clinical tool.

Clinical Results

MCG waveforms are remarkably similar to ECG's and no special nomenclature is required. In retrospect, this observation is not very surprising. As electromotive forces in the heart increase in amplitude, resulting currents in the torso will also increase. The larger current amplitudes will cause larger potential differences (ECG) as well as larger magnetic fields (MCG). Thus, the typical MCG shows a *P* wave, *QRS* complex, and *T* wave. *S-T* segment displacements may be present as well as *U* waves [14].

There is no reason to expect relative wave amplitudes to be the same in the MCG as in the ECG. The configurations of both the ECG and the MCG will vary with the site where they are recorded. In addition, the shape of the MCG will depend on the component of H that is selected. Cohen and Lepeschkin [14] have reported a comparison of ECG's and MCG's from seven normal males. Standard 12 lead ECG's were recorded together with scalar orthogonal Frank leads. A total of 34 MCG's were recorded from each subject on a 5 X 5 cm grid over the chest such as shown in Fig. 3. Only the component of the magnetic field normal to the chest was taken [15].

Cohen and Lepeschkin report that the *P* wave showed the most striking difference, being greatest in amplitude opposite the atria, and decreasing rapidly in amplitude as the detector was moved in any direction. The relative amplitude of the *P* wave at this site was higher than in the ECG, leading the authors to speculate that the MCG might be useful for analyzing complex arrythmias when the *P* wave is very low or hidden in the ventricular complex of the ECG.

The *QRS* complex often appeared to be more triphasic in

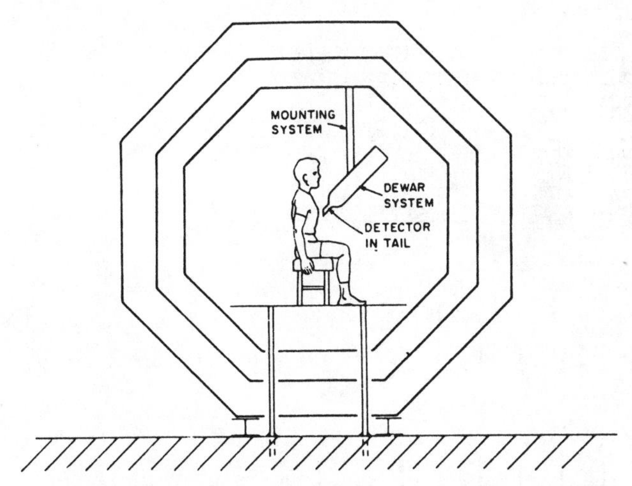

Fig. 2. Diagram of shielded enclosure at the National Magnet Laboratory, Massachusetts Institute of Technology. The room is octagonally shaped with five wall layers of shielding (three shown). The SQUID detector is situated in the tail of the Dewar containing liquid helium. All magnetic material is removed from the room. The subject is shown in orientation appropriate for recording the component of the magnetic field normal to the chest. (Illustration is courtesy of D. Cohen.)

the MCG than in the ECG. Also, maximal *S-T* displacements were relatively greater in the MCG. *T* waves tended to be of opposite polarity to *QRS* more frequently in the MCG, and were sometimes diphasic.

Baule and McFee [3] reported results from a group of 20 normal subjects and 20 subjects with heart disease. They used a two coil arrangement with one coil over the heart and a second displaced laterally by 20 cm [16] (see Fig. 1). The heart signal in the second coil was shown to be very small. Fifteen of the normal subjects had an entirely upright *QRS* deflection, with five having a small negative terminal deflection. Four of the abnormal subjects had strikingly abnormal deflections. A subject with right ventricular enlargement showed a large negative terminal deflection in addition to a larger than normal upright deflection. An infarct subject showed an entirely negative deflection

dc MCG

Recording of the MCG does not require any electrodes. The detector is not even in contact with the body. The ECG, on the other hand, is a record of the potentials appearing at electrodes attached to the skin. One characteristic of these electrodes is the appearance of dc potentials at the electrode-skin interface. These potentials are generally much larger than any dc potential of cardiac origin, they are usually highly variable from subject to subject and electrode site, they can depend strongly on skin preparation and the type of paste, and they often will drift in amplitude slowly with time. For these reasons, the very low-frequency components of the ECG are normally filtered out and do not appear in the tracing.

Since the MCG does not involve electrodes, and the SQUID magnetometer responds to dc fields, the MCG is capable of registering dc and very low-frequency components of the heart's electrical activity. Injury currents associated with ischemia and acute infarcts may well involve dc currents. Hence, the MCG could be an important tool for assessing damage to the heart under these conditions

S-T segment displacements are diagnostic indicators of ischemia and acute infarct in the ECG. Are dc currents present? Does the *S-T* event reflect some primary activity during repolarization, or is it rather a consequence of a baseline shift during the remainder of the heart cycle associated with a dc injury current? Conclusive answers to these questions have not been obtained from electrograms, in part due to the difficulty of making dc measurements. The MCG, however, may provide an answer.

Cohen and Kaufman [17] studied the dc MCG in the presence of experimental myocardial infarction in intact dogs. Their equipment had a frequency response from dc to about 40 Hz. The level of the baseline (e.g., *T-P* interval) in the MCG was determined by taking the dog out of the range of the magnetometer, and then wheeling him up to it. The difference between the MCG baseline and the out of range level was then a measure of the absolute dc level of the dog's MCG.

Occlusions were created by inflating a cuff which had previously been surgically implanted around a coronary artery. Baseline shifts and *S-T* segment shifts appeared, simultaneously, within 20 s after occlusion. The shifts were approximately equal and opposite (see Fig. 4). Their opposition was maintained for at least 15 min. When premature ventricular beats occurred in one dog, baseline shifts were also equal and opposite to *S-T* segment shifts. *S-T* segment shifts were only apparent; the *S-T* level remained constant while the baseline level on the remainder of the cycle changed. Both *S-T* segment and baseline shifts slowly decreased after 15 min.

An earlier study by Cohen *et al.* [18] had not shown a one-to-one correlation between baseline shifts and *S-T* segment shifts. In the earlier study, scans were concentrated in the period when occlusion was judged to be complete. The authors speculate that as irreversible infarction begins a new primary *S-T* component may appear. In the very early stages, however, their data are consistent with the hypothesis that the *S-T* segment shift is a secondary event which is a consequence of a periodically interrupted steady current. It remains to be seen whether the dc MCG will provide important new diagnostic information not available in the conventional ECG.

It should be noted that Cohen and Kaufman observed small steady fields at certain chest positions for some dogs. The origin of these fields is unknown.

Magnetic Susceptibility Plethysmography

If magnetic material is placed in a magnetic field, magnetic dipoles will be induced in the material, and the field will be distorted. Tissues are very weakly magnetic, and their magnetic properties can generally be neglected in analyses of the MCG. It is the conductive properties of body tissues which determine the currents in the body, and hence, the external magnetic field.

Wikswo *et al.* [19] have pointed out that the magnetic susceptibility of body tissue may have a perceptible influence on the MCG in the following way. There is a very small distortion of the earth's magnetic field due to the fact that body tissues are not completely nonmagnetic. Since the susceptibility of blood is somewhat greater than that of other tissues,

500 IEEE TRANSACTIONS ON BIOMEDICAL ENGINEERING, VOL. BME-26, NO. 9, SEPTEMBER 1979

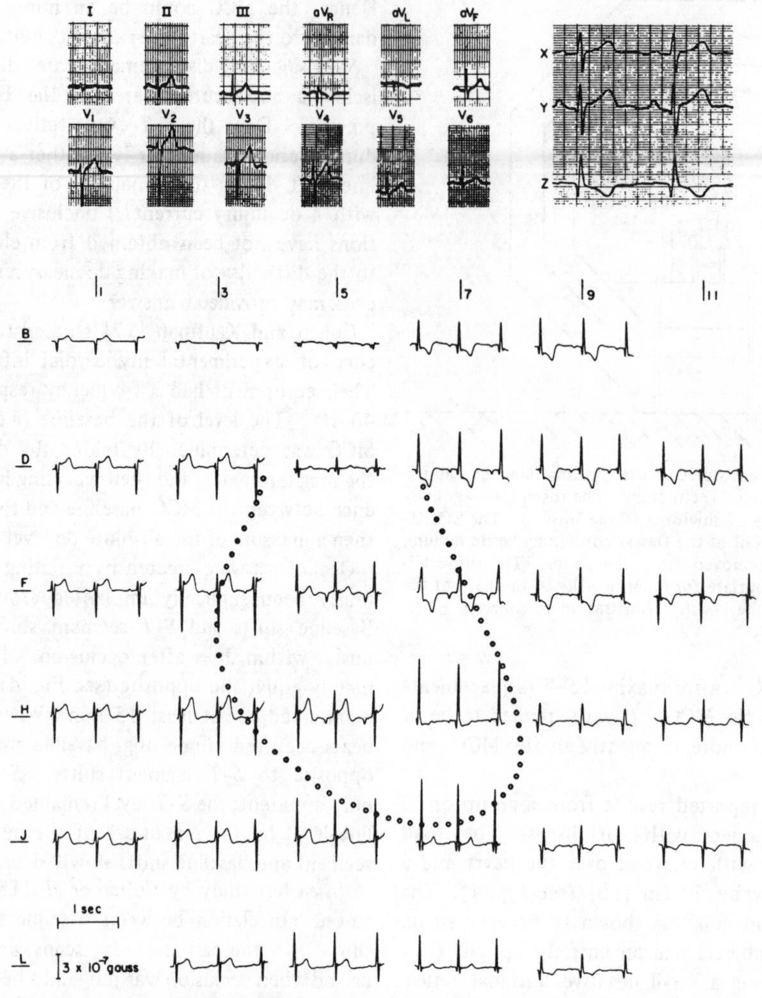

Fig. 3. MCG's recorded in the precordial region of a human subject. Component of magnetic field normal to the chest is shown. Small circles indicate X-ray outline of the heart. Conventional ECG leads and scalar Frank vector leads are shown at top. (Illustration is courtesy of D. Cohen.)

this distortion will vary with time during the cardiac cycle as the heart fills and then ejects blood.

This time varying magnetic field is extremely minute. The earth's magnetic field, however, is a million times the MCG intensity, and the resulting distortion may be an appreciable fraction of the MCG amplitude. On the basis of a simple model study, Wikswo *et al.* estimated that a 1 G field would produce a field change of 5×10^{-7} G. Hence, the earth's magnetic field of 0.5 G could produce a noticeable effect, and all MCG's recorded without shielding may be subject to this artifact. Furthermore, Wikswo *et al.* propose that the artifact itself could be used to obtain a measure of stroke volume, somewhat analogous to the technique of impedance plethysmography, where the electrical impedance across the chest varies with intracavitary blood volume (as well as air in the lungs). The effect can be greatly enhanced by using a large external constant magnetic field, and preliminary results have been obtained in this manner.

Theory

The following development of the expressions for the magnetic field arising from sources in a bounded volume conductor is adapted from Geselowitz [7] and Grynszpan and Geselowitz [9]. As with the ECG, the theory of the MCG can also be approached from the perspective of reciprocity and lead fields [2]. While the two approaches are equivalent, they can provide different insights into the relation between sources and fields. The lead field approach is often useful in developing leads, i.e., coil arrangements, which reflect particular weightings of the source distribution.

Let us represent bioelectric sources by an impressed current density J^i. Then in a region of conductivity σ the current density is

$$J = -\sigma \nabla V + J^i \tag{2}$$

where V is the electric potential. Let H be the magnetic field

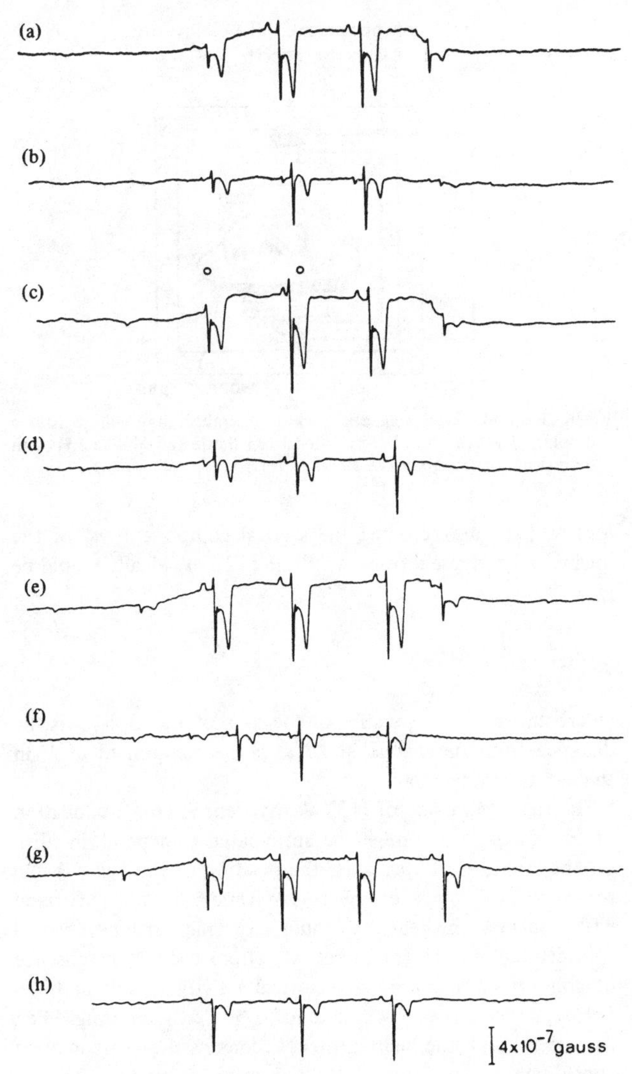

Fig. 4. dc MCG's recorded from the chest of a dog during a sequence of coronary occlusions and releases. MCG increases in amplitude as the dog is wheeled in toward the magnetometer, and then disappears as the dog is wheeled away. Baseline shifts tend to be equal and opposite to *S-T* segment shifts. (a) During first 2-min trial occlusion. (b) After release of second 2-min trial occlusion. (c) During 5-min trial occlusion. The two dots indicate PVB's. (d) Just after release of 5-min occlusion. (e) 15 min into 28-min occlusion. (f) 30 s after release of 28-min occlusion. (g) During second short occlusion following release of 28-min occlusion. (h) Just after release of this second short occlusion. (Illustration courtesy of D. Cohen.)

intensity. Since the bioelectric problem of interest is a quasi-static one,

$$\nabla \times H = J = -\sigma \nabla V + J^i. \tag{3}$$

Let

$$H = \nabla \times A. \tag{4}$$

Then

$$\nabla \times \nabla \times A = \nabla(\nabla \cdot A) - \nabla^2 A = -\sigma \nabla V + J^i, \tag{5}$$

where

$$\nabla \cdot A = -\sigma V, \tag{6}$$

$$\nabla^2 A = -J^i. \tag{7}$$

These equations may be interpreted to indicate that the electric field is related to the divergence of J^i while the magnetic field is related to the curl of J^i. This point has also been made by Plonsey [8].

For simplicity, we will assume that the torso is a homogeneous volume conductor. The results can be generalized for the case of an inhomogeneous conductor. If dS_0 is an element of area of the boundary surface of the body, and dv is an element of the volume conductor, then

$$4\pi H = \int J^i \times \nabla\left(\frac{1}{r}\right) dv + \int \sigma V \nabla\left(\frac{1}{r}\right) \times dS_0 \tag{8}$$

where r is the distance from the field observation point to dv or dS_0 and V is the electric potential (ECG) on the body surface.

A very important construct in ECG is the heart dipole or vector p which, for a homogeneous conductor, is given by

$$p = \int J^i \, dv = \int \sigma V \, dS_0. \tag{9}$$

Similarly, the magnetic dipole m is given by

$$m = \tfrac{1}{2} \int r \times [J^i \, dv - \sigma V \, dS_0] \tag{10}$$

where r is the radius vector from an arbitrary origin inside the volume conductor to dv or dS_0. Both m and p are independent of the origin.

Currents outside the volume conductor are zero, and therefore H in this region can be expressed as the gradient of a scalar potential U, i.e.,

$$H = -\nabla U. \tag{11}$$

If the magnetic scalar potential U could be readily measured, then the MCG could be presented in terms of a scalar rather than a vector field, resulting in the elimination of the problem of recording and displaying a vector quantity. Magnetometers in general will respond to H and we are faced with the problem in magnetocardiography of how best to handle the vector nature of the magnetic field intensity.

Insight into the magnetic field generated by electromotive forces in volume conductors can be achieved by considering a spherical volume conductor. Caution must be observed in interpreting the calculations since the symmetry of the sphere leads to special properties of its magnetic field. For example, a radially oriented current dipole source in a sphere will give rise to no external magnetic field. Fig. 5 shows the magnetic field of a current dipole in a homogeneous conducting sphere. It will be noted that the locus of turnover points of the field lines tends to correspond to the projection of the dipole on the surface of the sphere [6].

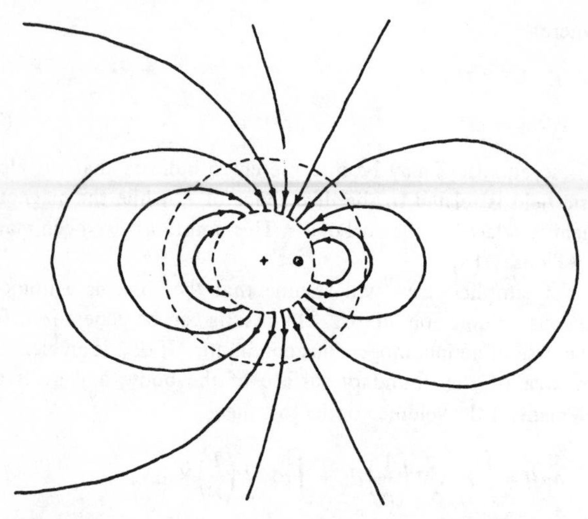

Fig. 5. Magnetic field of current dipole in a conducting sphere. Dipole indicated by ⊙ is pointing out of the plane of the drawing. Broken lines show two spheres. Pattern of magnetic field lines *outside* the surface of the sphere is independent of the size of the sphere.

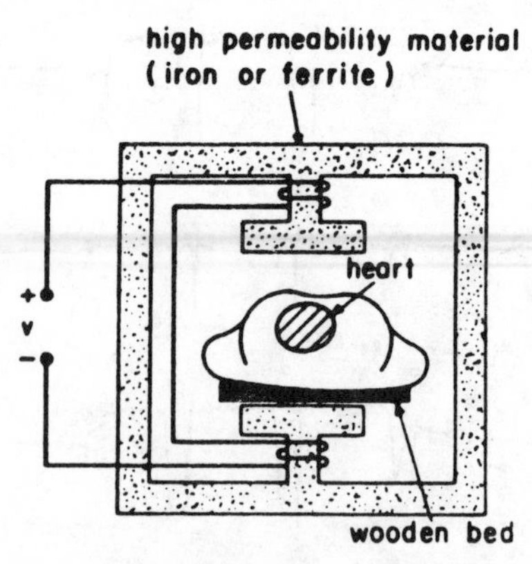

Fig. 6. Example of a magnetic pickup assembly that will register a dipolar magnetic field. (Reprinted from Baule and McFee [3] with permission of the *American Heart Journal*.)

Electric and Magnetic Vector

Equations (9) and (10) provide expressions for the electric and magnetic heart vectors, respectively (assuming a homogeneous torso). An important distinction is immediately evident. The electric dipole or vector $\boldsymbol{p}$ is the resultant of the electromotive forces in the heart, and does not depend on the volume conductor. The magnetic dipole, or vector $\boldsymbol{m}$ by contrast, does depend on the geometry of the volume. The geometrical dependence involves the ECG potential V, and in principle this term could be evaluated and "stripped away" leaving just the term involving the sources. In practice, it would seem to be entirely impractical to do this [at least starting from (10)] since evaluation of the integral requires a knowledge of the entire body surface ECG and shape of the torso. If the influence of this term could be predicted for various torso shapes, then a direct relationship between the magnetic vector and the cardiac sources could be achieved.

For convenience, let us call the first term in (10) the magnetic heart "pseudodipole" or "pseudovector." Note that this vector, which we will designate $\boldsymbol{m}'$, depends on the origin and is given by

$$\boldsymbol{m}' = \frac{1}{2}\int \boldsymbol{r} \times \boldsymbol{J}^i \, dv. \tag{12}$$

If the origin is approximately in the center of the heart, then $\boldsymbol{m}'$ would be dominated by electromotive forces perpendicular to the radius vector $\boldsymbol{r}$, or "tangential" forces. Since the predominant direction of the spread of excitation in the normal heart is from endocardium to epicardium, or in a "radial" direction, alterations in the excitation pattern in the heart might be expected to affect tangential forces disproportionately. Hence, the magnetic heart vector might possibly be more sensitive to certain pathologies than the electric heart vector [3].

Fig. 6 shows a coil arrangement proposed by Baule and McFee [3] for recording the sagittal component m_z' of the magnetic heart pseudovector. From (12), m_z' ideally would be given by

$$m_z' = \frac{1}{2}\int \rho \cdot J^i_{xy} \, dv \tag{13}$$

where the z axis is parallel to the axis of the coils, ρ is the distance from the z axis, and J^i_{xy} is the component of J^i in the x-y or frontal plane.

The right hand side of (13) clearly depends on the location of the z axis, which might be anticipated to depend, in turn, on the position of the coils, the position of the heart in the torso, and the shape of the torso. Unless the z axis is fixed with respect to the heart, m_z' could vary widely from individual to individual due to anatomical variations even in the absence of abnormalities in cardiac electrical activity. Baule and McFee argue that the axis will tend to be "self-centering," i.e., to pass through the heart center, so long as the heart may be considered to be spherically homogeneous and to be surrounded by lungs of negligible conductivity.

In practice, these assumptions hold only very approximately. Further studies are necessary to clarify the relation between the lead system of Fig. 6 and the magnetic dipole component as defined by (13). Note that Baule and McFee have also proposed coil arrangements for obtaining the other two components of the magnetic heart vector. On the basis of a model study, Malmivuo and Wikswo have suggested that the magnetic heart pseudovector could be obtained with reasonable accuracy from three mutually orthogonal coils at a single precordial site, i.e., directly from the measurement of the magnetic field intensity vector at a single point [20].

Model Studies of the Inverse Problem

On theoretical grounds, there is new information in the magnetic field not available in the electric field. Can the MCG then provide additional information which could be used to characterize the cardiac electromotive forces quantitatively with greater accuracy? This question can be approached from several directions. Rush has argued that it is unlikely that electromotive forces in the heart will give rise to currents which contribute to the magnetic field, but not to the electric field at the body surface [21]. His arguments are mildly strengthened by computer simulations in which the same electromotive forces have been shown to generate realistic normal ECG's and MCG's [22], [23]. It would appear important to extend these studies to include pathologies.

A more significant approach to the question of added information in the MCG is in the context of the inverse problem, i.e., the attempt to reconstruct the cardiac sources from available data at the surface of the torso. A promising approach to the inverse problem involves resolving the cardiac sources into a limited number of discrete dipoles, each of which ideally represents the electrical activity in a segment of myocardium [24]. Most commonly, the dipoles are fixed in orientation and location, but vary in strength to simulate the electrical behavior of the heart through its cycle [25]. Designate the strength of the jth dipole by P_j. If the inverse problem can be successfully solved, i.e., the time varying dipole moments $P_j(t)$ accurately calculated, then a quantitative characterization of the electrical activity in the heart during the cardiac cycle will be available.

Model studies to test the inverse solution have generally taken the following form. A known source is placed in a volume conductor and the forward problem is solved to give potentials at points on the surface. Next, the location and orientations of the equivalent dipoles are selected and transfer coefficients which relate each unit equivalent dipole to each surface site are determined. A set of equations is then formed which can be inverted to give the dipole strengths.

Let C_{ij}^E be the transfer coefficient which gives the contributions of the jth unit dipole to the electric potential V_i at site i on the surface of the volume conductor. If there are N equivalent dipoles, then

$$V_i = \sum_{j=1}^{N} C_{ij}^E P_j \tag{13}$$

where P_j is the strength of the jth dipole. Values of V_i for $i = 1$ to M are available from the solution to the forward problem. It is preferable to take M much larger than N. The M equations in the N unknowns can then be inverted to give the N dipole strengths. The question is how closely the inverse solution resembles the original source distribution, especially in the presence of noise. Noise can be either signal noise superimposed on the potential V, or "modeling noise" associated with variations in the assumed positions, orientations of the equivalent dipoles, or variations in the assumed shape of the volume conductor.

When the method of least squares is used to invert the overdetermined set of equations, the validity of the solution deteriorates drastically in the presence of very small levels of noise [26]. Lynn *et al.* [27] proposed an alternative form of solution in which each dipole moment is constrained to be nonnegative, and Brody and Hight [28] have shown encouraging results with this method in a model study.

The rationale for the nonnegative constraint is that the equivalent dipoles are selected so that it is unrealistic from electrophysiological considerations for their moments to be negative. It appears reasonable to be able to do this for normal hearts and a wide range of abnormalities, but perhaps not for all cardiac pathologies. The constrained solution is obtained by the method of quadratic programming [29].

Miller and Geselowitz [30] studied the effect of adding magnetic information to the solution of the inverse problem. Then in addition to (13) there is the relationship

$$U_i = \sum_{j=1}^{N} C_{ij}^M P_j \tag{14}$$

where U is the scalar magnetic potential and C^M is the corresponding magnetic transfer coefficient. The inverse solution was studied using electric potentials alone, magnetic potentials alone, and a combination of electric and magnetic potentials. A realistic cardiac generator was used and the volume conductor was taken to be a sphere. It was found that including magnetic data improved the solution in the presence of additive potential noise, but did not improve the solution in the presence of modeling noise.

In Conclusion

High-quality MCG's can be obtained with relative ease in adequately shielded chambers. Recording of MCG's in the hospital still poses problems which require solution if one wants to avoid the expense of constructing an elaborate shielded enclosure.

Clinical studies of MCG's to date are very limited. There are yet no firm data to indicate that the MCG will provide new diagnostic information not readily available from other procedures, especially the ECG. Experimental studies relating to steady currents of injury indicate considerable promise for the MCG in evaluating ischemia and acute infarcts. The sensitivity of the MCG to atrial activity may be useful in special cases.

It is not yet clear what new information is available in the magnetic field, especially when considered from the standpoint of signals in the presence of noise. Model studies have indicated some improvement when using both magnetic and electric data in the presence of some types of noise, but not in the presence of other types. One particular theoretical conclusion of interest is the enhanced sensitivity of MCG leads to tangential components of cardiac electromotive forces. Clinical studies of this effect remain to be undertaken.

Further clinical investigation of the MCG will require a decision concerning which "leads" to use. A complicating factor is the vector nature of the magnetic field intensity. It would appear worthwhile to consider a magnetic vector lead system. It should be noted that the magnetic heart vector or dipole is not the same as the electric heart vector or dipole. A further question to be resolved is the significance of the torso boundary effect on the magnetic field.

At present, the MCG is much more difficult and expensive to obtain than the ECG. No situations where the MCG provides new diagnostic information in humans have yet been demonstrated, but experimental results indicate some promising possibilities. Further research is required.

Acknowledgment

The generous assistance of Dr. D. Cohen in providing illustrations used in this article is gratefully appreciated.

References

[1] G. M. Baule and R. McFee, "Detection of the magnetic field of the heart," *Amer. Heart J.*, vol. 66, pp. 95-96, 1963.

[2] —, "Theory of magnetic detection of the heart's electrical activity," *J. Appl. Phys.*, vol. 36, pp. 2066-2074, 1965.

[3] —, "The magnetic heart vector," *Amer. Heart J.*, vol. 79, pp. 223-236, 1970.

[4] Y. D. Safonov, V. M. Pravotorov, V. M. Lube, and L. J. Yakimenkov, "Method of recording the magnetic field of the heart (magnetocardiography)," *Bull. Exp. Biol. Med.*, vol. 64, pp. 1022-1024, 1967.

[5] D. Cohen, "Magnetic fields around the torso: Production by electrical activity of the human heart," *Science*, vol. 156, pp. 652-654, 1967.

[6] D. Cohen and L. Chandler, "Measurements and a simplified interpretation of magnetocardiograms from humans," *Circulation*, vol. 39, pp. 395-402, 1969.

[7] D. Geselowitz, "On the magnetic field generated outside an inhomogeneous volume conductor by internal current sources," *IEEE Trans. Magn.*, vol. MAG-6, pp. 346-347, June 1970.

[8] R. Plonsey, "Capability and limitations of electrocardiography and magnetocardiography," *IEEE Trans. Biomed. Eng.*, vol. BME-19, pp. 239-244, May 1972.

[9] F. Grynszpan and D. B. Geselowitz, "Model studies of the magnetocardiogram," *Biophys. J.*, vol. 13, pp. 911-925, 1973.

[10] J. Zimmerman, P. Thiene, and J. Harding, "Design and operation of stable *r-f* biased superconducting point-contact quantum devices," *J. Appl. Phys.*, vol. 41, pp. 1572-1580, 1970.

[11] D. Cohen, E. Edelsack, and J. Zimmerman, "Magnetocardiograms taken inside a shielded room with a superconducting point-contact magnetometer," *Appl. Phys. Lett.*, vol. 16, pp. 278-280, 1970.

[12] M. Saarinen, P. Siltanen, P. J. Karp, and T. E. Katila, "The normal magnetocardiogram: I Morphology," *Ann. Clin. Res.*, vol. 10, suppl. 21, pp. 1-43, 1978.

[13] B. Denis, D. Matelin, C. Favier, M. Tanche, and P. Martin-Noel, "L'enregistrement du champ magnetique cardiaque–considerations techniques et premiers resultats en milieu hospitalier," *Arch. Mal Coeur*, vol. 69, p. 299, 1976.

[14] D. Cohen and E. Lepeschkin, "Review of magnetocardiography," paper for the *Proc. 12th Int. Colloq. Vectorcardiographicum*, Brussels, Belgium, Aug. 1971.

[15]. D. Cohen and D. McCaughan, "Magnetocardiograms and their variation over the chest for normal subjects," *Amer. J. Card.*, vol. 29, pp. 678-685, 1972.

[16] G. M. Baule, "Instrumentation for measuring the heart's magnetic field," *Trans. N.Y. Acad. Sci.*, vol. 27, pp. 689-700, 1965.

[17] D. Cohen and L. A. Kaufman, "Magnetic determination of the relationship between the *S-T* segment shift and the injury current produced by coronary artery occlusion," *Circ. Res.*, vol. 36, pp. 414-424, 1975.

[18] D. Cohen, J. Norman, F. Molokhia, and W. Hood, Jr., "Magnetocardiography of direct currents: *S-T* segment and baseline shifts during experimental myocardial infarction," *Science*, vol. 172, pp. 1329-1333, 1971.

[19] J. P. Wikswo, J. E. Opfer, and W. M. Fairbank, "Observation of human cardiac blood flow by non-invasive measurement of magnetic susceptibility changes," in *AIP Conf. Proc.*, vol. 18, pp. 1335-1339, 1974.

[20] J. A. V. Malmivuo and J. P. Wikswo, Jr., "A new practical lead system for vector magnetocardiography," *Proc. IEEE*, vol. 65, pp. 809-811, May 1977.

[21] S. Rush, "On the independence of magnetic and electric body surface recordings," *IEEE Trans. Biomed. Eng.*, vol. BME-22, pp. 157-167, May 1975.

[22] M. Horacek, "Digital model for studies in magnetocardiography," *IEEE Trans. Magn.*, vol. MAG-9, pp. 440-444, Sept. 1973.

[23] B. N. Cuffin and D. B. Geselowitz, "Computer model studies of the magnetocardiogram," *Ann. Biomed. Eng.*, vol. 5, pp. 164-178, 1977.

[24] E. J. Fischmann and M. R. Barber, "'Aimed' electrocardiography, model studies using a heart consisting of 6 electrically isolated areas," *Amer. Heart J.*, vol. 65, pp. 628-637, 1963.

[25] R. H. Selvester, R. Kalaba, C. R. Collier, R. Bellman, and H. Kagiwada, "A digital computer model of the vector-cardiogram with distance and boundary effects: Simulated myocardial infarction," *Amer. Heart J.*, vol. 74, pp. 792-808, 1967.

[26] C. L. Rogers and T. C. Pilkington, "Free moment current dipoles in inverse electrocardiography," *IEEE Trans. Biomed. Eng.*, vol. BME-15, pp. 312-323, Oct. 1968.

[27] M. S. Lynn, A. C. L. Barnard, J. H. Holt, Jr., and L. T. Sheffield, "A proposed method for the inverse problem in electrocardiography," *Biophys. J.*, vol. 7, pp. 925-945, 1967.

[28] D. A. Brody and J. A. Hight, "Test of an inverse electrocardiographic solution based on accurately determined model data," *IEEE Trans. Biomed. Eng.*, vol. BME-19, pp. 221-228, May 1972.

[29] P. Wolfe, "The simplex method for quadratic programming," *Econometrica*, vol. 27, pp. 382-398, 1959.

[30] W. T. Miller, III, and D. B. Geselowitz, "Use of electric and magnetic data to obtain a multiple dipole inverse cardiac generator: A spherical model study," *Ann. Biomed. Eng.*, vol. 2, pp. 343-360, 1974.

David B. Geselowitz (S'51–A'54–M'61–SM'72–F'78) was born in Philadelphia, PA, on May 18, 1930. He received the B.S., M.S., and Ph.D. degrees in electrical engineering from the University of Pennsylvania, Philadelphia, in 1951, 1954, and 1958, respectively.

He was on the faculty of the University of Pennsylvania from 1951 to 1971. Since 1957 he has been active in biomedical engineering education and research, specializing in electrocardiography. Since 1971 he has been Professor and Head of the Department of Bioengineering, The Pennsylvania State University, University Park.

Dr. Geselowitz spent 1978–1979 as a Guggenheim Fellow at Duke University, Durham, NC. He is a member of Tau Beta Pi, Eta Kappa Nu, AAPT, and the Biophysical Society. He is a Director of the Biomedical Engineering Society and a Fellow of the American College of Cardiology. He is currently Chairman of the American Heart Association Committee on Electrocardiography, and is a former editor of the IEEE Transactions on Biomedical Engineering. He currently serves on the editorial boards of the Proceedings of the IEEE and the *Journal of Electrocardiography*.

An Engineering Overview of Cardiac Pacing

By PETER P. TARJAN, Ph.D.
Chief Scientist
Cordis Research Corporation
and
ALAN D. BERSTEIN, Eng.Sc.D.
Director of Technical Research
Department of Surgery
and Technical Director
Pacemaker Center
Newark (N.J.) Beth Israel Medical Center

The history of electrical stimulation of the heart for reviving and maintaining its rhythm has been covered by several excellent reviews.[1] The roots of cardiac pacing with implantable devices may be traced back over a quarter of a century to the late 1950's, with passive implants powered by externally-worn stimulators coupled by inductive means.[2,3] Not much later, the idea of recharging the battery in an implanted stimulator through the intact skin was introduced into clinical practice. The promise of freedom from the external transmitter was very appealing, but remained unfulfilled until a suitable non-rechargeable battery was employed to power the implant.

Cardiac pacing became a form of true rehabilitation, albeit for too-short uninterrupted periods, once self-contained implants became available for clinical use in 1960. The mercury-zinc "Rubin" cells, together with advances in low-drain junction transistors and in encapsulating materials, made the implants possible. The primary technical problems remaining were those of excessive current drain in the pacemaker circuit itself, and an unsatisfactorily high battery self-discharge rate. Further development was also clearly needed to reduce the incidence of mechanical lead failure, and to optimize the electrical performance of the stimulating electrodes.

The problem of battery self-discharge was initially believed solved by radioisotopic power sources, but by the early 1970's, the availability of high-energy-density batteries using various lithium chemistries offered an irresistible alternative. Greatbatch traces the development of pacemaker power sources in another article in this issue.

Lead breakage due to mechanical fatigue or wear began to disappear with the use of multifilar conductors and polyurethane insulation. Current drain was steadily reduced, mainly by advances in CMOS integrated circuits, until the formation of the stimulus became the major burden on the battery. This aspect of pacing has received much attention in recent years through improvements in electrodes, including reductions in size, modifications in shape, and the use of different materials. For example, the substitution of porous for mirror-smooth surfaces, and the use of carbon rather than more polarizable metals have helped to reduce the current required for stimulation.

This article deals with several major issues of cardiac pacing:

1. An overview of cardiac stimulation;
2. Functional improvements in pacing;
3. The problems of sensing spontaneous cardiac events; and
4. Some technical aspects of pacemaker follow-up.

Stimulation of the heart

Cells in the heart, as elsewhere, are separated by membranes from their environment. The 10nm-thick membrane of a living cell exhibits fascinating voltage-sensitive permeability characteristics toward the ions in its fluid surroundings. These characteristics are responsible for a resting membrane potential difference of about 90mV (outside positive) on account of an electrical dipole layer supported by the membrane. When the resting state of the cell is disturbed, an **action potential (AP)** evolves that is characteristic of the cell and its physiologic state. In cardiac and skeletal muscle cells, which form fibers, the AP is followed by their shortening. Cardiac muscle fibers are also distinguished by their ability to propagate the AP from one to another, which results in neatly organized contractions of the heart.

The "all or nothing" principle in cellular electrophysiology (EP) is based on the observation that resting cells either respond to a stimulus with a fully developed AP, or they do not respond at all. The stimulus intensity at which a response occurs is the threshold for the prevailing situation.

The "all or nothing" principle also applies to systems, or syncytia, of cells, such as the muscle mass of the heart. Fortunately, not all the cells need be depolarized at once by a stimulus to obtain an organized contraction of the heart. This reaction is attributed to propagation of APs along the fibers, and to the associated current flow that induces depolarizations in adjacent fibers. At least in principle, a single fiber could be depolarized and then cause the entire organ to follow. There are some theories based on empirical data, however, that associate propagated APs with the initial depolarization of a **critical mass or volume of excitable heart muscle.**[4] These theories explain some of the difficulties in obtaining an organized contraction in response to stimulation with an intracellular microelectrode.

Depolarization of the heart can be stimulated through an electrode in either epicardial or endocardial contact with heart muscle. (Transthoracic stimulation is possible, but requires much larger amounts of energy and usually causes severe discomfort because of induced skeletal-muscle contractions. Esophageal stimulation is also possible.) The size and shape of the electrode is of importance. Other significant factors are the electrode material, surface finish, site, and the time elapsed since implantation.

Stimulation can be provided either via two electrodes, not necessarily of equal size, but both in contact with or close to heart muscle, or by a small electrode in contact with the heart and a much larger return electrode situated elsewhere in the body. The former is called **bipolar** and the latter **unipolar** stimulation. The stimulation threshold is characterized at any instant by two parameters: the intensity and the duration of the stimulus required to produce a cardiac contraction. The intensity may be specified in terms of either the voltage applied between the two electrodes or the current passing through them. In either case, the relationship between intensity and duration is characteristically hyperbolic, provided the amplitude is averaged over the duration of the stimulus.

This hyperbolic relationship was described by Lapicque[5] at the turn of the century:

$$\text{Intensity} = \frac{A}{T} + B$$

where T is the stimulus duration and A and B are system constants.

For infinitely long stimuli, the intensity

Dr. Tarjan is Chief Scientist for the Cordis Research Corportation in Miami. He received the B.S.E.E. in 1959 from Purdue University, the S.M.E.E. in 1960 from the Massachusetts Institute of Technology, and the Ph.D. in Biomedical Engineering in 1968 from Syracuse University. Dr. Berstein is Director of Technical Research, Department of Surgery, and Technical Director of the Pacemaker Center, Newark Beth Israel Medical Center in Newark, New Jersey. He received the B.A. in Music History and Theory in 1960 from Bowdoin College, Brunswick, Maine, the M.S.E.E. in 1975 and the Eng.Sc.D. in 1979, both from the New Jersey Institute of Technology, Newark.

0278-0054/84/0200-0010$1.00 © 1984 IEEE

becomes B, which Lapicque called the rheobase. The duration, T_c, at an intensity of twice the rheobase is defined as the chronaxie:

$$T_c = \frac{A}{B}$$

If the intensity is given in terms of current, I, then a second relationship between charge, Q, and duration can be used:

$$Q = \int_0^T i(t)dt = \bar{I}T = A + BT$$

where I is the mean intensity.

This relationship is linear, and is easy to test using rectangular current pulses for stimulation, so that the mean intensity and the instantaneous intensity are equal. Empirical data indicate that the linear relationship does not hold for very short pulses, substantially below the chronaxie. There the relationship tends toward a constant value. For the sake of conserving the power source, one would like to use stimuli of short duration with minimal charge consumption. Naturally, the intensity must maintain a margin of safety above the threshold of stimulation.

Human clinical data has revealed, for specific electrode types, a relationship between threshold values and time following implantation. It appears that over a period of 7 to 14 days, the threshold intensity increases to a peak of possibly 10-times the initial value, although more recent studies[6] have shown significantly smaller increases. The threshold intensity gradually declines following the peak, and usually stabilizes somewhere between the initial value and perhaps twice that value. This threshold evolution is attributed to tissue irritation followed by the development of an organized fibrous capsule of non-excitable tissue that separates the surface of the electrode from the excitable tissue (see Figure 1). Several sets of clinical data have allowed the establishment of a relation between threshold parameter values and the surface area of the active electrode. This relationship is based on statistical data for certain electrodes at implantation and after maturation of the electrode-tissue interface. It has also been suggested[9] that a linear relation exists between the square root of the surface area and the chronaxie. This relation, taken together with the hyperbolic threshold approximation described above, offers a systematic approach to the establishment of an adequate safety margin in clinical pacing.[10] Initial threshold currents, at a 1 ms stimulus duration, decrease linearly with increasing surface area of the smaller of the two electrodes.

This brings forth the concept of current density at the site of excitation. For 1 ms pulses, the threshold current density, J, is about 50 μA/mm².

Over the long term, larger electrodes were found to be associated with smaller increases in threshold intensity than were smaller electrodes. This finding is explained on the basis of the separation between the electrode and the excitable tissue. To achieve a response, it is necessary for the critical current density to be reached at the nearest excitable tissue site. The distance from the electrode's surface to this tissue is determined by the thickness of the fibrous capsule that forms around the electrode.

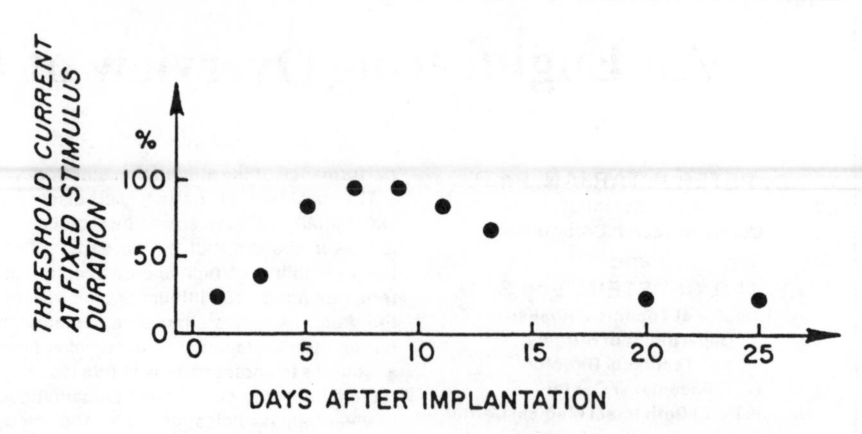

Figure 1: The evolution of the threshold of cardiac stimulation during the first several weeks after electrode implantation. The changes in threshold current are associated with maturation of the electrode-tissue interface.

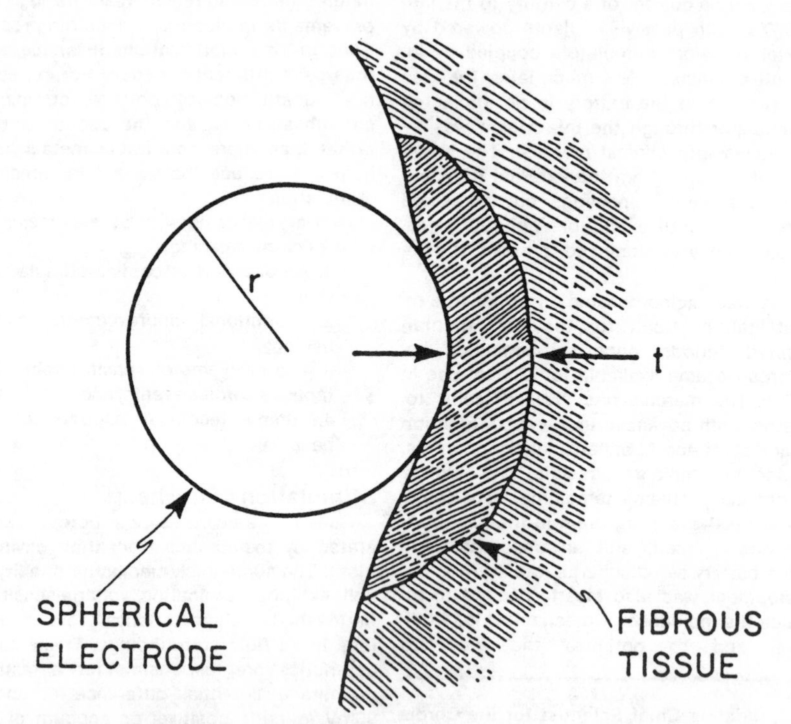

Figure 2: A schematic representation of a spherical electrode in contact with the heart, used in discussing threshold current.

Consider a spherical electrode of radius r, in contact with the heart, and connected to a power source by a thin insulated wire that is assumed not to disturb the spherical symmetry (see Figure 2).

The initial (acute) threshold current is related to the electrode surface area, S, by:

$$I_a = SJ_{th} \simeq 4\pi r^2 J_{th}$$

The eventual (chronic) value is determined by the thickness, t, of the capsule:

$$I = 4\pi (r + t)^2 J_{th}$$

It should be noted that the ratio of chronic to acute threshold current:

$$\frac{I_{ch}}{I_a} = \frac{(r + t)^2}{(r)^2}$$

is strongly dependent on the tissue reaction to the electrode as expressed by t.

Shortly after implantation of the electrode, the surrounding tissue swells in response to irritation, and the separation between the electrode and excitable tissue is large. A ten-fold rise in threshold intensity at the peak value indicates that the affected tissue thickness is about the diameter of the spherical electrode, but when the interface matures the separation produced by the encapsulation is much smaller.

There are two important considerations for choosing the size or surface area of an electrode. The smaller the size, the less current is needed for stimulation. But with smaller size, the early rise in threshold can be significant, and the chronic reaction becomes important.

A quantitative approach is offered by considering the effects of electrode resistance and size on the voltage required to achieve the desired current flow.

The resistance, R, between a small spherical electrode, of radius r, concentric with a much larger spherical electrode, can be show to be:

$$R = \frac{\varrho}{4\pi r}$$

where ϱ is the resistivity of the medium between the electrodes.

The voltage requirement (by Ohm's Law) is:

$$V_{th,ch} = \left[4\pi(r + t)^2 J_{th}\right]\left[\frac{\varrho}{4\pi r}\right] = \frac{(r + t)^2}{r} \varrho J_{th}$$

$V_{th,ch}$ has a minimum value when $r = t$, the consequences of which are fairly obvious.

The power source may also be optimized for its size on the basis of energy expenditure, E_{th}, per stimulus at threshold:

$$E_{th} = (I_{th})^2 R$$

The optimal size from this viewpoint is $r = t/3$.

These considerations have ignored the phenomenon of electrode polarization, which occurs when current flows from a metal electrode into an electrolyte. The polarization voltage is current-density dependent. The relation is non-linear, but for a first approximation one may assume that:

$$V = CJ$$

where C is a constant dependent on the choice of the electrode material and on the electrolyte. Thus, for the acute case:

$$V_{th} = J_{th} SR + CJ_{th}$$

In general:

$$V_{th,ch} = \frac{J_{th} \varrho (r + t)^2}{r} + C \frac{J_{th} (r + t)^2}{r^2}$$

Polarization complicates optimization, but the above equation shows the importance of low-polarization electrodes for the chronic case. The polarization loss depends on the current density at the electrode surface, rather than at the excitation site.

The optimal size of our spherical electrode is then

$$r = \tfrac{1}{2} t (1 + \sqrt{1 + 8C/t})$$

The conclusions from the analysis of the spherical unipolar system can be generalized to hemispherical and other electrode shapes; the less the tissue response and polarization, the better the performance. The smaller the electrode, the lower the acute threshold. Small electrodes, however, are associated with high impedance. Hence, the selection of the power source must take into consideration not only the current, but also the voltage requirement for stimulation.

Functional developments

The functional development of clinical pacing has progressed through several stages from asynchronous ventricular stimulation (the same rate of stimulation regardless of the spontaneous activity of the heart), to the presently most advanced dual-chamber demand pacing.

It was recognized in the early 1960's that a pacemaker that synchronized ventricular contractions with spontaneous atrial activity would provide a natural (physiologic) pattern of cardiac contraction and thereby greatly improve hemodynamic function. This seemingly simple concept required significant changes in pacemaker pulse generator design. A sensing system was added to detect atrial depolarizations, whose amplitude and frequency characteristics overlapped significantly with those of power-line interference and skeletal-muscle potentials. Also, the atrial electrodes were not, and still are not, sufficiently immune to sensing spontaneous ventricular activity occurring several centimeters from the atrial sensing electrodes, but produced by a much larger "generator." The signals are typically a few mV in amplitude and 10 to 20 ms in duration.

Another avenue of development was toward "non-competing" or "on demand" pacemakers for patients with intermittent need for pacing support. To keep the lead simple, the stimulating and sensing functions were both assigned to the same electrode set. After delivery of a stimulus, the sensing circuit is rendered inoperative for several hundred milliseconds due to polarization of the electrodes. Fortunately, this effect did not impede progress, as the natural refractory period of the heart muscle (its inability to respond to stimulation while in the state of depolarization or recovery from a previous depolarization) is of about the same duration. After an effective stimulus, therefore, there is no spontaneous activity to be sensed, apart from the response to that stimulus.

Electromagnetic interference from microwave ovens prompted significant changes in pacemaker design and encapsulation in the early 1970's; all pulse generators are now shielded by metallic enclosures, fitted with frequency-selective feedthrough connectors, and designed with filtering techniques related to time and frequency. When the level of interference is able to overcome all of these protective mechanisms, a final electronic safeguard comes into play: while the interference is present, the system operates in the asynchronous mode.

The desire to change the stimulus intensity noninvasively, in order to save battery power and solve some electrode problems, and the need to change the rate of stimulation for the sake of higher cardiac output led to the development of noninvasively "programmable" pacemakers. For the most part, these pacemakers communicate with the external programming device by means of pulsed magnetic fields or pulsed radio-frequency transmission. Earlier systems used simple pulse-counting methods, but contemporary systems employ binary-coded trains, thereby increasing transmission speed. The noninvasively programmable "states" have increased from 24 (6 rates and 4 intensities) to tens of millions of potentially useful and permissible combinations of parameter values. In addition to rate and intensity, contemporary pacing systems permit the alteration of sensitivity, stimulus duration, and refractory period for each channel. They also provide for changes of stimulation mode (ranging from asynchronous single-chamber pacing to dual-chamber "on-demand" stimulation), for changes in electrode configuration from unipolar to bipolar, and for alterations in certain timing functions, such as the delay between atrial sensing and the subsequent ventricular stimulus.

"Physiologic" pacing today stands for the concept of an adaptive system that will provide the appropriate stimulation pattern, regardless of the reason for an otherwise slow heart rate. If the atrial rate is too slow, then only the atrium is stimulated; if heart block develops, then synchronization of the ventricle with the atrium is maintained; if both sinus bradycardia and heart block are present, then the atrium and ventricle are stimulated in the physiologically proper sequence. Physiologic pacing is generally accepted as being able to improve cardiac output by 10 to 20 percent in the typical patient, as compared with ventricular pacing alone.

There is only one form of bradycardia that cannot be treated adequately by presently available pacemakers. This is where a higher cardiac output is needed temporarily by a patient with sinus bradycardia whose spontaneous atrial rate does not increase substantially with exercise or stress. One example is an athlete with sinus bradycardia who needs a faster heart rate to increase his cardiac output during exercise. Another is a patient in atrial fibrillation. Such patients might benefit from new pacemakers that are equipped to detect changes in some biological parameter that is linked to increased load on the heart. The theoretical range of such parameters is broad, but presently there are only investigational prototypes in clinical testing. The sensors range from blood pH,[11] through mixed venous oxygen saturation,[12] and body motion sensed by accelerometers,[13] to those dependent on electrical parameter changes such as the intracardiac stimulus-to-T-wave interval.[14] (Details on this topic are contained in the article by Fearnot, Smith, and Geddes in this issue.)

Additional capabilities afforded by some new cardiac pacemakers include the selection of certain special stimulation patterns for the automatic termination of tachyarrhythmias. This news field requires an understanding, gained through electrophysiologic testing methods, of the abnormalities in the heart's conduction system during tachycardia. (Mehra's article in this issue describes this application.)

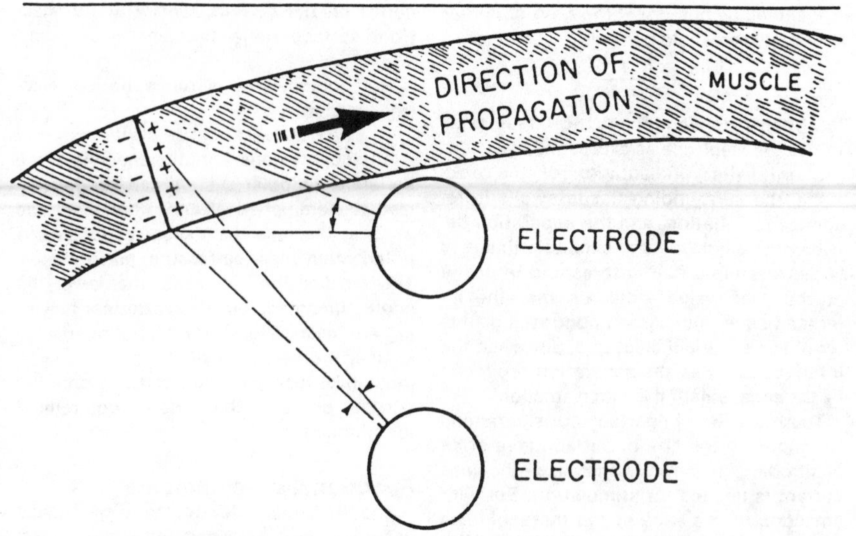

Figure 3: A schematic representation showing response to a moving dipole layer as a function of electrode position relative to the layer. This model is used in explaining the polarity and amplitude of signals detected by the electrode.

Sensing spontaneous events

Depolarization of the heart is a highly organized series of events — action potentials — in space and time. The detection, by a pacemaker, of the spontaneous activity of the heart calls for electrical sensing circuits that are able to detect minute (1-10 mV) signals having spectra in the 10 to 100 Hz range, and maximum instantaneous rates of change of about 1 V/sec. The spectrum of a QRS complex (associated with ventricular depolarization) is significantly different from that of a T-wave (which occurs during ventricular repolarization), hence the two can be separated by a high-pass filter. The same is not quite true for differentiating between premature ventricular beats and normal ventricular events. Successfully identifying spontaneous atrial activity is often problematic when large ventricular complexes are also sensed by the atrial channel of a dual-chamber or an atrial pacemaker.

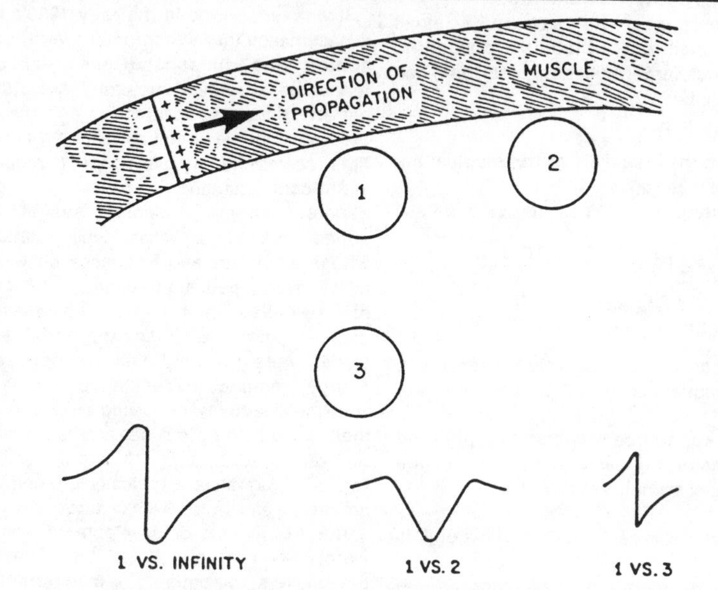

Figure 4: A comparison of unipolar (1 vs. infinity) and bipolar (1 vs. 2, 1 vs. 3) sensing of a moving dipole layer. Bipolar sensing is seen to be highly dependent on electrode orientation.

The nature of the problem might be easier to understand by employing a simplified model for depolarization. The propagation of the activity can be depicted as an electrical dipole layer moving through the heart in a direction perpendicular to the surface to which an electrode is attached (see Figure 3). With respect to infinity, the electrode's potential is proportional to the solid angle between it and the perimeter of the traveling dipole layer, as described. As the dipole approaches the electrode, this angle becomes larger, until the electrode begins to view the oppositely-charged side of the dipole surface. This charge reversal causes a change in the signal's polarity; if the dipole layer is planar, the polarity change is abrupt, with a resultant large time derivative occurring at that moment. A warped, non-planar dipole layer's passage causes a more gradual reversal of the signal's polarity. Also, as the electrode becomes separated from the surface of the tissue that contains the moving dipole surface, the solid angle is reduced for each position of the dipole surface, thus resulting in a signal of smaller amplitude.

This model also allows construction of signals from the moving dipole layer that would be received by pairs of electrodes with perpendicular or parallel orientation with respect to the tissue, as illustrated in Figure 4. The bipolar electrode pairs can be seen to be much more sensitive to nearby events than are the unipolar electrodes. The sensitivity to local activity is enhanced by reducing the separation between the electrodes, but at the same time the intensity of the signal is diminished.

This simple model should serve to show why a bipolar electrode pair is suitable for rejecting "far-field" signals from the other heart chambers and interference from sources outside the body, while selectively detecting the activity in the chamber where it is situated. The model also shows the relationship between the propagation velocity of the dipole layer and the rate of change in the detected signal, as well as the importance of the orientation of the axis of the electrodes with respect to the direction of propagation of the dipolar surface.

As the atrial wall is much thinner than the apical region of the right ventricle, it is suggested by the model that a dipolar surface propagating through the atrium would produce a smaller signal, in a given pair of electrodes, than would be detected by the same electrodes placed in the right ventricle. This is indeed the case.

In summary, the detection circuit must be so designed that, for the expected range of electrode configurations, it should be able to detect signals that represent the depolarization activity in the adjacent chamber and reject all other signals. This goal is accomplished by using bandpass filters with carefully chosen cutoff frequencies and "skirts." If the low cutoff frequency is too low, then problems arise because of T-wave sensing and the detection of the electrodes' polarization after a stimulus. At the same time, there is an increase of the amplitude of the filtered signal of interest because its spectrum reaches below the proper cutoff frequency. The choice of the upper cutoff frequency is also critical, especially for unipolar systems where it is necessary to reject muscle potentials generated by tissues near the metal case of the implant. This spectrum often overlaps that of the cardiac signal, but has a much broader range.

In addition to filtering in the frequency domain, there is a certain amount of sorting of signals based on the timing of events. For example, two ventricular signals cannot occur within 200 ms because of the refractoriness of the muscle. Also, the timing of atrial and ventricular events allows some gating of signals from the sensing circuit so as to prevent the system from interpreting stimulus-related artifacts as cardiac signals.

In the future, the waveshape of the intracardiac signal also may be analyzed to determine its validity and origin. It is not yet feasible, in a low-power implantable system, to digitize and process the signals fast enough to attain the necessary, timely pacemaker response.

Pacemaker follow-up

Certain aspects, such as monitoring rate and pulse duration of follow-up, have become simpler with improvements in the quality of implants and the implantation process. Other aspects became more complex as the complexity of the pacing systems themselves has increased. Acute problems are often caused by improper lead placement and superficial (or omitted) testing of lead performance during implantation.

The information collected during implantation also serves as a baseline for later troubleshooting. It is useful to characterize each electrode and electrode pair by obtaining "strength-duration" relations (stimulation-threshold measurements) at several different stimulus durations, determining the apparent electrode impedance (typically by taking the ratio of simultaneous, instantaneous voltage and current). It is also useful to record the waveforms of spontaneous events, as small electrode displacements can be diagnosed by changes in these waveforms. Once the stimulator is connected to the leads, a record of the programmed stimulus waveform should be obtained, as changes in this waveform can be deduced later from body-surface recordings, and observed directly during pacemaker replacement. Changes in the stimulus-artifact waveform, detected during routine follow-up, often provide the first clue to malfunction in the pulse generator or leads, permitting appropriate action to be taken before clinical problems arise. Naturally, the chosen settings of the programmable parameters should be added to the implantation records.

Pacemaker follow-up is performed either through a telephone line, in person, or a combination of both. (Details of transtelephone follow-up systems can be found in the article by Herzeler in this issue.)

Office visits allow the recording of stimulus-artifact waveforms and high-quality electrocardiograms. Time measurements can be performed but, in addition, the stimulus duration can be measured. Computerized data processing facilitates the identification of trends in changing parameters. In older pulse generators, battery depletion is heralded by a change in rate. In newer models, where timing is controlled by a very stable crystal clock, battery depletion or system malfunction typically causes a back-up system to continue stimulating the heart, but at a significantly different rate. This rate change can be easily detected during follow-up so as to allow timely replacement of the pulse generator. During office visits, stimulation and sensitivity thresholds can be estimated using the appropriate external programming device; new values may be chosen, if needed, to maintain appropriate safety margins.

In addition to the use of computers for management of follow-up data, it is already possible to automate some of the measurements and programming maneuvers, in order to save time and make the tests more uniform and comprehensive. Techniques are now under development whereby the computer can assess follow-up measurement results in terms of earlier values, to aid in detecting possible measurement errors or actual system changes that may require intervention.[15]

REFERENCES:

1. D.C. Schechter, Background of clinical cardiac electrostimulation. I. Responsiveness of quiescent, bare heart to electricity. **N.Y. State J. Med.**, 71:2575, 1971.
2. W.D. Widemann, W.L. Glenn, L. Eisenberg, and A. Mauro, Radio-frequency cardiac pacemaker. **Annals of the N.Y. Acad. of Sciences**, 111:992, 1964.
3. L. Cammilli, R. Pozzi, G. Pizzichi, and G. De Saint-Pierre, Radio-frequency pacemaker with receiver coil implanted on the heart. **Annals of the N.Y. Acad. of Sciences**, 111:1007, 1964.
4. F.W. Lindemans and Jan J. Denier Van Der Gon, Current thresholds and liminal size in excitation of heart muscle. **Cardiovascular Research**, 12:477, 1978.
5. L. Lapicque, Definition experimentale de l'excitabilite. **Soc. Biol.**, 77:280, 1909.
6. G.C. Timmis, J. Helland, D.C. Westveer, J. Stewart, and S. Gordon, The evolution of low threshold leads. **Clin. Prog. in Pacing and Electrophys.**, I:326, 1983.
7. S. Furman, B. Parker, and D.J.W. Escher, Decreasing electrode size and increasing efficiency of cardiac stimulation. **J. Surg. Res.**, 11:105, 1971.
8. N.P.D. Smyth, P.P. Tarjan, E. Chernoff, and N. Baker, The significance of electrode surface area and stimulating thresholds in permanent cardiac pacing. **J. of Thorac. Cardiovasc. Surg.**, 71:559, 1976.
9. W. Irnich, The chronaxie time and its practical importance. **Presentations of the 1st Europ. Symp. on Card. Pacing, London**, 69, 1978.
10. A.D. Bernstein, V. Parsonnet, S. Saksena, and E. Shilling, Threshold curve approximations for pacemaker output programming. **Proceed. of 2nd Europ. Symp. on Card. Pacing, Florence**, 115, 1982.
11. L. Cammilli, L. Alcidi, G. Papeschi, V. Wiechmann, G. Grassi, and L. Padeletti, Clinical evaluation of the pH-triggered pacemaker one year after implantation in man. **Presentations of the 1st Europ. Symp. on Card. Pacing, London**, 39, 1978.
12. A. Wirtzfeld, K. Stangl, R. Heinze, Th. Bock, and H.D. Liess, Mixed venous oxygen saturation for rate control of an implantable pacing system. **Proceed. of the VIIth World Symp. on Card. Pac.**, 271, 1983.
13. D.P. Humen, K. Anderson, D. Brumwell, S. Huntley, and G.J. Klein, A pacemaker which automatically increases its rate with physical activity. **Proceed. of the VIIth World Symp. of Card. Pac.**, 259, 1983.
14. A.F. Rickards and R.M. Donaldson, Rate responsive pacing suing the TX pacemaker. **Proceed. of the VIIth World Symp. on Card. Pac.**, 253, 1983.
15. A.D. Bernstein and V. Parsonnet, Microcomputer and microprocessor applications in cardiac pacing. **Medical Instrumentation**, 17:329, 1983.

A new convention at the convention

Attendees at most conventions that I have attended wear their nametag pinned to their left lapel. For right-handed people this is the easiest position to reach. I would like to propose a new convention, to start at the annual IEEE/EMBS Conference. That is —wear your nametag on the right lapel.

When meeting an old acquaintance whose name you might just have forgotten, it is more graceful to read the nametag than to try to make excuses for a faulty memory. In this situation you try to sneak a look at the name while energetically shaking hands with this almost remembered person. For those of us approaching 50, farsightedness and forgetfulness, a quick downward glance at the nametag is a neater gesture, and usually more successful than the awkward right-to-left, swivel-neck manuever. So let's all try something new when we meet in Los Angeles!

— ALVIN WALD

Transcutaneous Electrical Nerve Stimulation for Pain Control

By ANDREW Y.J. SZETO, Ph.D.
San Diego State University
San Diego, California
and JUDITH K. NYQUIST, Ph.D.
Pain Treatment Center
Scripps Clinic & Research Foundation
La Jolla, California

Pain is a sensory experience common to nearly all of us and one we avoid or minimize. This negative connotation obscures the fact that an organism's capacity for survival as an individual, as well as survival as a species, depends on neural circuitry that features both a muscular reflex withdrawal mechanism in response to a potentially damaging stimulus and a perceptual mechanism to allow the organism to formulate and carry out more complex escape/withdrawal behavioral strategies. Individuals with congenital insensitivity to pain bear the scars of a lifetime of damaging environmental stimuli that go unperceived and unheeded. Neurological and neuropathological studies on some of these pain-insensitive individuals suggest abnormalities of the sensory input fibers and/or the preliminary processing neurons in the spinal cord. The ability to respond to pain, then, is an asset when it causes us to pull our burned finger away from a hot stove or motivates us to consult a physician when severe abdominal pain signals an inflamed appendix.

There are instances, however, when pain persists beyond the time of usefulness as a protective, warning system or beyond the normal healing process. There is no longer any offending stimulus, yet the neural circuitry has gone awry and continues to signal "pain" to the individual. Pain of this type is pathological, whether physical or psychological or both. If this pain persists beyond a period of six months, it is called chronic pain. Examples of chronic pain include low back pain, phantom limb pain, and peripheral neuralgias. Whereas acute pain usually is associated with tissue damage and is more amenable to treatment, chronic pain can be stressful and socially debilitating, often accompanied by depression. Disturbance in sleep and appetite, preoccupation with the body, and general irritability are common.

The idea of electrical stimulation for pain relief is not a new one. Ancient Greek records report that an electric fish could produce numbness. Later refinements included placing the fish directly on the painful site resulting in relief, for example, from gout and headache. The development in non-Western cultures of acupuncture based on mechanical sensory stimulation with needles has evolved into electroacupuncture — a combination of an ancient technique and electrical stimulation. However, the current therapy of transcutaneous electrical nerve stimulation for pain relief is based on a different rationale that developed from neuroanatomical and neurophysiological observations, as well as neurosurgical reports.

Specificity theory

One of the early theories of sensation was the specificity theory, which was based on the assumption that a direct pathway exists through the nervous system from a receptor in the skin to the brain for each sensation — touch, warmth, cold, pain. This view of a specific pain pathway probably dates from Descartes (1644) who used the analogy of pulling at the end of a rope (the skin receptor) to ring a bell at the other end (the pain center in the brain). In 1894-95, von Frey proposed a theory of cutaneous senses based more on inference and deduction than on scientific data. He assigned a type of specialized anatomical structure in the skin to each particular sensation. According to his theory, Meissner corpuscles subserved the sensation of touch, Krause end-bulbs were responsible for cold sensation, and Ruffini end-organs were warmth receptors. Since pain spots in the skin were found everywhere, and since free nerve endings (with unmyelinated nerve fiber terminals and undifferentiated receptor structures) were likewise nearly ubiquitous, von Frey deduced that the free nerve endings were pain receptors.

During the last 50 years, the findings of neuroanatomists and neurophysiologists were incorporated into the specificity theory. On the basis of fiber diameter, specific nerve fibers were assigned specific sensory modalities, determined by the receptor to which they were attached. Touch fibers are large diameter (10-20 μm) and myelinated; cold fibers are the smaller (1-6 μm) myelinated fibers. Pain sensibility is carried by both small myelinated A-delta fibers, and warming activates even smaller (0.5-1 μm) unmyelinated C fibers. These pain fibers synapse onto specific pain neurons in the upper portion of the spinal cord gray matter. The axons of these pain neurons cross the midline and ascend in the anterolateral portion of the spinal cord to the thalamus.

For most specifists, the pain center in the thalamus was the end point of the pain pathway in the central nervous system. This premise was based primarily on the observation that cerebral cortex damage or removal did not result in the abolition of pain sensation. Even the specificity theorists, however, were uncertain about the role of the cerebral cortex in pain processing.

Pattern/convergence theory

The specificity concept of the pain pathway outlined above provided the rationale for pain treatment by sectioning nerves or making destructive lesions to interrupt the transmission of pain signals. Unfortunately for the patient, the pain relief following such procedures was often temporary with a return of the pain in six to nine months. Sometimes the pain was worse than before the surgical procedure. Thus, the specificity theory was inadequate to explain pain mechanisms.

Furthermore, it was known that non-painful stimulation of the skin could, in some cases, evoke pain and that the pain could outlast the period of actual stimulation. Such observations suggested that it was the pattern of activity that is important in pain processing and not simply activation of an element in a specific pain pathway. Neurophysiological experimentation has borne this out by showing that even at the receptor/peripheral nerve level, there are single receptors that respond not exclusively to painful (or nociceptive) stimulation but also to non-nociceptive stimulation (touch, temperature). Such multimodal receptor properties refute the idea of absolute receptor specificity.

Dr. Nyquist is Clinical Research Coordinator at the Pain Treatment Center, Scripps Clinic and Research Foundation. She received the B.A. in Biology and Chemistry in 1963 from St. Olaf George College, Northfield, Minnesota, and the Ph.D. in Physiology and Biophysics in 1969 from the University of Washington. Dr. Szeto is Associate Professor in the Department of Electrical and Computer Engineering at San Diego State University. He received the B.S. in Electronics from UCLA in 1971, the M.S. and M.Engr. in Bioelectronics and Control Systems from the University of California-Berkeley in 1973 and 1974, and the Ph.D. in Man-Machine Systems and Rehabilitation Engineering from UCLA in 1977.

0278-0054/83/0400-0014$1.00 © 1983 IEEE

Melzack-Wall theory of pain

Other experimental evidence has shown that although some spinal cord neurons respond exclusively to nociceptive stimulation, others respond to a convergence of both nociceptive and non-nociceptive inputs. Such neurophysiological data culminated in a new theory of pain proposed in 1965 by Psychologist Ronald Melzack and Neurophysiologist Patrick Wall.[1] This theory of pain is based on the interactions between different peripheral nerve inputs to spinal cord neurons. These spinal cord neurons, when activated, give rise to the experience of "pain."

There are several important features of this model, one of which has led to the development of electrical stimulation for pain relief. Melzack and Wall's model showed that whereas both nociceptive (painful) and non-nociceptive inputs converge onto the same spinal cord neuron, the inputs themselves interact in a special way. The model proposed that non-nociceptive input fibers can inhibit nociceptive input fibers before they synapse on the spinal cord neuron. The model predicted, then, that activating the non-nociceptive input fibers (naturally, by mechanical rubbing of the skin, or by low-intensity, repetitive electrical stimulation of appropriate fibers) would inhibit the nociceptive input and thereby reduce pain. The development of electrical stimulation for pain relief will be discussed in the next section of this paper.

Another feature of the Melzack-Wall model of pain, which was not very well developed at the time, was a black box labelled "central control." This central control was intended to account for the broad range of pain modulating influences such as placebo response, hypnosis, medication, stress, and distraction. Since the model was originally proposed, a great deal of research has focused on descending inhibitory modulation by endogenous opiate (endorphin, enkephalin) and non-opiate (serotonin) systems. These descending systems are implicated in some later refinements of Transcutaneous Electrical Nerve Stimulation (TENS) and will be discussed in the third section of this paper.

Development of TENS

Because Melzack and Wall's gate theory of pain[1] was amenable to experimental verification, many experiments were conducted to verify its basic tenets. Wall and Sweet[2] first demonstrated in the human that stimulating primary afferent neurons could relieve pain. Since that demonstration, various invasive and noninvasive electrical stimulators have been devised for pain control in man.[3,4,5]

The four general types of electrical stimulation in common use are: 1) TENS (transcutaneous electrical nerve stimulation) employing skin surface electrodes; 2) implantable peripheral nerve stimulation; 3) deep brain stimulation; and 4) dorsal column stimulation (DCS) in which electrodes are placed either into the subdural space or outside the dura at the thoracic or cervical level of the spinal column. As indicated by the title, this paper discusses the development, usage, and theory of transcutaneous electrical nerve stimulation. Implanted peripheral nerve stimulators and deep brain stimulators have been used with varying degrees of success. The latter appears especially effective in controlling the widespread and severe pain of advanced cancer.

Transcutaneous electrical nerve stimulation was initially used as part of a comprehensive screening program to establish patient candidacy for dorsal column stimulation.[3,6]Electrical nerve stimulation from battery-powered stimulators was used to test the patient's response to nerve stimulation and to document the extent of pain relief. The pain relief produced by transcutaneous electrical nerve stimulation was so impressive that studies were subsequently conducted to determine its role as a separate therapeutic modality. Since then, improved stimulators and skin surface electrodes have been commercially developed and used by numerous patients with chronic and acute pain.

TENS for chronic pain

The ability of TENS to reduce pain was initially examined in chronic pain patients; many of the patients had not obtained relief through analgesics, physical therapy, prescribed exercises, and ablative surgery. In these early clinical applications of TENS, purely empirical means were tried in terms of electrode placement, type of stimulation used, and the tenacity with which efficacious stimulation sites were sought. Accordingly, the percentage of chronic pain patients benefiting from TENS varied from study to study, but most investigators found that about 33-66 percent of the patients benefited significantly using this form of stimulation.[4,5,7,8] The pain problems treated by TENS in these early studies included peripheral neuropathy, stump pain, phantom limb pain, chronic lumbar syndrome (low back pain), spinal cord injury, sciatica, post-incisional chest pain, and post-herpetic neuralgia.

During the early days of TENS, when the details of its mechanism were fuzzy, clinicians and researchers raised questions regarding the placebo effect. But several lines of evidence indicate that TENS has a very real effect on pain. In a double-blinded, crossover trial using genuine and sham transcutaneous nerve stimulators, Thorsteinsson et al.[9] found the placebo effect of TENS to be 32 percent and quite similar to the placebo effect associated with medications in general. Forty-eight percent of the patients who used the genuine TENS unit obtained complete or partial pain relief. The largest difference in analgesic effect between the real and sham stimulators occurred in patients with neuropathies.

Another line of evidence that argues against the supposition that TENS acts only as a placebo comes from the long-term study of TENS efficacy by Eriksson et al.[10] Of the 123 chronic pain patients studied, they found that 41 and 31 percent of them continued to use TENS 12 and 24 months, respectively, after their initial trial period. Three-quarters of the successfully relieved patients reported more than 50 percent pain relief (as measured by visual analog scales) together with increased levels of social activity and decreased levels of analgesic drug intake. Placebo effects rarely persist beyond one or two weeks.

Some of the early difficulties associated with TENS therapy included the lack of information regarding stimulation parameters and electrode placements that most effectively ameliorate chronic pain. Most clinicians followed the general guidelines provided by TENS manufacturers in placing the electrodes near the pain site and then setting the stimulator to produce a moderate pulse width, a pulse current that did not evoke direct motor responses, and a pulse frequency varied to achieve discernible paresthesia. Though helpful, these guidelines lacked precision, scientific rationale, and optimized stimulation parameters for various chronic pain states. Clinicians often were frustrated by this mode of therapy, and some doubted its efficacy.

In light of these problems, pain researchers endeavored to define TENS therapy more fully. Studies of TENS since 1975 have examined its long-term efficacy,[10] its relationship to acupuncture[8,11] and release of endogenous opiates,[12] optimal electrode placement,[13] more efficacious stimulating parameters,[12,13] and the types of pain syndromes responseive to TENS. The scientific basis for TENS and its success rate have been increased somewhat by these studies but, unfortunately, not enough to make TENS a universally accepted treatment for pain. While there is now less guesswork involved, the experience, knowledge, and tenacity exercised by the clinician still contribute heavily to the successful application of TENS.

Despite its impreciseness, TENS remains a very attractive way of treating pain. Except for possible skin irritation under the electrodes, there are no negative physiological side effects from TENS. Pain patients who respond to TENS greatly prefer it to analgesics. For some patients, TENS has provided the only acceptable form of pain relief.

TENS for acute pain

Because of these significant attributes, TENS is being used for more types of pain, particularly acute pain. In the pioneering study by Hymes et al.,[14] TENS was found to reduce the intensity of postoperative pain by 80 percent, the incidence of respiratory and gastrointestinal tract complications, and the length of stay in the postoperative intensive care unit. A number of other studies have examined the efficacy of TENS for post-op pain following upper abdominal surgery,[15,16] hip or knee joint replacements,[17] spine surgery,[18] and cesarean births.[19] Postoperative TENS is now available as an option to certain patients prior to surgery at a number of hospitals around the country.

Because using narcotics for pain may carry some undesirable side effects (e.g., sedation, nausea, dizziness, antidiuresis, addiction, etc.), transcutaneous nerve stimulation also has been tried for acute pain arising from advanced cases of cancer,[20] dental procedures,[21] neuromuscular strains of "whiplash,"[22] and labor.[23,24] The success rates for these applications of TENS range from 33 to 88 percent. For a labor or postcesarean patient, TENS therapy has a special appeal because it does not affect the infant and allows the mother to be alert early in the postpartum period. Another strong advantage of TENS usage is that pain medication is not secreted through the breast milk and thus cannot affect the infant during the first few critical days of life.

Despite the non-universality of TENS, its non-narcotic approach and few undesirable side effects have enlarged the market for TENS to $50-70 million in annual retail sales in the United States. The top nine manufacturers account for about 95 percent of the sales, with the remainder being scattered among numerous regional companies.[25] Manufacturers and dealers of TENS agree that its market will not dramatically escalate until TENS therapy is accepted by a greater number of physicians in private practice outside major medical centers. Efforts at physician education appear to be the goal of many manufacturers for 1983-84.

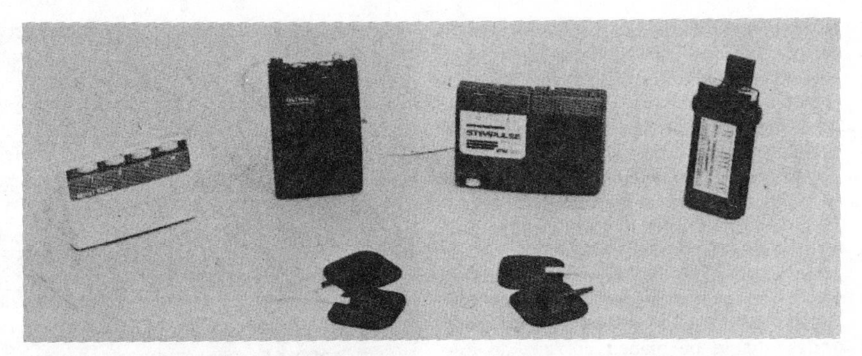

Figure 1: Typical TENS units on the market today. Tenzcare (3M, Inc.); DYNEX II (La Jolla Technology, Inc.); STIMPULSE (Stimtech, Inc.); and NEUROMOD (Medtronic, Inc.).

Stimulation parameters

In conjunction with the continuing research efforts to understand the mechanism by which TENS operates, stimulation parameters for it have undergone evolutionary developments. Some early surface electrical stimulators yielded a complex, inconsistent, and electronically "dirty" output.[26] Other and most recent stimulators use battery power and solid state circuits to produce "cleaner" asymmetrical, biphasic pulses operating into a skin-like impedance load.

Research efforts of biomedical engineers, coupled with expanding knowledge about the neural mechanisms of TENS, gradually resolved key issues such as the electrical characteristics of the skin/electrode interface, constant voltage versus constant current output, and the desired ranges for frequency, voltage, and current amplitude. Due to the nonlinear, time-varying, and complex nature of the skin/electrode interface[27,28] and the need to selectively stimulate large diameter afferent nerve fibers, the stimulator should have a high impedance output (i.e., constant current output). To avoid possible stinging sensations associated with electrocutaneous stimulation, voltage compliance should be limited to about 50 volts for pulses greater than 50 μsec.[29] To provide sufficient strength to excite the targeted nerves, the output stimulus should be able to yield 60 to 80 mA of peak current with pulse widths from 40 to 250 μsec. To avoid iontophoresis and any net ionic changes within the stimulated tissue, a zero net DC stimulus waveform is required. To provide the full range of transcutaneous nerve stimulation,[7,10] the stimulator should produce frequencies of 1 to 130 Hz.

Despite energetic attempts to define an optimal output waveform, disagreements about it still remain. The majority of the TENS units available today (see Figure 1) produce asymmetrical, biphasic rectangular pulses. Some units feature a spiked waveform or a rectangular pulse of one polarity followed by a spike or rectangular pulse of the opposite polarity. Marketing claims that one waveform is superior in efficacy or comfort to some other waveform are usually unsubstantiated clinically and cannot be satisfactorily explained physiologically. Part of the difficulty in the search for an optimal waveshape lies in the highly subjective nature of pain and pain relief, the impact of personality differences on pain reports, variability of pain syndromes, and the lack of a good animal model for chronic pain.

The latest models of TENS tend to modulate the various stimulation parameters. State-of-the-art stimulators offer a choice of output waveforms and modulation modes so that clinicians and their patients can choose the best stimulation for them. Some of the more common modulation schemes include a low frequency burst mode (1-2 burst/sec with 7-10 pulses/burst at 100-150 pulses/sec), alternating frequency mode, and intensity modulation mode. The rationale for these approaches include reduced sensory adaptation to TENS, greater user comfort, and/or wider spectrum of nerves excited without causing muscle contractions. These explanations, appealing and physiologically reasonable as they might be, generally lack direct experimental verification. Marketing and sales data indicate, however, that many patients are receptive to these newer TENS units and believe them to be at least as effective for their pain as their older units.

Electrode placement

The issue of optimal electrode placement has been the subject of numerous research projects [8,13] and TENS application manuals. [30,31] Appropriate placement of electrodes is one of the most critical factors influencing the success of TENS. For an area of the body to be an optimal stimulation site, (1) it should be physiologically related to the source of pain; (2) it must be readily accessible to a TENS electrode; (3) it must provide a direct link to the CNS; and (4) it should be anatomically distinct.

For conventional transcutaneous electrical nerve stimulation (10-100 pps), the electrodes should be placed over the pain site or in the same dermatome that includes or is immediately proximal to the painful area.[30] Stimulation intensity should be set to achieve a tingling sensation or paraesthesia in the pain area and applied continuously for up to eight hours. Both electrodes should be placed proximal to the pain site with the negative electrode more centrally located.

In cases when patients do not receive sufficient analgesia from conventional TENS, the use of "acupuncture-like TENS" and electrode placements over specific points (acupuncture, motor, and/or trigger) may be indicated. Clinicians are often guided in their search for these points using charts and tables distributed by TENS manufacturers. This particular mode of stimulation is characterized by 1-2 Hz pulses or by low frequency (1-2 Hz) bursts of 100 Hz pulses at sufficient intensity to cause muscle contraction (See Figure 2). The typical treatment regimen of acupuncture-like TENS is 30-45 minutes of strong stimulation administered at regular intervals three or four times a day.

If a patient is still not getting adequate pain relief, or if his/her particular lifestyle necessitates electrode placements away from the painful region, then the electrodes may be placed at the spinal segmental level corresponding to the area of pain, or contralateral or proximal to the area of pain. Bilateral stimulation using two sets of electrodes is another placement. Given the myriad pain syndromes and choices for electrode placement, it is clear that maximal pain relief from TENS rests on the skill and knowledge of the clinician applying this treatment modality.

Research since Melzack-Wall

The Melzack-Wall theory of pain spawned significant advances not only in non-surgical, non-destructive treatment for pain, but also provided the impetus for new avenues of investigation into the basic neurophysiological and neurochemical mechanisms of mammalian pain. Some of the details of the original theory have been modified as a result of these investigations, primarily because the original theory was too simple and was incomplete, not because it was incorrect.[32] The significant neurophysiological findings can perhaps best be put into perspective by looking at the inputs to the spinal cord processing unit as either segmental (at the level of the spinal cord) or supraspinal (descending from more central structures above the level of the spinal cord.

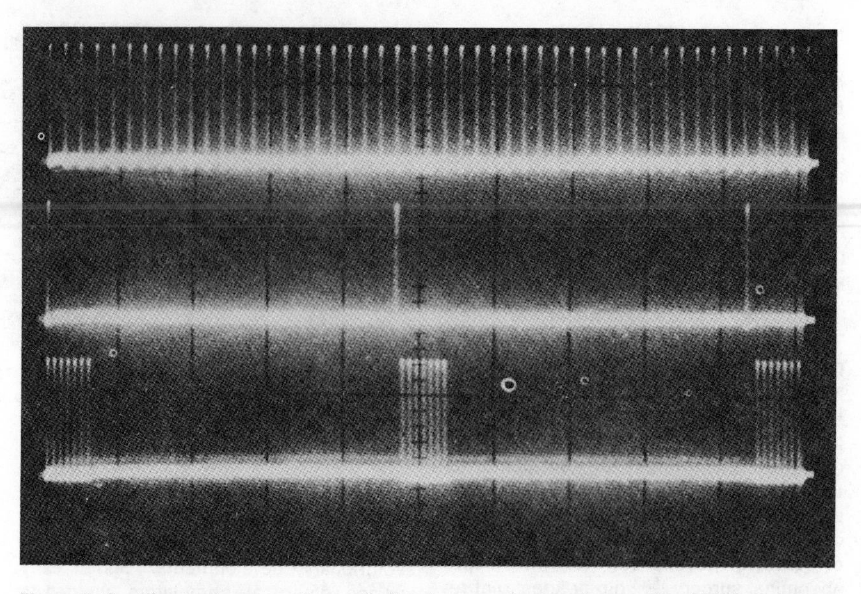

Figure 2: Oscilloscopic comparison of stimulus waveforms: Conventional high frequency TENS (top), low frequency TENS (middle), and low frequency burst TENS (bottom). Time scale is 0.1 sec/div.

Segmental influences

The original Melzack-Wall pain model emphasized presynaptic inhibition of nociceptive input via small fibers by non-nociceptive input via large fibers acting through small excitatory substantia gelatinosa neurons in the spinal cord. Research subsequent to 1965 has shown that these small interneurons may be either excitatory or inhibitory to the postsynaptic transmission cell. To make the situation even more complicated, the input from these excitatory and inhibitory interneurons can be modulated presynatically. So the neuronal "computational unit" in the spinal cord sees a variety of segmental inputs, both nociceptive and non-nociceptive, excitatory and inhibitory. Each of these inputs can be modulated either presynaptically or postsynaptically or both.

In addition to the interactions and modulation occuring at the snyapse in the spinal cord, there is an experimental evidence that repetitive electrical stimulation (such as TENS) produces excitability changes in the peripheral nerve fiber itself, before the level of the spinal cord.[33,34,35] These excitability changes include increased electrical threshold, decreased conduction velocity, and, in some cases, block conduction in the fiber during the period of stimulation and for a short time thereafter.

Supraspinal modulation

Perhaps the most exciting findings in the area of pain research since 1965 have been the discovery and isolation of endogenous opioid peptides, the endorphins and enkephalins,[36,37] and discovery and localization of opiate receptors in the central nervous system.[38] It has been shown that there is a powerful pain-inhibiting system originating in the periaqueductal gray substance surrounding the cerebral aqueduct in the midbrain. Neurons in this region can be activated by morphine or by electrical stimulation. These neurons activate other neurons in the brainstem that project to the spinal cord and release serotonin — an inhibitory, non-opioid neurotransmitter. The interactions of these endogenous opioid and non-opioid systems are currently being worked out in a a variety of animal models in both acute experiments on anesthetized and in unanesthetized, behaving animals. It now appears that there are a number of these opioid peptides (met-enkephalin, leu-eukephalin, dynorphin, alpha-endorphin, beta-endorphin) and perhaps an equal number of specific opioid receptors.

With regard to how transcutaneous nerve stimulation influences segmental and supraspinal mechenisms, it is hypothesized that the descending inhibitory modulating systems are activated by high-intensity, low-frequency (1-2 Hz), acupuncture-like TENS.[7] In contrast, low-intensity, high-frequency TENS (conventional TENS) likely modulates pain by acting at the segmental level without the participation of the descending systems.[39]

Prognosis

The 60s and 70s encompassed the development of the technology of electrical stimulation for relief of pain and basic research in the area of the body's endogenous opioid systems. Melzack and Wall's heuristic model of pain processing in the spinal cord, proposed in 1965, represented a synthesis of then current neurophysiology, neuroanatomy, and clinical observations. A decade later, in 1975, the story of the endogenous opioids began to unfold with the initial research emphasis on identification and localization of these substances. Recent research has been directed toward discovering the physiological significance and how these systems are naturally activated in the behaving organism. Perhaps the decade of the '80s will bring a better understanding of to what degree electrical stimulation for pain relief utilizes these endogenous systems for pain modulation. Certainly the immediate need in the area of therapeutic electrical stimulation is for the precise "tailoring" of stimulus parameters and electrode placement to various types of pain.

References:

1. R. Melzack and P.D. Wall, "Pain Mechanisms: A New Theory," **Science**, vol.150, pages 971-979, 1965.
2. P.D. Wall and W.H. Sweet, "Temporary Abolition of Pain in Man," **Science**, vol. 155, pages 103-104, 1967.
3. C.N. Shealy, J.T. Mortimer, and N.R. Hagfors, "Dorsal Column Electroanalgesia," **J. Neurosurg.**, vol. 32, pages 560-564. 1970.
4. D.M. Long, "Cutaneous Afferent Stimulation for Relief of Chronic Pain," **Congr. Neural Surg.**, vol. 21, pages 257-268, 1974.
5. W.M. Wolff, J.A. Lewis, and R.H. Simon, "Expereince with Electrical Stimulation Devices for the Control of Chronic Pain," **Medical Instrumentation**, vol. 9 (5), pages 217-220, 1975.
6. J.A. Picaza, B. Cannon, S. Hunter, A. Boyd, J. Guma, and D. Maurer, "Pain Supression by Peripheral Nerve Stimulation," **Surg. Neurol.**, vol. 4, pages 105-124, July 1975.
7. R. Melzack, "Prolonged Relief of Pain by Brief Intense Transcutaneous Somatic Stimulation," **Pain**, vol. 1, pages 357-373, 1975.
8. E.J. Fox and R. Melzack, "Transcutaneous Electrical Stimulation and Acupuncture: Comparison of Treatment for Low-Back Pain," **Pain**, vol. 2, pages 141-148, 1976.
9. G. Thorsteinsson, H.H. Stonnington, G.K. Stillwell and L.R. Elveback, "The Placebo Effect of Transcutaneous Electrical Stimulation,"**Pain**, vol. 5, pages 31-41, 1978.
10. M.B.E. Eriksson, B.H. Sjölund, and S. Nielzen, "Long Term Results of Peripheral Conditioning Stimulation as an Analgesic Measure in Chronic Pain," **Pain**, vol. 6, pages 335-347, 1979.
11. J.C. Willer, A Roby, P. Boulu, and F. Boureau "Comparative Effects of Electroacupuncture and Transcutaneous Nerve Stimulation on Human Blink Reflex," **Pain**, vol. 14, pages 267-278, 1982.
12. B.H. Sjölund and M.B.E. Eriksson, "Endorphins and Analgesia Produced by Peripheral Conditioning Stimulation," **Advances in Pain Research and Therapy**, vol. 3, J.J. Bonica et al. (editors) Raven Press, N.Y., 1979, pages 587-592.
13. S.L. Wolf, M.R. Gersh, and V.R. Rao, "Examination of Electrode Placements and Stimulating Parameters in Treating Chronic Pain with Conventional Transcutaneous Electrical Nerve Stimulation (TENS)," **Pain**, vol. 11, pages 37-47, 1981.
14. A.C. Hymes, D. Raab, E, Yonehiro, G. Nelson, and A. Printy, "Acute Pain Control by Electrostimulation: A Preliminary Report,"**Adv. Neurol.**, vol. 4, pages 761-767, 1974.
15. A.M. Cooperman, B. Hall, K. Mikalacki, R. Hardy, and E. Sadar, "Use of Transcutaneous Electrical Stimualtion in the Control of Postoperative Pain,"**Amer. J. Surg.**, vol. 133, pages 185-187, 1977.
16. J. Ali, C.S. Yaffe, and C. Serrette, "The Effect of Transcutaneous Electric Nerve Stimualtion on Postoperative Pain and Pulmonary Function," **Surgery**, vol. 89 (4), pages 507-512, 1981.
17. M. Stabile anf T. Mallory, "The Management of Postoperative Pain in Total Joint Replacement: Transcutaneous Electrical Nerve Stimulation is Evaluated in Total Hip and Knee Patients," **Orthopaedic Rev.**, vol. 7, pages 121-123, 1978.
18. R.R. Richardson and E.B. Siqueria, "Transcutaneous Electrical Neurostimualtion in Postlaminectomy Pain"**Spine**, vol. 5 (4), pages 361-365, 1980.
19. J.E. Riley, "The Impact of Transcutaneous Electrical Nerve Stimulation on the Postcesarean Patient," **Jour. of Gyn. Nursing**, pages 325-329, September/October 1982.
20. M.J. Ostrowski and V.A. Dodd, "Transcutaneous Nerve Stimulation for Relief of Pain in Advanced Malignant Disease," **Nursing Times**, vol. 73, pages 1233-1238, August 1977.
21. A. Pertovaara, P. Kemppainen, G. Johansson,and S.L. Karonen, "Dental Analgesia Produced by Non-painful, Low-frequency Stimualtion is not Influenced by Stress or Reversed by Naloxone," **Pain**, vol 13, pages 379-384, 1982.
22. R.R. Richardson and E.B. Siqueria, "Transcutaneous Electrcal Neurostimulation in Acute Cervical Hyperextension-hyperflexion Injuries," **Illinois Medical Journal**, vol. 159 (4), pages 227-230, 1981.
23. R. Erkkola, P. Pikkola, and J. Kanto, "Transcutaneous Nerve Stimulation for Pain Relief During Labor: A Controlled Study," **Annales Chirurgiae of Gynaecologiae**, vol. 69, pages 273-277, 1980.
24. L.E. Augustinsson, P. Bohlin, P. Bundsen, C.A. Carlsson, L, Forssman, P. Sjöberg, and N.O. Tyreman, "Pain Relief During Delivery by Transcutaneous Electrical Nerve Stimulation,"**Pain**, vol 4, pages 59-65, 1977.
25. K. Brown, "TENS Market Update," **Homecare**, pages 40-58, January 1983.
26. C. Burton and D.D. Maurer, "Pain Suppression by Transcutaneous Electronic Stimulation," **IEEE Trans. Biomed. Eng.**, vol. 21 (2), pages 81-88, 1974.
27. K.R. Brennen, "The Characterization of Transcutaneous Stimulating Electrodes," **IEEE Trans. Biomed. Eng.**, vol. 23 (4), pages 337-340, 1976.
28. A. van Boxtel, "Skin Resistance During Square-wave Electrical Pulses of 1 to 10 mA," **Med. & Biol. Eng. and Computing**, vol. 15, pages 679-687, 1977.
29. A.Y.J. Szeto and F.A. Saunders, "Electrocutaneous Stimulation for Sensory Communication in Rehabilitation Engineering," **IEEE Trans. Biomed. Eng.**, vol. 29 (4), pages 300-308, 1982.
30. J.S. Mannheimer, **Optimal Stimulation Sites for TENS Electrodes**, La Jolla Technology, Inc., CA, 1980.
31. J. Yu and W. Carroll, **Electrode Placement Manual for TENS**, Medtronic, Inc., Minneapolis MN, 1982.
32. R. Melzack and P.D. Wall, **The Challenge of Pain**, New York: Basic Books, 1982.
33. R.j. Ignelzi and J.K. Nyquist, "Direct Effect of Electrical Stimulation on Peripheral Nerve Evoked Activity: Implications in Pain Relief," **J. Neurosurg.**, vol. 45, pages 159-165, 1976
34. R.J. Ignelzi and J.K. Nyquist, "Excitability Changes in Peripheral Nerve After Repetitive Electrical Stimualtion. Implications in Pain Modulation," **J. Neurosurg.**, vol. 51, pages 824-833, 1979.
35. J.N. Campbell and A. Taub, "Local Analesia from Percutaneous Electrical Stimulation: A Peripheral Mechanism," **Arch. Neurol. (Chic.)**, vol. 28, pages 347-350, 1973.
36. J. Hughes, "Isolation of an Endogenous Compound from the Brain with Pharmacological Properties Similar to Morphine," **Brain Res.**, vol. 88, pages 295-308, 1975.
37. L. Terenius and A Wahlstrom, "Search for an Endogenous Ligand for the Opiate Receptor," **Acta Physiol. Scand.**, vol. 94, pages 74-81, 1975.
38. C.B. Pert, M.J. Kuhar, and S.H. Snyder, "Opiate Receptor: Autoradiographic Localization in Rat Brain," **Proc. Natl. Acad. Sci.**, vol. 73, pages 3729-3733, 1976.
39. S.E. Abram, A.C. Reynolds, and J.F. Cusick, "Failure of Naloxone to Reverse Analgesia from Transcutaneous Electrical Stimulation in Patients with Chronic Pain," **Anesth. Analg.**, vol. 60, pages 81-84, 1981.

0278-0054/83/0400-0018$1.00